Sustainable Technologies for Food Waste Management

This book comprehensively explores various aspects of food waste management, ranging from understanding the issue to innovative solutions that convert waste into valuable resources. The initial chapters provide an in-depth analysis of advanced methods for characterizing food waste. The book examines cutting-edge methodologies, including physico-chemical, thermal, and biological approaches, for converting food waste into valuable resources such as biohydrogen, biopolymers, biogas, organic acids, bioactive compounds, biochar, compost, and vermicompost. Additionally, it delves into the utilization of advanced technologies like machine learning, the Internet of Things (IoT), sensors, and robotics for food waste treatment and management, thus contributing to a sustainable future. It underscores the conversion of food waste into renewable energy sources, sustainable materials, and valuable compounds. This book will be useful for researchers, practitioners, and policymakers seeking sustainable approaches to tackle the global challenge of food waste.

- Offers in-depth insights into advanced methods for characterizing food waste, enabling precise and targeted waste reduction strategies
- Examines physico-chemical, thermal, and biological approaches for converting food waste into valuable resources
- Presents machine learning, IoT, sensors, and robotics for sustainable food waste management and treatment
- Highlights the conversion of food waste into valuable resources, including biohydrogen, biopolymers, bioactive compounds, compost, and vermicompost
- Provides specific details about the food waste supply chain and associated policies

Sustainable Industrial and Environmental Bioprocesses

This book series aims to provide a comprehensive collection of books focusing on bioprocesses in industrial and environmental biotechnology. The multidisciplinary content encompasses chemical and biochemical engineering, industrial microbiology, and energy biosciences, all with a central theme of sustainable development and circular economy principles. The books align with the Sustainable Development Goals (SDGs) and offer state-of-the-art information and in-depth knowledge on the subject matter. The books in this series also emphasizes the application of emerging tools, such as machine learning and artificial intelligence, for the advancement of bioprocesses. While primarily targeting academicians and researchers, the series is valuable for policy planners and industry professionals, with carefully tailored contents to cater to their specific needs and interests.

Series Editor: Dr Ashok Pandey

Professor Ashok Pandey is currently Executive Director at the Center for IEnergy and Environmental Sustainability, Lucknow, India. His major Research and technological development interests are industrial and environmental biotechnology and energy biosciences, focusing on biomass, biofuels, chemicals, waste, wealth, and energy, etc. He has 16 patents and has published 120 books and more than 1000 papers and book chapters. Professor Pandey is the recipient of many national and international awards and honors, which include Fellow, The World Academy of Sciences (TWAS), Academician of European Academy of Sciences and Arts, etc.

Decentralized Sanitation and Water Treatment: Treatment in Cold Environments and Techno-Economic Aspects
R.D. Tyagi, Ashok Pandey, Patrick Drogui, Bhoomika Yadav, Sridhar Pilli and Jonathan W.C. Wong

Biodegradation of Toxic and Hazardous Chemicals: Detection and Mineralization
Kashyap K. Dubey, Kamal. K. Pant, Ashok Pandey and Maria Angeles Sanromán

Biodegradation of Toxic and Hazardous Chemicals: Remediation and Resource Recovery
Kashyap K. Dubey, Kamal K. Pant, Ashok Pandey and Maria Angeles Sanromán

Waste Management in Climate Change and Sustainability- Lignocellulosic Waste
Sunita Varjani, Izharul Haq, Ashok Pandey, Vijai Kumar Gupta, and Xuan-Thanh Bui

Ashwagandha for Quality of Life: Scientific Evidence
Sunil C Kaul and Renu Wadhwa

Microfluidics in Food Processing: Technologies and Applications
Ayon Tarafdar, Ranjna Sirohi, Barjinder Pal Kaur, Ashok Pandey, and Claude-Gilles Dussap

Sustainable Technologies in Food Waste Management
Ranjna Sirohi, Ayon Tarafdar, Luciana Porto de Souza Vandenberghe, Mohammad J. Taherzadeh, and Ashok Pandey

https://www.routledge.com/Sustainable-Industrial-and-Environmental-Bioprocesses/book-series/SIEB

Sustainable Technologies for Food Waste Management

Edited by

Ranjna Sirohi
Sri Karan Narendra Agriculture University, Jobner, Rajasthan, India

Ayon Tarafdar
ICAR-Indian Veterinary Research Institute, Bareilly, India

Luciana Porto de Souza Vandenberghe
Federal University of Paraná, Curitiba, Brazil

Mohammad J. Taherzadeh
University of Borås, Borås, Sweden

Ashok Pandey
Centre for Energy and Environmental Sustainability, Lucknow, India

CRC Press
Taylor & Francis Group
Boca Raton London New York

CRC Press is an imprint of the
Taylor & Francis Group, an **informa** business

Designed cover image: Shutterstock

First edition published 2025

by CRC Press
2385 NW Executive Center Drive, Suite 320, Boca Raton FL 33431

and by CRC Press
4 Park Square, Milton Park, Abingdon, Oxon, OX14 4RN

CRC Press is an imprint of Taylor & Francis Group, LLC

ISBN: 9781032705941 (hbk)
ISBN: 9781032705965 (pbk)
ISBN: 9781032706030 (ebk)

DOI: 10.1201/9781032706030

Typeset in Times
by Deanta Global Publishing Services, Chennai, India

Contents

About the Editors

Dr. Ranjna Sirohi is currently working as an assistant professor at the SKN Agricultural University, Jaipur, India. Previously, she worked at the University of Petroleum and Energy Studies, Dehradun, Uttarakhand, India. She obtained her B. Tech. degree in Biochemical Engineering, her M. Tech. degree in Agricultural Engineering, and her Ph.D. degree in Process and Food Engineering from G.B. Pant University of Agriculture and Technology, Pantnagar, India. She worked as a research professor at Korea University in Seoul, Republic of Korea, before joining UPES. She has also worked as a researcher at the École Polytechnique Fédérale de Lausanne, Switzerland. Her major research interests are in bioprocess technology, food and food waste valuation, waste to wealth, and biofuels. She has more than 150 publications, including seven books, 21 book chapters, popular articles, 21 conference proceedings, and two filed patents. Dr. Sirohi has won several awards and is a life member of several professional societies.

Dr. Ayon Tarafdar is currently working as an ARS Scientist at the ICAR, Indian Veterinary Research Institute under the Ministry of Agriculture and Farmers Welfare. He obtained his B. Tech. degree from IIT Kharagpur (India), his M. Tech. from GBPUA&T, Pantnagar (India), and his Ph.D. from NIFTEM (Kundli, India) in Food Engineering. He has worked as a fully funded researcher at Washington University in St. Louis, USA. His main research focus has been on non-thermal food processing operations, bioprocess modeling and optimization, machine learning, food waste valorization, advanced food characterization tools, and precision agriculture. He has published more than 93 scientific papers and reviews, in addition to 13 book chapters, two books, and one filed patent. Dr. Tarafdar has been awarded several meritorious fellowships during his academic journey, including the TEQIP-II fellowship, the NIFTEM research fellowship, the UGC-JRF fellowship, the CSIR direct SRF fellowship, and the SERB-DST international travel grant. In addition to this, he has been conferred with the Young Scientist Award and best poster and presentation awards for his work at several conferences both in India and abroad. He serves on the editorial board and is a guest editor for various journals.

Dr. Luciana Porto de Souza Vandenberghe is a Full Professor at the Department of Bioprocess Engineering and Biotechnology, Federal University of Paraná, Curitiba, Brazil. Dr. Vandenberghe obtained her Ph.D. degree in Génie de Procédés Industriels Biotechnologie from the Université de Technologie de Compiègne (2000), France. Her areas of interest include bioprocess engineering and biotechnology, and industrial microbiology, with a focus on the valorization of solid and liquid agro-industrial products through submerged and solid-state fermentation for biomolecule production, including biofuels, industrial enzymes, organic acids, bioplastics, and plant growth hormones. She has published 140 papers, 2 books, 60 book chapters, and 25 patents. She is an editor and an editorial board member of various journals.

Professor Mohammad J. Taherzadeh has been a Professor of Bioprocess Technology since 2004 at the University of Borås in Sweden and chairman of the Swedish Centre for Resource Recovery. Prof. Taherzadeh has a Ph.D. in Bioscience from Chalmers University of Technology in Sweden (1999) and MSc from Sharif University of Technology (1991) and BSc from Isfahan University of Technology (1989), both in Chemical Engineering from Iran. He is developing processes to convert wastes and residuals to value-added products, with a focus on pre-treatment and fermentation using filamentous fungi and anaerobic bacteria. He has conducted extensive research on food waste and its valorization in a variety of products, and also developed several industrial processes in this field. He has +440 publications in scientific peer-reviewed journals, +40 book chapters, and eight books published by different publishers.

Professor Ashok Pandey is currently Executive Director at the Center for IEnergy and Environmental Sustainability, Lucknow, India. His major Research and technological development interests are industrial and environmental biotechnology and energy biosciences, focusing on biomass, biofuels, chemicals, waste, wealth, and energy, etc. He has 16 patents and has published 120 books and more than 1000 papers and book chapters. Professor Pandey is the recipient of many national and international awards and honors, which include Fellow, The World Academy of Sciences (TWAS), Academician of European Academy of Sciences and Arts, etc.

List of Contributors

Marcela Moreira Albuquerque
Department of Bioprocess Engineering and Biotechnology
Federal University of Paraná
Curitiba, Paraná, Brazil

Anupama Anand
Department of Food Science and Technology
Dr. YS Parmar University of Horticulture and Forestry
Nauni, Himachal Pradesh, India

Mukesh Kumar Awasthi
College of Natural Resources and Environment
Northwest A&F University
Shaanxi, China

Kunal Balhara
Department of Food Engineering
National Institute of Food Technology Entrepreneurship and Management
Sonipat, Haryana, India

Rafael Novaes Barros
Department of Bioprocess Engineering and Biotechnology
Federal University of Paraná
Curitiba, Paraná, Brazil

Mehak Bhardwaj
Amity Institute of Biotechnology
Amity University Chhattisgarh
Raipur, Chhattisgarh, India

Shiqing Chen
College of Natural Resources and Environment
Northwest A&F University
Shaanxi, China

Yuzhe Chen
College of Natural Resources and Environment
Northwest A&F University
Shaanxi, China

Swarnima Dey
Department of Food Technology, Faculty of Science and Humanities
SRM University, Delhi-NCR
Sonepat, Haryana, India

Prachi Gaur
Amity Institute of Biotechnology
Amity University
Lucknow, Uttar Pradesh, India

Vivek Kumar Gaur
Centre for Energy and Environmental Sustainability
Lucknow, Uttar Pradesh, India
School of Energy and Chemical Engineering, UNIST
Ulsan, Republic of Korea

Vartika Gupta
Department of Botany
University of Lucknow
Lucknow, Uttar Pradesh, India

Suyue Hao
College of Natural Resources and Environment
Northwest A&F University
Shaanxi, China

Sharareh Harirchi
Swedish Centre for Resource Recovery
University of Borås
Borås, Sweden
Department of Biotechnology
Iranian Research Organization for Science and Technology
Tehran, Iran

Ida Eyi Heathcote-Fumador
Chalmers University of Technology
Gothenburg, Sweden

Ranjan Kaushik
Center of Food Science and Technology
CCSHAU
Hisar, Haryana, India

Varaprasad Kolla
Amity Institute of Biotechnology
Amity University Chhattisgarh
Raipur, Chhattisgarh, India

Yogesh Kumar
Department of Food Technology
Faculty of Science and Humanities
SRM University, Delhi-NCR
Sonepat, Haryana, India

Komal Kumari
Central University of South Bihar
Fatehpur, Bihar, India

Elaine M'Nkubitu
Independent researcher
Nairobi, Kenya

Clara Matte Borges Machado
Department of Bioprocess Engineering and Biotechnology
Federal University of Paraná
Curitiba, Paraná, Brazil

Walter José Martinez-Burgos
Department of Bioprocess Engineering and Biotechnology
Federal University of Paraná
Curitiba, Paraná, Brazil

Isaac Arnold McSey
Group Technology - Infrastructure
Volvo Autonomous Solutions
Gothenburg, Sweden

Adriane Bianchi Pedroni Medeiros
Department of Bioprocess Engineering and Biotechnology
Federal University of Paraná
Curitiba, Paraná, Brazil

Ariane Fátima Murawski de Mello
Department of Bioprocess Engineering and Biotechnology
Federal University of Paraná
Curitiba, Paraná, Brazil

Mojdeh Mirshafiei
Department of Biotechnology
School of Chemical Engineering, College of Engineering
University of Tehran
Tehran, Iran

Taru Negi
Department of Food Science and Technology
GBPUAT
Pantnagar, Uttarakhand, India

Juliana de Oliveira
Unicuritiba University Center
Reboucas, Curitiba, Brazil

Priscilla Zwiercheczewski de Oliveira
Department of Bioprocess Engineering and Biotechnology
Federal University of Paraná
Curitiba, Paraná, Brazil

Abdullah Bilal Öztürk
Department of Chemical Engineering
Yildiz Technical University
Esenler, Istanbul, Türkiye
Swedish Centre for Resource Recovery
University of Borås
Borås, Sweden

Ashok Pandey
Centre for Energy and Environmental Sustainability
Lucknow, Uttar Pradesh, India

Mohsen Parchami
Department of Forest Biomaterials and Technology
Swedish University of Agricultural Sciences
Umeå, Sweden

Cristine Rodrigues
Department of Bioprocess Engineering and Biotechnology
Federal University of Paraná
Curitiba, Paraná, Brazil

Shubham Samkaria
Department of Agrotechnology
CSIR-IHBT
Palampur, Himachal Pradesh, India

Gabriela de Bona Sartor
Department of Bioprocess Engineering and Biotechnology
Federal University of Paraná
Curitiba, Paraná, Brazil

Thamarys Scapini
Department of Bioprocess Engineering and Biotechnology
Federal University of Paraná
Curitiba, Paraná, Brazil

Poonam Sharma
Department of Bioengineering
Integral University
Lucknow, Uttar Pradesh, India

Uadal Singh
SKN Agriculture University
Jobner, Rajasthan, India
Rajasthan Agricultural Research Institute
Jaipur, Rajasthan, India

Ranjna Sirohi
Sri Karan Narendra Agriculture University
Jobner, Rajasthan, India

Payal Sahu
Amity Institute of Biotechnology
Amity University Chhattisgarh
Raipur, Chhattisgarh, India

Shubham Samkaria
CSIR-IHBT
Palampur, Himachal Pradesh, India

Amita Shakya
Amity Institute of Biotechnology
Amity University Chhattisgarh
Raipur, Chhattisgarh, India

Carlos Ricardo Soccol
Department of Bioprocess Engineering and Biotechnology
Federal University of Paraná
Curitiba, Paraná, Brazil

Arjun Suresh
Save Vibrant Earth Foundation
Noida, Uttar Pradesh, India

Mohammad J. Taherzadeh
Swedish Centre for Resource Recovery
University of Borås
Borås, Sweden

Ayon Tarafdar
Livestock Production and Management Section
ICAR-Indian Veterinary Research Institute
Bareilly, Uttar Pradesh, India

Daniel Thomi
Independent Researcher
Cyber Risk and Control
Nairobi, Kenya

Luciana Porto de Souza Vandenberghe
Department of Bioprocess Engineering and Biotechnology
Federal University of Paraná
Curitiba, Paraná, Brazil

Piyush Verma
School of Health Science and Technology, UPES
Dehradun, Uttarakhand, India

Steven Wainaina
Swedish Centre for Resource Recovery
University of Borås
Borås, Sweden

Adenise Lorenci Woiciechowski
Department of Bioprocess Engineering and Biotechnology
Federal University of Paraná
Curitiba, Paraná, Brazil

Fatemeh Yazdian
Department of Life Science Engineering
Faculty of New Sciences and Technologies
University of Tehran
Tehran, Iran

Haoyuan Zhang
College of Natural Resources and Environment
Northwest A&F University
Shaanxi, China

1 Food Waste Management – An Introduction

Piyush Verma, Ranjna Sirohi, Ayon Tarafdar, and Ashok Pandey

1.1 INTRODUCTION

A fundamentally important balance between the world's economic expansion and the depletion of natural resources must be maintained through sustainable production and consumption habits (Ananno et al., 2021). The challenge of sustainability begs the question of whether future generations will be able to access the same or a greater range of resources as the current generation. For this reason, it is crucial to address the sustainability issue. It would be unjust if our generation used up more of our resources and left less for future generations (Saijo, 2020). The resources of the world are greatly impacted by the methods in which goods are made and utilized (Pellegrini et al., 2019).

Food waste (FW) is a serious global problem that is being seen from more public, governmental, and corporate viewpoints (Issock et al., 2020) in addition to being an environmental one for both developed and developing countries. It arises throughout the food chain, from production to over-harvesting, processing, transportation, storage, and consumption by consumers and the food service industry. Every year, about 1.3 billion metric tons (MT) of edible food are wasted worldwide (HLPE, 2014; Wang et al., 2021). Ecological and socioeconomic factors suffer when up to one-third of food that is intentionally produced for human nutrition is wasted and never eaten. Therefore, managing FW is one of the biggest environmental concerns in the world. FW includes incorrect use of water, land, and energy resources in addition to the underutilization of edibles (Tsang et al., 2019).

From an environmental perspective, FW accounts for 8–10% of the world's annual greenhouse gas (GHG) emissions (3.3 billion tons) (IPCC, 2019). The agricultural phase of FW has an annual (blue) water footprint of approximately 250 km^3, which is five times the volume of Lake Garda and greater than any country's annual (blue) water footprint for food consumption (Fattibene et al., 2020). About 22% of GHG emissions and 30% of global energy use are attributed to the food industry only (Grosso & Falasconi, 2018). Food loss or waste (FLW) has a worldwide carbon footprint that accounts for almost 8% of all anthropogenic GHG emissions, which leads to climate change. It is an example of how limited resources, such as land, water, and energy, are wasted over a product's life cycle. About 4.5 kilograms of CO_2 are discharged into the atmosphere for every kilogram of food produced (Ferdes et al., 2022). Thyberg and Tonjes (2016) claim that FW in landfills turns into GHG-like methane, which has a 25 times larger potential to cause global warming than carbon dioxide over 100 years.

A thorough understanding of waste patterns and food consumption is necessary for the creation of structured waste management systems, as FW poses significant risks to economic, social, and environmental issues (Attiq et al., 2021). To achieve sustainability and satisfy the Sustainable Development Goals (SDG) of the United Nations (UN), the Farm-to-Fork strategy focuses on addressing food loss and waste (FLW) throughout the food supply chain (FSC) (EU, 2020). The FW hierarchy supports the idea that efforts should be made to reduce FW as well as use it as a resource (Albizzati et al., 2021). FW can be minimized to have a minimal negative impact on waste management systems, reduce costs, and promote food security through appropriate waste management. Consideration should also be given to the notable decline in environmental quality in addition to

DOI: 10.1201/9781032706030-1

these losses (Mishra et al., 2020). Precautionary actions can aid in reducing the quantity of FW that can be prevented. Only with effective waste recycling and management techniques can the amount of inevitable FW be reduced (Dahiya et al., 2018).

1.2 DEFINITION AND TYPES OF FOOD WASTE

The standards used to define FW vary. Food wastage was defined by Parfitt et al., (2010) as "spoilt food arising at the end of the food cycle, which refers to retailers' and consumers' practice," while Lipinski et al., (2013) defined it as "food which is of acceptable quality and qualified for human consumption but is not consumed because it is squandered either before or after it deteriorates." FW is any edible food that is discarded regardless of whether it has expired or has been allowed to go bad (Sahoo et al., 2023). The European FUSIONS "Food Use for Social Innovation by Optimizing Waste Prevention Strategies" project defines FW as "any food, and inedible parts of food, removed from the food supply chain to be recovered or disposed of (including composting, crops ploughed in/not harvested, anaerobic digestion, bio-energy production, co-generation, incineration, disposal to sewer, landfill, or discarded to sea)" (Goossens et al., 2019).

To select the optimal management strategy for FW, categorization is a crucial first step. Gaining more knowledge about the various waste products produced in the FSC can help identify a management strategy that minimizes negative effects on the environment while maximizing positive social and economic effects (Garcia et al., 2015). The forms of FW that exist today vary greatly in terms of where they come from, their chemical makeup, their physical characteristics, and whether or not they contain bacteria. Depending on the best recovery and value-adding technology selected, there are numerous ways to categorize this waste. The easiest way to categorize FW is by the kind of industry sector that produces it. This classification method's benefit is that it makes it simple and quick to set up a recovery route and handle waste from a similar area with shared features, all while drastically lowering expenses (Ferdes et al., 2022). FW is therefore divided into categories, including waste from the dairy, meat, fish, poultry, cereal, bakery, fruits, vegetables, brewing, and wine industries. Food source, edibility, the presence of animal products, complexity, treatment, packaged versus unpackaged (critical for sorting and further processing), packaging biodegradability, supply chain stage, and other factors could all be considered when categorizing FW (Garcia et al., 2015).

The European Commission (2014) has divided FW sources in the literature into three categories: "Food losses are defined as follows: (i) food products lost during the production phase; (ii) food products lost during the consumption phase, such as fruit cores and banana peels; (iii) food products (Figure 1.1) that could have been eaten but were lost during the consumption phase." Furthermore, FW can be categorized according to its chemical composition, such as C-H-N-O-S-Cl content, or

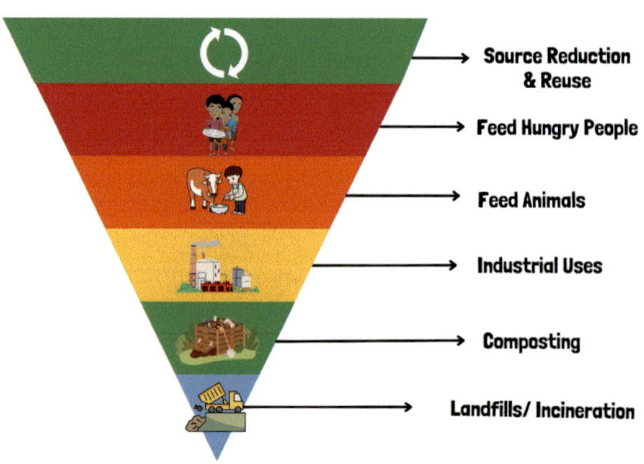

FIGURE 1.1 European Union food waste hierarchy.

nutrient composition, such as carbohydrate and fat content, and storage temperature, such as ambient, chilled, or frozen. However, the data these examples provide is insufficient to rank certain waste management options higher than others (Garcia et al., 2017).

Various researchers have additionally divided FW into categories at the household level, such as cooked or uncooked, packaged or unpackaged (waste that is packaged is further divided into opened and unopened packaging), and based on why it is being disposed of (Schott et al., 2015). Lin et al., (2013) conducted an alternative thorough classification effort, and FW is classified into the following categories: fruits and vegetables leftover from organic crops, packaging, animal byproducts, mixed FW, and household garbage (Garcia et al., 2017). FW is also divided into two categories: edible FW that can be avoided and non-edible FW that cannot be avoided.

1.3 FOOD WASTE MANAGEMENT IN CIRCULAR BIOECONOMY

In the present discourse on sustainability, the effective utilization of natural resources in circular production systems is a compelling argument. To attain a greater degree of sustainability, businesses and scientists are becoming more involved in the process of valorizing agri-FW. Concepts like the "bioeconomy" and the "circular economy" (CE) are at the forefront of recent progress. The CE Action Plan (European Commission, 2014) defines the CE as the long-term preservation of materials, resources, and product value in the economy combined with a decrease in waste output (Mak et al., 2020). The SDGs of the UN, as well as pledges to minimize GHG emissions and promote sustainable consumption, also refer to the "bioeconomy" (Ferdes et al., 2022).

Concurrently, there is a push for a transition from a linear to a CE because of the increasing demand for resources and energy to meet the needs of a rapidly expanding and resource-intensive population. Waste is viewed as a secondary resource in a CE (Teigiserova et al., 2020). Including the circular bioeconomy in FSC management could be a good way to deal with waste management problems. By transforming "waste" into valuable feedstock via its regenerative cycle, the circular bioeconomy mimics the natural ecosystem and creates circular supply networks. It can be costly and time-consuming to deploy a circular bioeconomy. Furthermore, there are technological obstacles to incorporating the circular bioeconomy's restorative and regenerative concepts into current FSC systems (Kharola et al., 2022).

The section titled "environment and CE Action Plan" in the list of primary criteria centers on the idea of producing zero waste by substituting methods like composting for disposal and incineration or using alternative energy sources to ensure resource sustainability. It is a crucial component of sustainability that can be applied to the problem of FW (Kazancoglu et al., 2018). Within the developing circular bioeconomy, there are several obstacles to preventing FW and developing a sustainable system for managing FW. The first one takes into account the various definitions of FW and, as a result, explores the potential for creating consistent regulatory criteria for quantification (accounting). The inability to classify FW into relevant groups that may be connected to a corresponding waste hierarchy is the second problem. The nomenclature used in the waste hierarchy currently in use does not accurately correspond to sustainable management (Teigiserova et al., 2020).

1.4 CURRENT SCENARIO OF FOOD WASTE

Forty-six million more people experienced hunger in 2021 than in 2020, and since 2019, before the COVID-19 pandemic, 150 million more people have experienced hunger. The COVID-19 pandemic and severe climate change have harmed global food production in recent years (Zhong et al., 2020). Moreover, the global food supply has been impacted by the Russian-Ukrainian conflict (Ben Hassen & El Bilali, 2022). The susceptibility of agricultural food systems has also been brought to light by several incidents. Therefore, addressing FW is a global issue and a crucial component (Alamar et al., 2018). Because FW harms the environment and the economy, managing it is one of the major worldwide concerns of the twenty-first century. The ongoing increase in FW has resulted in a substantial shift in waste management strategies, including legislation and government policies. Rather than being disposed of in landfills, waste is now treated through reduction, recycling, and reuse, with an emphasis on energy and resource recovery from waste (Tamasiga et al., 2022).

As stated by the State of Food Security and Nutrition in the World 2022, 828 million people worldwide experienced hunger in 2021, and 29.3% of people worldwide experienced moderate to severe food insecurity (Bravi et al., 2020; UN report, 2022). In 2019, 931 MT of food was wasted worldwide according to statistics, and 17% of the food produced worldwide may go to waste. Reducing FW is crucial to achieving the SDG set forth by the United Nations. Publications about FW have been steadily increasing over the past few decades. Reducing FW, promoting sustainability, and providing balanced nutrition are ways to address concerns with food security (Galli et al., 2019).

For people, societies, and countries to be healthy, there must be access to and availability of wholesome food (Thyberg & Tonjes, 2016). The 2030 Agenda highlights the need for better waste management systems in light of the growing number of undernourished individuals and the depletion of natural resources. Studies reveal that the amount of FW produced annually in developed and developing nations is roughly 107 kg/capita and 56 kg/capita, respectively. This indicates that waste production is directly correlated with living standards (Ferdes et al., 2022).

UN SDG 12, which addresses the environmental impact that all three processes (production, processing, and distribution) have on the environment, turned its attention to food-related issues. FW by food group was analyzed, and researchers discovered that fruit and vegetables made up 24% and 22% of the waste, respectively, followed by cereals (12%), meat (11%), and oil crops (10%). The food categories that comprise the smallest segments of the FSC, fish and eggs, likewise produce the least amount of FW in absolute terms, even though a large portion of such food groups – 50% and 31%, respectively – go to waste (Caldeira et al., 2019).

The issue of FW is widespread around the world, but no one solution works for everyone. Instead, community-based approaches to FW management are developed, and area-specific actions are implemented following the characteristics of the waste, such as the varying composition of regional waste. FW can be turned into wealth if it is managed, disposed of, and recycled in a holistic manner that is tailored at the micro level based on the type of waste generated, local regulations, as well as people's attitudes and environmental consciousness (Sinha & Tripathi, 2021). From the standpoint of a CE, it is evident from ongoing trends that there is still a long way to go before meeting international standards for the disposal of FW. However, some case study examples offer hope and should be widely implemented to reduce FW and maximize its value for both environmental sustainability and economic viability.

1.5 STATISTICS AND DATA ON GLOBAL FOOD WASTE GENERATION

The worldwide human population is expected to reach over 10 billion by 2050, accompanied by a significant increase in global food consumption, crippling the world's food supply structure (Haldar et al., 2022). Every year, a massive volume of food and kitchen waste accumulates as a result of conventional FW management procedures (Sharma et al., 2021). The 1.3 billion tonnes of FW produced each year occupy approximately 28% of total agricultural land, equivalent to 1.4 billion hectares of usable cultivable space (Sharma et al., 2021; Sahoo et al., 2023). The total amount of food loss and waste in developed and developing countries is similar, at 671 million metric tons (MMT) and 623 MMT, respectively, but per capita FLW differs substantially, with developed countries having much higher FLW, implying that improved living standards may result in increased food loss and waste output (Wang et al., 2021). According to recent estimates, there is one trillion dollars' worth of FW globally annually, which not only causes financial problems but also food insecurity (Septianto et al., 2020).

FW in underdeveloped nations is thought to cost USD 310 billion a year, wasting significant amounts of energy and capital and continuously taxing natural resources like land and water (Aamir et al., 2018). Approximately 40% of FW in Europe happens during the stages of production, handling/storage, and processing, and 60% happens during the stages of distribution, retail, and consumption (Moshtaghian et al., 2021). FW is produced in large quantities in India by supermarkets, restaurants, lodging facilities, grocery stores, apartments, airport cafeterias, and the food processing sector. According to the UN Environment Program (UNEP) FW Index Report 2021 (Chaudhary et al., 2021), FW per capita in India was 90 kg per year in the high-income group, 68 kg in the middle-income group, and 63 in the low-income group.

Caldeira et al., (2019) estimate that for every 638 million tons of primary food commodities that are input, 129 million tons of FW are produced along the FSC. The majority of this waste is produced at the consumption level (46%) and is then followed by primary production (25%) and processing (24%), and distribution and retail accounts 5% of the food waste generated in the supply chain. The production of FW per capita in Pakistan is 93, 74, and 118 kg/year; in India, it is 63, 68, and 90 kg/year; in South Africa, it is 27, 30, and 45 kg/year; and in Ghana, it is 80, 86, and 86 kg/year, according to the low, medium, and high-income groups in the specific study area (Sinha & Tripathi 2021; Sahoo et al., 2023). Indian households squander the least amount of food in South Asia – 50 kilograms of food per capita per year. FW in 2019 totaled 931 million tonnes worldwide, with homes squandering the most (570 million tonnes), followed by the retail and food service industries (Pal & Bhatia, 2022). The amount of FW generated by households in India was estimated to be approximately 68,760,163 tonnes annually with a medium degree of confidence. This figure was greater than the amounts in Bhutan (60,000 tonnes/year), Bangladesh (10,618,233 tonnes/year), and Afghanistan (3,109,153 tonnes/year) (Sinha & Tripathi, 2021).

According to estimates made by Nguyen Hao et al., (2015), 1.3 billion tonnes of FW are currently disposed of in landfills each year. About 30% of food produced for human consumption is lost or wasted globally, costing the economy more than USD 900 billion. This amounts to 1.32 billion tonnes of food (Mak et al., 2020). According to Lawrence and Friel (2019), FW is predicted to cost the world $936 billion yearly. Another report estimates that 25% of the food and beverages that American families purchase are thrown away, resulting in a capital waste of $2,277 a year for a family of four (Gunders and Bloom, 2017). In the UK, 61% of the food that is purchased is thrown away by customers, while 61% of it is still appropriate for human consumption (Sharma et al., 2021). See Tables 1.1 and 1.2.

TABLE 1.1
Major Food Groups' Percentage Contributions to the Overall FW (by Weight) in Various Global Areas

Region	Cereals	Fruits	Vegetables	Dairy	Meat & Fish
East Asia & Pacific	27	11	37	2	5
Europe & Central Asia	21	11	17	12	5
Latin America	25	18	10	11	7
Middle East & N. Africa	35	17	28	3	4
North America	13	121	142	22	6
South Asia	33	17	28	6	1
Sub-Saharan Africa	142	23	20	0.5	4

TABLE 1.2
Household FW Estimates (from Measured Data Points or Extrapolation) for Various Regions (UNEP, 2021)

Region	Country	Household FW Estimate (kg/capita/year)	Household FW Estimate (tonnes/year)
Northern Africa	Algeria	91	913,918,529
	Egypt	91	9,136,941
	Sudan	97	4,162,396
Sub-Saharan Africa	Angola	100	3,169,523
	Botswana	92	211,802
	Ethiopia	92	10,327,236
	Congo	100	535,851
	Kenya	99	5,217,367
	South Africa	40	2,329,228
	Zambia	78	1,391,729
	Zimbabwe	100	1,458,564
Latin America and the Caribbean	Argentina	72	3,243,563
	Bahamas	74	28,794
	Brazil	60	12,578,308
	Chile	74	1,401,043
	Jamaica	72	213,552
	Mexico	94	11,979,364
	Paraguay	72	510,256
	Uruguay	74	255,909
Northern America	Bermuda	74	744,606
	Canada	79	2,938,321
	Greenland	74	744,178
	United States of America	59	19,359,951
Central Asia	Kazakhstan	76	1,404,584
	Kyrgyzstan	91	583,951
	Tajikistan	97	906,209
	Turkmenistan	76	449,895
	Uzbekistan	91	3,001,868
South-Eastern Asia	Cambodia	86	1,423,397
	Malaysia	91	2,921,577
	Myanmar	86	4,666,125
	Singapore	80	465,385
	Viet Nam	76	7,346,717
Southern Asia	Afghanistan	82	3,109,153
	Bangladesh	65	10,618,233
	India	50	68,760,163
	Maldives	71	37,688
	Nepal	79	2,249,412
	Pakistan	74	15,947,645
	Sri Lanka	76	1,617,738
Europe	Poland	56	2,119,455
	Ukraine	76	3,344,904
	Russian Federation	33	4,868,564

(*Continued*)

TABLE 1.2 (CONTINUED)
Household FW Estimates (from Measured Data Points or Extrapolation) for Various Regions (UNEP, 2021)

Region	Country	Household FW Estimate (kg/capita/year)	Household FW Estimate (tonnes/year)
	Denmark	81	469,449
	Finland	65	361,937
	Norway	79	423,857
	United Kingdom	77	5,199,825
	Italy	67	4,059,806
	France	85	5,522,358
	Germany	75	6,263,775
	Switzerland	72	6,167

1.6 FACTORS CONTRIBUTING TO FOOD WASTE AT DIFFERENT STAGES OF THE SUPPLY CHAIN (PRODUCTION, DISTRIBUTION, CONSUMPTION)

1.6.1 Consumer Behaviour and Food Consumption Patterns

A sizable portion of FW is caused by consumers, both directly and indirectly. For the most part, this waste could be prevented if consumers were ready to accept unacceptable food. Since most FW happens at the household level, it would be more helpful if the sources of consumer-generated FW were better understood in order to avoid FW. According to an analysis by Dung et al., (2014), the annual per capita FW in developed and developing countries is 107 kg and 56 kg, respectively. According to these figures, there is a clear difference in FW generation between developed and developing nations, with better living conditions leading to more FW creation. Low living standards are associated with lower quality expectations for food production and, thus, lower levels of FW per person. However, it is believed that the total quantity of FW in emerging nations is not significantly smaller than that in developed ones due to the influence of growing populations and increasing economic constraints (Thi et al., 2015).

According to reports, there is a positive correlation between a nation's level of prosperity and the amount of FW (Xue et al., 2017). As more people move into the middle class, they become less concerned with food to set themselves apart from those in lower socioeconomic groups (Aschemann-Witzel, 2018). FW rises in developing economies together with a spike in GDP and the expansion of the middle class (Aydin & Yildirim, 2021). The idea of sustainable consumption is gaining popularity these days, and people are choosing their food more thoughtfully (Nagarajan et al., 2022). Society's establishment of sustainable purchasing patterns was prompted by growing awareness of environmental challenges and climate change (Venghaus et al., 2022). According to Block et al., (2022), consumers frequently underestimate how much food they will consume, which is the root cause of rising FW.

Moral beliefs are a crucial component of FW behavior. FW behaviour is strongly influenced by moral values, which also have an impact on eating, shopping, and understanding food conservation. Food conservation awareness and eating habits do not directly affect FW; however, shopping behaviors are found to be a significant predictor of FW behavior. Aydin and Yildirim (2021) in their study reported the relationship between consumers' purchasing, eating, and eating habits, as well as their moral attitudes and knowledge about food conservation, and their behavior related to FW. They collected data from 328 people in Turkey surveyed using structural equation modeling. The study's findings showed that moral beliefs directly influence eating, purchasing, and FW behaviors as well as awareness of food conservation. Additionally, it has been demonstrated that FW behavior is significantly influenced by buying patterns.

Porpino et al., (2016) state that a key factor in understanding food provision and FW is abundance. Large amounts and overpreparation of food are widespread practices in various cultures, such as Latin American culture. The degree to which consumers understand proper food handling practices influences the amount of FW they produce (Aschemann-Witzel et al., 2015). There is less waste produced by customers who are more informed about the ideal storage conditions for food (Bravi et al., 2019). According to Pearson et al., (2013), one of the main causes of FW is a lack of awareness and understanding regarding food safety. McCarthy and Liu (2017) claim that consumers who are more concerned with sustainability and who experience guilt and anxiety when they toss food away waste less food. The majority of FW research involves consumer purchasing routines, which can be a significant driver in identifying their FW behavior (Schanes et al., 2018). According to multiple studies, shoppers who create shopping lists and only buy what they need waste less food (Bravi et al., 2019). Notably, the COVID-19 pandemic has severely affected food purchasing decisions, consumption, and management by consumers, which has decreased household waste (Scacchi et al., 2021). Almost twice as many consumers as before the pandemic, 81%, now make lists before shopping and check the product's expiration date (Nicewicz & Bilska, 2021). Even though purchases increased during the COVID-19 pandemic, there has been a decrease in the number of spoiled and tossed products due to increasing awareness about FW and its effects on the environment (Šimanskienė et al., 2022).

1.6.2 Role of Food Industry Practices and Regulations in Waste Generation

One of the biggest sectors in the world, the food industry plays an important role in the economies of many countries. Food processing industries are considered to be rapidly expanding sectors due to the increase in the need for food to prevent hunger and the need for nourishing food to support the development of a healthy country. However, in the next 50 years, there will be a significant rise in food production due to world population growth and demands on the FSC. Due to these factors, there is a growing national and worldwide interest in the social, political, and scientific aspects of the large amounts of food sector waste (Otles et al., 2015). There is a lot of FW (dry or wet) produced by the food industry, and this is gaining attention due to its effects on the environment, the economy, and the growing concerns about food arising from the world's population growth, demographic shifts, and the impact of climate change on food production (Amicarelli & Bux, 2021).

"Food industrial wastes" often refer to portions of leftovers created as a result of food manufacturing procedures (Figure 1.2). They are made up of a variety of organic materials that are thrown away after manufacture. Industries that process food, generate a variety of wastes as fruit and vegetables, meat, oil, dairy industry, etc. It is a possible substrate for the bioprocessing of many different materials. These materials, which have uses in both the industrial and medicinal sectors, include pigments, acids, enzymes, biofuels, bioactive agents, and biopolymers (Sarker et al., 2024). Nowadays, around one-third of fresh crop is squandered during the food processing chain and is dumped in landfills and the ocean (Kaur et al., 2023). However, it could be utilized to produce various value-added products.

In the food industry, a significant quantity of waste is produced during the receiving, processing, and packing stages in both solid and liquid forms. FWs can be used as food, feed, or fodder because they are extremely rich in nutrients and energy sources. The food processing industries have disposed of a significant volume of FW continually throughout the past few decades. According to the FAO, 11–23%, 17–19%, 8–17%, and >50% of FW is attributed to retail, industrial processing, agriculture, and the final consuming phases (Amicarelli & Bux, 2021). Fourteen million metric tons of waste are produced by the food processing sectors, according to reports (Narasimmalu & Ramasamy, 2020). Food loss and waste were 34%, 13%, and 4%, respectively, at Canada's primary processing, secondary processing (additional processing), and distribution stages. About 30–35% of whole fruits produced by the fruit industry are considered FW; however, the reported FW for durian was 60% (Gowman et al., 2019).

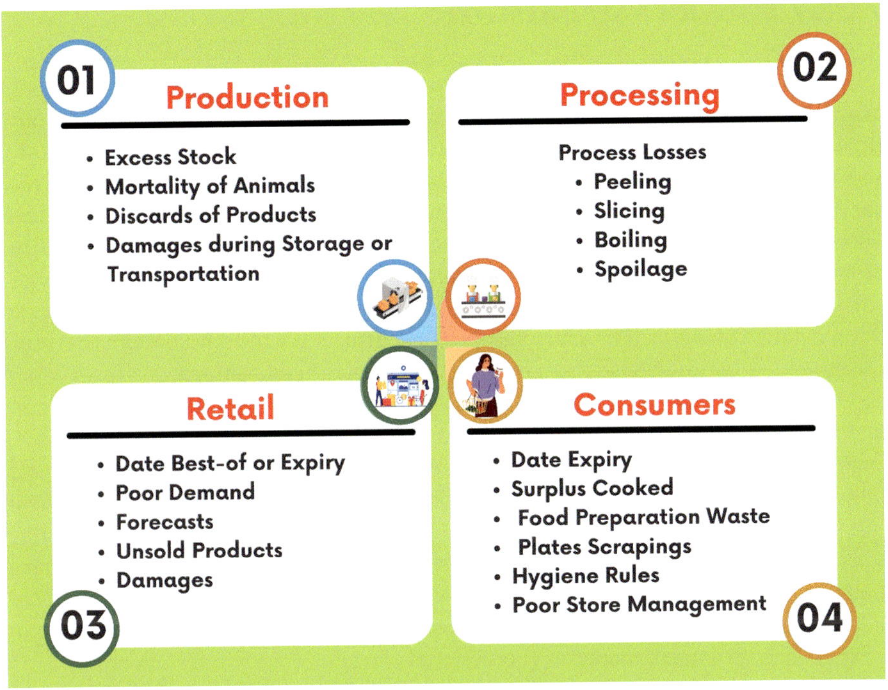

FIGURE 1.2 Food waste situations at various stages of the production chain.

As per the 2019 report by the FAO, the Philippines, USA, India, and China are the top global producers of FW (FAO, 2019). International regulatory frameworks, national laws, and directives about waste management designate waste prevention/minimization and by-product valuation as critical tactics for enhancing food security and establishing an efficient management system within the food industry (Otles et al., 2015). Numerous studies indicate that certain technologies can result in an efficient reduction of waste. Installing cold chain equipment, for instance, close to farms in isolated locations might save the stocks of perishable goods, such as fruits and vegetables, which may be in excess during particular seasons (Negi & Anand, 2015).

Emissions of damaging GHGs result from waste. Specifically, 26% of GHG emissions worldwide are attributed to the food industry, with FW accounting for 6% of this total. FW decomposes in landfills under anaerobic circumstances, producing methane gas (CH_4), which has a global warming tendency 25 times greater than that of CO_2 (Martin-Rios et al., 2020). The process of separating desirable goods from unwanted by-products creates waste in the food processing industry (Kwan et al., 2018).

Utilizing food-processing waste to create high-value bioproducts has been the subject of extensive research efforts in recent years. Eliminating or drastically reducing FW has been a primary priority for governments and the food industry. In a CE, waste is valued when it is turned into commodities like chemicals, materials, and fuels. It enhances the economy, society, and environment in addition to creating value-added products like organic fertilizers, animal feed, biofuels, and power from FW (Sarker et al., 2024). Sustainable bioconversion of FW into marketable goods reduces the annoyances caused by FW decomposition in landfills and the surrounding environment in addition to providing economic benefits (Bilal et al., 2019). Value addition is essential to the CE and the control of FW. Developments in integrated biological processes for the simultaneous recovery of resources and production of energy with zero solid waste disposal should offer a promising future for the management of waste from the food processing sector. Increased regulation in conjunction with incentive schemes such as tax breaks for individuals, businesses, industries, or organizations will be more important in lowering carbon credits through FW valorization (Sinha & Tripathi, 2021).

1.7 GLOBAL POLICIES AND INITIATIVES

1.7.1 International Efforts to Tackle Food Waste

Combating FW is a global issue, and numerous nations have implemented laws, procedures, and policies to address the issue. Reducing FW is mostly dependent on policies (Fesenfeld et al., 2022). Reducing FW would help with interconnected sustainability issues like limits of natural resources, food security, and climate change. Accordingly, one of the crucial concerns pertaining to sustainable development is creating a suitable plan for decreasing FLW (Ishangulyyev et al., 2019). Many strategies have been proposed to decrease or prevent FW, with many of them already being implemented. To determine which initiatives provide the best opportunities and which activities are most promising, a detailed study of FW interventions is necessary. It's becoming more and more obvious that minimizing the food systems' numerous detrimental environmental effects is necessary to guarantee that there will be enough food to sustainably feed the world's expanding population (Thyberg and Tonjes, 2016).

It is imperative and necessary for food systems to transition to more sustainable ones. Specifically, demand must be moderated, more food must be produced, governance must be strengthened, and waste must be reduced. One of the 17 objectives to be accomplished by 2030 is waste management, which is a task that can contribute to environmental conservation. Ineffective waste management has seriously damaged the ecology and harmed human health. Waste production is predicted to rise in tandem with economic expansion and population growth; lower-middle-income nations are most likely to experience the largest increases (Ferdes et al., 2022).

One of the main objectives of UN-SDG 12, "Ensure sustainable consumption and production patterns," addresses the problem of FW. Similar to other products, food has an influence on the environment during its manufacturing, processing, and distribution stages, as well as during the "use phase," which is when it is consumed at home (Grosso & Falasconi, 2018). Food loss and waste management are strongly tied to the 12th goal, which deals with "Responsible Production and Consumption." SDG 12 has set ambitious goals, such as implementing a ten-year framework of sustainable consumption and production initiatives aimed at reducing food losses throughout supply chains, including post-harvest losses, and halving global food waste per person at the retail and consumer levels by 2030 (Grosso & Falasconi, 2018). Minimizing food loss and waste also helps achieve other SDGs, such as SDG-15 on biodiversity and terrestrial ecosystems, SDG-13 on climate change, SDG-2 on zero hunger, and SDG-6 on sustainable water management (Jaouhari et al., 2023). The UN (FAO) provides several important insights about the state of FW overall. The differences between industrialized and emerging economies are fascinating (Ferdes et al., 2022). Policymakers will undoubtedly be drawn to the laborious and demanding process of cumulatively estimating the capital loss suffered due to FW, which will prompt stricter legislation aimed at reducing FW worldwide (Sharma et al., 2021).

As the Paris Agreement and the 2030 SDG are being implemented, policymakers, practitioners, and researchers have come to realize how urgent it is to address the problem of FW in emerging economies' urban areas. For instance, in France, retailers are required by law to donate food that would otherwise go bad. This rule was passed in 2016 (de Los Mozos et al., 2020). The food recovery hierarchy was unveiled by the US Environmental Protection Agency (USEPA) in 2015. To reduce FW, several priority tiers were put forth, including source reduction, industrial usage, feeding the hungry, feeding animals, composting as a substitute, and incineration and landfill disposal. This food recovery hierarchy was established by the EPA, which addressed the main problem before addressing other options. Since prevention is where waste occurs most frequently, such as from residential sources, it is ranked first in the first tier of source reduction (Zamri et al., 2020).

1.7.2 Case Studies of Countries with Successful Food Waste Reduction Policies

The FLW behavior during the COVID-19 outbreak in Spain was subjected to a multivariable analysis (Aldaco et al., 2020). Variations in eating habits over a short period affect patterns of food loss and waste. The methodology, which used Spain as a case study, comprised a life cycle thinking strategy that included an economic and thorough nutritional assessment, a material flow analysis of the supply chain, and an examination of the inputs and outputs of the Spanish food basket. Based on the data, it could be concluded that during the initial weeks of the COVID-19 lockdown, there was a partial reallocation of extra-domestic consumption to households (a 12% rise in household food loss and waste) but no substantial change in overall food loss and waste generation. To round out the multivariate impact profile of the COVID-19 epidemic on food loss and waste generation and management, the economic impact (+11%), GHG emissions (+10%), and nutritional content (–8%) are included. As a result, this study emphasizes once more how important it is to take action to reduce FW, especially in the household sector, to better utilize food surpluses, prevent and control food loss and waste, and prepare for unforeseen future events.

A novel concept called "Farm to Farm" has been created in an attempt to lessen the quantity of FW produced by restaurants in India, which is estimated to be worth nearly 10 billion USD annually and amounts to over 67 million metric tonnes (Dabas et al., 2021). The city of Hyderabad, which generates 400–500 tonnes of FW daily, was the study's location. It was anticipated that by 2022, 6,000 MT of waste would be handled daily in 8–10 cities around India. The system was designed to process 10 tonnes of FW per day or 300 tonnes of FW per month. The released gas will be improved, bottled, and sold as compressed biogas – a quarter less expensive than regular LPG used for cooking in Indian homes. A discounted rate of manure can be offered to farmers to increase soil fertility, as the same crop that was grown on the farm is now being cultivated on another farm.

This method involves gathering food scraps from different lodging establishments and turning them into nutrient-rich organic biofertilizers through fermentation. Because restaurant food is cooked with grease and spices, the leftover food is indigestible. Pilot research involved incubating food with microbial consortia to convert it to organic manure and gas. This system was upgraded to handle one ton of FW every day, after receiving a government subsidy, from handling 25 kg of garbage each day.

From a life cycle viewpoint, Ogunmoroti et al., (2022) evaluated the management of FW, its present and future conditions, and the related environmental effects in all Chinese provinces. Based on the findings, FW in China was projected to be 61.37 million tons in 2020 and, depending on different socioeconomic scenarios, may reach 100.95–139.82 million tons by 2040. Tianjin was determined to have the largest per capita FW, with an average rise of 29.67% from 2016. Accordingly, FW-related GHG emissions in China were projected to reach 149.25 $MTCO_2$-eq. in 2020, with a significant influence from the types of FW in individual provinces. Different socioeconomic growth routes are expected to result in a 70%–148% rise in the environmental burden of FW. By shifting FW to anaerobic digestion, GHG emissions from FW can be cut by 74% by 2040. If many Chinese provinces carry on with their current FW management tendency, China's transition to sustainable waste management may be hindered and unsustainable economically. These results may prove crucial in establishing objectives for reducing FW at the provincial level and formulating plans to guarantee community participation in such initiatives.

In a lab setting, the feasibility of producing bioethanol from FW produced in Greater Noida, Uttar Pradesh, was evaluated. Food trash was gathered and shredded for the generation of bioethanol after drying (Thapa et al., 2019). Studies on fermentation and hydrolysis were carried out in order, with control variables. A correlation was found between the development of yeast and an ethanol output of 13.78 g/100 g of dry FW. It is theoretically possible to manufacture 329,756 L of bioethanol per day from municipal solid waste (MSW) in Delhi. It was shown that the organic component of MSW may be successfully converted to ethanol by controlled fermentation with the

Saccharomyces cerevisiae strain. This is a financially and environmentally advantageous alternative for valorization (Sahoo et al., 2023).

1.7.2.1 Food Waste Management Policies

The worldwide problem of managing FW has long been a source of concern for both industrialized and developing nations. There is an enormous amount of FW produced as a result of numerous socioeconomic reasons (Mu'azu et al., 2019). Climate change, environmental degradation, and economic difficulties associated with FW generation and disposal are addressed through a variety of initiatives and legislation. Taking into account the first level of decision-making, several initiatives, such as the SDGs, the European Green Deal, the CE Action Plan, and the Farm-to-Fork Strategy, were previously adopted in response to the issue. Furthermore, enterprises must engage voluntarily, make financial expenditures, and participate spontaneously to minimize FW or turn it into byproducts (Jaouhari et al., 2023). After realizing the benefits of converting FW into value-added materials (bioplastics and bioactive compounds) and energy-releasing products (biogas, biodiesel, and electricity), the French government adopted a policy of FW valorization and created punitive laws to ensure strict adherence by the public. This policy's implementation prevented an estimated 88 million tons of FW from decomposing, saving US $167 billion annually (Tsang et al., 2019; Sharma et al., 2021).

There is an obvious pattern on a worldwide scale. While food losses in developing nations are greater throughout the production and post-harvest stages, in affluent countries the largest single contribution comes from the consumption phase, with much of that occurring at the family level (FAO, 2014). The existing statistics in developed nations like Japan (Liu et al., 2016) and the EU member states (Stenmarck et al., 2016) identify the FW that occurs throughout several stages of the FSC, including manufacturing, processing, storage and transit, commercialization, and residential settings.

Initiatives from the UK to reduce FW can be an inspiration on a political level. In comparison to 2007, the UK was able to reduce household FW by approximately 1.44 million tonnes in 2018. As a result, private home FW dropped from 132 kg/(cap/year) to roughly 100 kg/(cap/year). The UK's Waste and Resources Action Program (WRAP) was founded to promote recycling and sustainable waste management. WRAP started the Love Food Hate Waste campaign in 2007 as a follow-up to their recycling activities and to raise public awareness of FW issues. The initiative aimed to raise consumer awareness about FW and motivate them to decrease it (Leverenz et al., 2021).

Industry consensus on reducing FW in Norway: In 2015, the Norwegian government and the food industry agreed to gradually reduce FW, with a target of 15% by 2020, 30% by 2025, and 50% by 2030. This was founded on the nation's pledge to adhere to the UN's SDG 12.3. As part of this plan, the industry must assess its standing, submit reports, carry out cost-cutting initiatives, and distribute extra food to charitable groups. In the meantime, the government is creating mechanisms to monitor the procedure, evaluate reports, and enable the contribution of excess inventory. Every year, at least one meeting is held by the parties involved to evaluate the progress made (Ananno et al., 2021).

The pay-as-you-throw mandate in South Korea: Among the nations with the strictest laws against FW is South Korea. The Waste Management Law was the first step taken by the government in 1986 to implement the 3R (reduce, reuse, recycle) program. The "polluter pay" concept was included in the nation's legal system via the 1992 Act on Resource Saving and Recycling Promotion. In 1995, a national volume-based trash system was implemented, and in 1997, it became mandatory to separate FW. FW recycling as animal feed or raw materials for fertilizers became required in 2005 when the nation outlawed the disposal of FW in landfills (Nguyen et al., 2017). The South Korean government implemented a pay-as-you-throw policy for FW in 2013. The policy required citizens to separate FW into biodegradable bags and dispose of it in RFID-enabled automated bins that weigh the quantity of trash disposed of (Sheldon, 2020). The amount that each home is fined depends on how much food is wasted. This strategy has worked well to encourage responsible consumption (Nguyen et al., 2020). Only 2% of the nation's FW was recycled in 1995; by 2019, that percentage had skyrocketed to 95% (Sheldon, 2020).

1.7.2.2 Analysis of Legislative Frameworks, Incentives, and Strategies

Reducing FW at the source, preventing overproduction, effectively distributing food supplies, and measuring portions correctly are only a few of the tactics covered by the policies designed to solve the problems associated with FW. Both domestic and global regulatory frameworks point out that preventing and minimizing waste and maximizing the value of byproducts are essential tactics for an efficient management system that supports the food industry's sustainability (Despoudi et al., 2021). By implementing strict rules, raising public awareness and engagement, and utilizing cutting-edge technical solutions that are currently accessible, several European nations have successfully reduced FW (Joshi & Visvanathan, 2019).

Various nations have used distinct approaches to tackle the problem of managing FW. These tactics encourage policies that avoid FW, encourage the segregation of FW, and ensure that it is disposed of appropriately to facilitate interventions across several phases. Certain nations have created applications for smartphones that notify users when food items are about to expire and may be bought at a discount. Also, policies such as using leftover food from hotels, restaurants, and department shops must be redistributed for human consumption rather than being disposed of as garbage in nations including Denmark, Norway, France, and Sweden (Principato & Principato, 2018). FW in Denmark decreased by 25% between 2010 and 2015, according to FUSION partner Stop Wasting Food Movement Denmark, the largest NGO addressing FW. Similarly, the voluntary effort was responsible for the UK's WRAP reporting a 21% countrywide reduction in avoidable FW (home FW decreased by 15%) between 2007 and 2012 (Liu et al., 2016).

Hong Kong sought to reduce its FW from 3600 t/d in 2014 to 2160 t/d in 2022 by boosting FW separation, recycling and treating separated FW, and treating non-separated FW before final disposal. To do this, it has started initiatives like the "Food Wise Hong Kong Campaign," which shares industry and commercial best practices to collaborate with government agencies, educational institutions, and non-profit organizations to prevent and minimize FW at the source. The "Green Lunch Charter" was developed as a result of collaboration between the public, academic institutions, and the Education Bureau of Hong Kong. It encourages schools to limit FW and use disposable lunch boxes (Joshi and Visvanathan, 2019).

The Municipal Solid Wastes (Management and Handling) Rule in India was first implemented in 2000 and then amended in 2013 and 2015. The rule emphasizes that waste must be sorted into three streams: hazardous, non-hazardous, and biodegradable, and it must be placed in color-coded bags before being given to collectors (Ong et al., 2018). Additionally, businesses that establish collection centers at farms and food retail establishments receive financial support. Mega Food Parks have also been developed to connect the food and agriculture sectors in order to maximize value and reduce waste. In addition to national development goals and policies, the country has committed to a number of international development goals, which have encouraged, if not required, it to create several FW management policies. For instance, given that landfills account for between 11 and 13% of global greenhouse gas emissions, the nations that have ratified the Kyoto Protocol are more likely to divert FW from them (Awasthi et al., 2019).

To combat the issue, European cities have collaborated closely to share best practices and lessons learned. They have established significant networks, such as the EUROCITIES Working Group on FW, and implemented programs and mechanisms to address it, including market incentives, awareness-raising campaigns, food donations, and integrated actions that also help create jobs, biofertilizers for urban agriculture, and renewable energy (Magarini et al., 2018).

1.8 CIRCULAR BIOECONOMY AND FOOD WASTE

The second decade of this century saw a rise in the use of the term "circular bioeconomy." It is the belief of certain scientists worldwide that CE and bioeconomy are two ideas that are similar in nature. Although the two ideas might support one another, there are several ways in which they might disagree. The bioeconomy goes beyond CE's primary focus of "efficient use of resources to increase end-of-life." The cascading use of resources is still absent from the bioeconomy, despite its wide variety of objectives. As a result, the term "circular bioeconomy" has been coined to describe a novel idea for the cascading use of bioresources and the valuation of biowaste (Negi & Kumar, 2021).

The main objective of the CE is to displace the prevalent linear economics model that has been in use for decades, which extracts raw materials from the environment to be processed into finished goods and discarded when their useful lives are over (Urbinati et al., 2017). The shift to degrowth – a process that combines reduced resource usage with greater well-being – seems imminent, especially in industrialized economies that have outgrown their ecological bounds. In this regard, the CE may be crucial (Fassio et al., 2023). The creation of a circular bioeconomy and boosting the value of material flows are receiving more and more attention in research. The goal of the circular bioeconomy is to reduce greenhouse gas emissions while achieving sustainable production and consumption (Mak et al., 2020). The growing tendency of FW footprints necessitates the creation of appropriate FW management plans to maintain environmental sustainability. FW is a multifaceted waste stream that affects the environment and the world economy. FW is a non-consumable, zero-value resource. In this case, using well-structured and effective valorization technologies to convert FW into various biofuels, such as biodiesel, bioethanol, biohydrogen, bio-oil, biochar, and biomethane, can be a desirable and workable strategy to combat the current global energy crisis and build a sustainable bioeconomy (Dhiman & Mukherjee, 2021). For the management of FW, the CE offers an appealing reference framework (Teigiserova et al., 2020). Stated differently, the CE paradigm's goals are supported by FW valorization alternatives like reduction, reuse, and recycling (Geueke et al., 2018). Like waste hierarchies in general, FW hierarchies offer practitioners and policymakers invaluable input on potential measures to be implemented.

Sustainable development based on a CE today has a prominent place on the international agenda. The CE is based on the principles of recycling, resource efficiency, waste minimization, and value maximization. Waste from the food sector is an untapped resource that can be turned into goods that are beneficial in the CE (Bilal & Iqbal, 2019). The fundamental ideas behind the CE are complementary to those of the bioeconomy and should make it easier to recycle and reuse materials that are used to build integrated sustainable approaches that elaborate on holistic resource utilization. It is anticipated that the transition to a circular bioeconomy would also provide a clear understanding of the importance of organizational, social, and technological innovation in fostering economic growth, job creation, and restoring competitiveness (Maina et al., 2017).

The CE urgently needs to replace the linear economy (based on fossil fuels), which calls for the creation of materials and chemicals using sustainable practices as well as sustainable resources. Here, FW is viewed as a possible feedstock for the circular bioeconomy's central tenet – the sustainable synthesis of minerals and chemicals. FW, therefore, has a tremendous amount of potential to strengthen the bioeconomy. The ability to turn FW into a wide range of goods, including biofuels, platform chemicals, biopolymers, proteins, enzymes, biofertilizers, and other bio-based materials and compounds, can guarantee production sustainability and address environmental concerns (Tsegaye et al., 2021).

FW typically contains too much moisture to be burned or treated thermally, and because it produces leachate and emits methane into the atmosphere when it is dumped in landfills, it can pose environmental risks (Mirabella et al., 2014). Consequently, this enormous number of bio-based products indicates a significant potential contribution from the bioeconomy to the CE, which may not only optimize resource utilization but also aid in addressing the issue of managing FW. As the circular use of bio-based resources gains momentum, numerous studies investigating the possibilities for circularity about FW and by-products have come to light. It is currently not possible to implement every solution that has been analyzed at an industrial scale (Girotto et al., 2015).

More and more people are becoming interested in the valorization of FW for the creation of bioethanol, biogas, hydrogen, and biodiesel (Kiran et al., 2014; Corrado and Sala, 2018).

Consumption-stage FW and processing waste can both be utilized to produce biofuels (Pham et al., 2015) or as supplies of nutrients for biotechnological processes (Pleissner et al., 2016). Future industrial advancements centered on the valorization of FW have to be grounded in a thorough accounting of the amounts that are already accessible, a techno-economic potential analysis, and a lifecycle-based environmental assessment of the benefits and burdens (Corrado & Sala, 2018).

1.9 FOOD WASTE MANAGEMENT

The goal of FW valorization methods is to transform FW into products with added value that can be included in different phases of the FSC. The following are some environmentally friendly methods for valorizing FW: recycling, composting, bioconversion, and upcycling (Tchonkouang et al., 2023). Recycling is the process of utilizing "food scraps," "organic resources," or "materials" that are unfit for human consumption. The FW hierarchy lists anaerobic digestion for energy production, animal feed, industrial uses such as chemicals or cosmetics, and composting as the four most environmentally friendly ways to use scraps.

Foods made from materials or products that would otherwise be regarded as waste or byproducts are often referred to as "upcycled" foods. As compared to alternatives like animal feed or composting, this strategy, which turns excess materials into value-added food items, presents a viable way to tackle FW. One of the newest concepts in the toolbox of techniques for decreasing and valuing FW is upcycled food (Tchonkouang et al., 2023). Foods that have been upcycled are made utilizing ingredients that are often not meant for human consumption, such as leftovers and waste from food processing and preparation, underutilized or damaged food components, residues, and byproducts. Therefore, by transforming waste materials and byproducts into useful and nutritious ingredients that can be reintroduced into the FSC as parts of new food items, upcycling increases the value of the waste material and byproducts. Foods and ingredients that have been recycled are frequently referred to as value-added excess items or waste-to-value food products (Moshtaghian et al., 2021).

By bioconverting FW, a wide range of value-added products are produced, including industrial enzymes, high fructose syrup, docosahexaenoic acid, pigments, vermicompost, activated carbon adsorbents, antioxidants, single-cell protein, bioactive molecules, and liquid fuels like biobutanol, bioethanol, and biodiesel. Additionally, bioelectricity, nutraceuticals, biogas, and biopolymers are produced, benefiting the larger food ecosystem (Bangar et al., 2023). A few manufacturers of repurposed foods and ingredients are putting their final goods on the market. Coffee Cherry Company (US) and Kaffe Bueno (Denmark) are two examples of businesses and goods that employ coffee grounds to make baked goods. While Planetarians (US) and NapiFeryn BioTech (Poland) turn the leftovers from refining sunflower or rapeseed oil into food-grade protein-rich powders, Remashed (UK) and Regrained (US) upcycle discarded grains from beer production to make granola bars (Grasso & Asioli, 2020). To make it simpler for manufacturers to produce these meals, the Upcycled Food Association launched a certification program for upcycled ingredients and foods (Moshtaghian et al., 2021). These foods provide an enticing solution to mitigate FW in terms of both the economy and the ecology.

Biotransformation, also known as bioconversion, is the process of using biological processes or agents, such as enzymes and microbial species, to change organic resources, like biomass from animal or vegetative waste, into marketable goods or renewable energy sources (Maurya et al., 2021). By converting FWs into value-added products like feed and food additives, animal feeds, enzymes, food-grade pigments, bio-based fuels, single-cell protein (SCP), fertilizers, and other helpful biochemicals or bioproducts, bioconversion of FWs promotes sustainable development and increases food security (Tropea, 2022). Many benefits arise with bioconversion, such as high biomaterial and energy recovery value, less landfill space, cheaper technology, and higher farmer revenue from the sale of horticultural and agricultural waste (Hashempour-Baltork et al., 2022).

1.9.1 Emerging Technologies

FW sustainability is an important field of study that has expanded rapidly in the last several years. There are many instances of research looking for sustainable ways to manage FW, but these studies have typically focused on one aspect of sustainability: the implications for the economy, environment, or society (Garcia et al., 2017). The field of FW research is always coming up with new methods, tools, strategies, and control frameworks to streamline processes. Furthermore, to achieve technical innovation, life cycle assessment, and sustainable development, it is desirable to have an improved technique for disposing of FW and to have integrated and sustainable ways for FW disposal (Ferdes et al., 2022). It has been discovered that the development of sophisticated methods for FW valuation is time- and environmentally beneficial. It also makes a substantial contribution to the circular bioeconomy and removes the mountains of decomposing waste sustainably (Sharma et al., 2021).

1.9.1.1 Bioreactor-Assisted Valorization

According to Abreu et al., (2019), the microbial-assisted valorization of organic waste streams into value-added products is of great importance because it greatly contributes to the circular bioeconomy and the sustainable management of FW. A possible opportunity that could turn FW into a profit is microbial cultivation in a bioreactor. According to Pino et al., (2018), a bioreactor is a mechanically stirred container or vessel that provides microbial cells with the biochemical and biomechanical requirements they need in a biologically active environment. This allows the microorganisms to grow at their optimal rate at the pilot scale in a controlled manner, resulting in a high yield (Sharma et al., 2021). Given the right pH, temperature, humidity, and water activity, FW can serve as a substrate for microorganisms to grow to their full potential and produce secondary metabolites (Singhania et al., 2009). A 45L continuous mode anaerobic bioreactor operating at 35°C and requiring around 40 days of hydraulic retention time was utilized to produce biogas from industrial coffee waste. Following the fermentation of coffee waste, cellulose and lignin were hydrolyzed to create roughly 18 NL/L of biogas with an approximate 80% v/v methane content (Battista et al., 2016).

1.9.1.2 Ultrasound-Assisted Extraction (UAE)

FW can be extracted using ultrasound technology to create high-end products while also lowering environmental impact. Ultrasound-assisted extraction (UAE) creates cavitation bubbles inside the material's cell wall through the use of ultrasonic waves. Cell wall destruction is caused by the production and collision of these massive bubbles, which also enhances the extraction process' mass transfer (Sharma et al., 2021). It has been shown that tobacco trash is a rich source of bioactive chemicals and antioxidants that would have been lost without the use of cutting-edge technologies. From the leaf and the tobacco waste, phenolic chemicals and solanesol were found, specifically rutin (93.7 and 11.56 µg/mL), caffeic acid (10.8 and 2.34 µg/mL), solanesol (598.9 and 294.9 µg/mL), and chlorogenic acid (804.2 and 3.64 µg/mL) (Banožić et al., 2019). Similar to the conventional extraction method, at comparable temperatures (60°C) and pH concentrations, ultrasound treatment is also said to produce pectin from tomato waste that is of higher quality in terms of its methoxy content, calcium pectate content, hydrouronic acid content, and degree of esterification (Grassino et al., 2016).

1.9.1.3 Microwave-Assisted Extraction (MAE)

Ionic conduction and internal molecular dipolar rotation cause heat to be produced inside the material during microwave-assisted extraction (MAE). The heat produced causes the cell wall to lyse, which facilitates the movement of bioactive chemicals from the area around the cell wall into the extraction system (Chuyen et al., 2018). Because MAE has a shorter extraction period than other methods, it has been used in a variety of waste streams to extract a wide range of bioactive chemicals. Garrido et al., (2019) used MAE to extract antioxidants and phenolic compounds from Chardonnay grape marc, utilizing 1.77 g of solid mass and 48% ethanol as the solvent for a 10-minute extraction period. Procyanidins, epicatechin, and catechins made up the majority of the 1.21 ± 0.04 mg galacturonic acid equivalent/mL total phenolic yield that was achieved. Using a 1:10 liquid-solid ratio and 80% (v/v) ethanol solvent, the lemon essential oil was extracted using microwave-assisted hydrodistillation in conjunction with a Clevenger device. Any leftover residue was then treated with MAE to extract lemon pigment at 80°C for 50 minutes. According to Martínez-Abad et al., (2020), the treatment resulted in the total production of 2 weight percent lemon essential oil and 6 weight percent lemon pigment.

1.9.1.4 Enzyme Immobilization-Assisted Valorization

FW serves as a substrate for the enzymatic conversion of important substances, including biodiesel, biofuel, pectin, oligosaccharides, food stabilizers, prebiotics, and biosurfactants, because it is rich in organic content (Sharma et al., 2021). Enzyme catalysts are employed these days to produce bioproducts. Enzyme catalysts provide several benefits over chemical catalysts, including high selectivity, low energy requirements, and catalytic power. Enzymatic conversion of FW into products like biodiesel, biofuel, sweeteners (higher sweetening index but lower calorie value), food supplements, prebiotics, thickeners, nutraceuticals, antioxidants, stabilizers, food colorants, and fat replacers has been achieved through processes like hydrolysis, deamination, oxidation, phosphorylation, acylation, glycosylation, and esterification (Chen et al., 2019).

1.9.2 Role of Data Analytics, IoT, and AI in Optimizing Supply Chains and Reducing Waste

The rise in disruptive events from both natural and man-made sources is making supply networks more vulnerable (Ali et al., 2018). Due to the perishable nature of food, FSCs are more intricate than other manufacturing/service supply chains (Ali and Nakade, 2017). To increase food security and slow down climate change, food loss and waste must be continuously monitored. Real-time sensors are a technology that can continually monitor food quality and assist in reducing FW by sensing quality characteristics like temperature and humidity. Systems for easily monitoring and maintaining food freshness and safety have to be developed in light of the expanding food industry and the rising demand for long-term food preservation (Mustafa & Andreescu, 2018).

Food and nutrition security, climate-controlled unpredictability, FW, power and governance concerns, and value distribution within FSCs are just a few of the many obstacles that FSCs have had to deal with over the years (Gokarn & Kuthambalayan, 2017; Ali et al., 2019). Digital tools have emerged as a practical means of preventing food loss and waste (Jagtap and Rahimifard, 2019). Digital tools like sensors, mobile barcode scanners, wireless laser scanners, global positioning systems, blockchain, and other information-sensing equipment can help achieve intelligent identification, tracking, monitoring, and management (Bux et al., 2022). These technologies can monitor many product kinds, including meat, milk, and other food goods, continually and can impact food loss and waste within the larger context of food security (Irani et al., 2018; da Costa et al., 2022).

The term "Internet of Things" (IoT) refers to scenarios where computing capability and network connectivity extend to objects, sensors, and everyday items not typically considered computers, allowing these devices to generate, exchange, and consume data with minimal human intervention, according to a 2015 report from the Internet Society (ISOC) (Rose et al., 2015). "RFID, Wireless Sensor Network (WSN), Long Term Evolution (LTE), Long Range (LoRa), Zensys Wave (Z-Wave), IEEE 802.15.4, Near-field communication (NFC), Machine to Machine (M2M), Ultra-wideband (UWB), and IPv6 Low-power Wireless Personal Area Network (6LoWPAN), etc." are some of the different types of Internet of Things devices (Yadav et al., 2022). To reduce FW, Jagtap and Rahimifard (2019) deployed IoT in the UK food manufacturing sector to track real-time FW data. Four distinct use cases in the context of the UK were used by the authors to demonstrate the uses of their suggested methodology.

Managing food nearing its expiration date, keeping an eye on food quality, and optimizing inventory and store design are just a few ways that the implementation of the Internet of Things (IoT) can help players minimize food loss and waste. The proper physical environment, particularly concerning temperature and humidity, can be provided using sensor technologies to help lessen food loss and waste (Kamble et al., 2019). The CE paradigm, which asserts that socially radical and technologically advanced solutions can build effective circular patterns, relies heavily on technology (Ghisellini et al., 2019). This is particularly true when it comes to converting the food industry's CE frameworks into concrete activities, which calls for technological help. Technology's ability to prevent and manage excess food is acknowledged as being essential to improving the FW hierarchy. As a result, it provides cutting-edge solutions to assist businesses in the highest and most important tiers of the hierarchy. Through online platforms or apps, information technologies can make it easier to share and redistribute food (Harvey et al., 2019). Furthermore, once surplus food is generated, strategies for managing it well are typically associated with transformation technologies, as these technologies deal with processing the surplus food to produce energy, fertilizers, or animal feed (Arshadi et al., 2016; Ciccullo et al., 2021).

Food businesses now have new chances to cut waste in their supply chains thanks to modern digital technology, sometimes known as Industry 4.0 technologies (da Costa et al., 2022). For instance, Internet of Things sensors can monitor humidity, temperature, and other pertinent factors in the food's storage area while it is being moved on trucks. The information is then sent to cloud storage for remote access. Decision-makers who remotely monitor the temperature can quickly adjust it if the food is kept in less-than-ideal conditions (for instance, if the storage temperature is above the maximum threshold). This will lessen the likelihood that the food will go to waste (Ramanathan et al., 2023).

Different digital technologies have been developed in the last few years to improve the efficiency of different FSC phases (Aramyan et al., 2021). Sensors that are connected to the Internet of Things (IoT) can monitor temperature and other factors to keep food fresher for longer. Owners of food can receive timely warning messages when sensor outputs are connected to the cloud and monitored through smartphone-based applications. If temperature control systems malfunction, this kind of warning system can prompt quick action to minimize food wastage. Real-time information sharing using data from IoT networks, optimizing food delivery locations based on real-time food quality monitoring, and extending food shelf life through real-time cold-chain monitoring and control are additional technological approaches to help reduce FW (Ramanathan et al., 2023).

1.9.3 Biotechnological Advancements for Converting Food Waste into Value-Added Products

The focus has now switched to investigating the renewable potential of various waste materials due to technological improvements and the creation of new policies based on the idea of waste to energy. A significant portion of waste produced worldwide is made up of agro-industrial residues that are left over after products derived from agriculture are industrially processed (Sodhi et al., 2022). FW is a valuable source of biological components that can be utilized in biotechnology applications such as enzymatic or fermentative processes, despite its underappreciated components (Torres-Leon et al., 2021). The technological feasibility of turning FW into a variety of value-added chemicals and biofuels is demonstrated by numerous scientific research investigations on FW valorization that are now being conducted. FW can be utilized to create high-value products, as well as a variety of biomaterials and biofuels (Teigiserova et al., 2020).

Organic fractions, proteins, lipids, carbohydrates, and inorganic components create the majority of FW. It is possible to create simpler chemical substances such as fatty acids, glucose, and amino acids (Mak et al., 2020). Using a variety of chemical and biological processes, FW can be transformed into bio-energy and bio-commodity chemicals (Xiong et al., 2019). FW has been highlighted as a source of value-added materials and chemicals by Esteban and Ladero (2018). An overview of enzymatic, chemical, and biotechnological processes that use FW as a raw source to produce chemicals and materials is given. A FW biorefinery, a viable approach to the circular bioeconomy, was reviewed by Dahiya et al., (2018). The several bioprocesses used to produce energy and different commodity molecules are summarized in this article. Kitchen waste and FW can be used to create a variety of value-added goods. This includes films, high fructose syrup, nutraceuticals, levulinic acid, mushroom cultivation, pigments, vermicompost, organic acids, xanthan gum, single-cell protein, sugars, bionanocomposite, chitosan, corrosion inhibitors, bioactive, activated carbon adsorbent, biobutanol, biodiesel, biogas, bioethanol, and wax esters.

1.10 CONCLUSIONS AND PERSPECTIVES

Several interconnected variables, including the growing global population, linear resource consumption, resource security, unabated waste generation, and social inequality, are placing extreme strain on the planet Earth and, if they continue, will result in an unsustainable twenty-first century. A total of 17 SDGs and 169 goals were outlined by the UN in September 2015, and 193 countries committed to the goal of making the world a better place to live. The goal of the process was for all UN member states, including all stakeholders, to agree on the SDGs to create sustainable environments. Making sure patterns of production and consumption are sustainable is the main goal of SDG 12. By 2030, Target 12.3 seeks to cut food loss throughout production, including post-harvest losses, and FW at the retail and consuming stages by 50%. It has been nine years since then, and there is still more to be done to lessen the growing food-related worries. By encouraging 3R practices, SDG target 12.3 seeks to minimize food loss during production (including post-harvest losses) and waste at retail and consuming stages by 50% by 2030. By 2050, there will likely be nine billion people on the planet, which will inevitably lead to a rise in the need for food. Food production will rise in response to the rise in demand, which will lead to an increase in FW. Sustainability throughout the supply chain is crucial to reducing the impact of FW.

Efforts to reduce FW are being made in many nations to strengthen food security and the food industry's sustainability. Depending on the process's final products and valorization, FW's effects on the economy and environment differ. It appears that the conversion and valorization of goods or co-products produced from FW can improve the sustainability of the food industries. Currently, biomaterials, biofertilizers, and energy recovery are the most common uses for FW. Rich and creative conversion of FW, however, may create more opportunities for high-value product production, foster greater industry collaboration in support of the industrial symbiosis concept, and support the CE drive. At all geographical levels, public authorities, institutions, and non-governmental organizations are becoming more focused on and committed to achieving food security objectives as well as effective and sustainable food systems. Gaining a greater understanding of the FSC can help identify FLW hotspots and important areas, enabling the development of customized policies and measures to increase sustainability. A comprehensive approach is necessary to address this issue, encompassing not only the social, economic, and environmental dimensions of sustainability but also the participation of all FSC actors – processors, sellers, farmers, distributors, and consumers – from production to consumption.

REFERENCES

Aamir, M., Ahmad, H., Javaid, Q., & Hasan, S. M. (2018). Waste not, want not: A case study on food waste in restaurants of Lahore, Pakistan. *Journal of Food Products Marketing*, *24*(5), 591–610.

Abreu, A. A., Tavares, F., Alves, M. M., Cavaleiro, A. J., & Pereira, M. A. (2019). Garden and food waste co-fermentation for biohydrogen and biomethane production in a two-step hyperthermophilic-mesophilic process. *Bioresource technology*, *278*, 180–186.

Alamar, M. D. C., Falagán, N., Aktas, E., & Terry, L. A. (2018). Minimising food waste: A call for multidisciplinary research. *Journal of the Science of Food and Agriculture*, *98*(1), 8–11.

Albizzati, P. F., Tonini, D., & Astrup, T. F. (2021). A quantitative sustainability assessment of food waste management in the European Union. *Environmental Science & Technology*, *55*(23), 16099–16109.

Aldaco, R., Hoehn, D., Laso, J., Margallo, M., Ruiz-Salmón, J., Cristobal, J., & Vazquez-Rowe, I. (2020). Food waste management during the COVID-19 outbreak: A holistic climate, economic and nutritional approach. *Science of the Total Environment*, *742*, 140524.

Ali, S. M., Moktadir, M. A., Kabir, G., Chakma, J., Rumi, M. J. U., & Islam, M. T. (2019). Framework for evaluating risks in food supply chain: Implications in food wastage reduction. *Journal of Cleaner Production*, *228*, 786–800.

Ali, S. M., & Nakade, K. (2017). Optimal ordering policies in a multi-sourcing supply chain with supply and demand disruptions-a CVaR approach. *International Journal of Logistics Systems and Management*, *28*(2), 180–199.

Ali, S. M., Rahman, M. H., Tumpa, T. J., Rifat, A. A. M., & Paul, S. K. (2018). Examining price and service competition among retailers in a supply chain under potential demand disruption. *Journal of Retailing and Consumer Services*, *40*, 40–47.

Amicarelli, V., & Bux, C. (2021). Food waste measurement toward a fair, healthy and environmental-friendly food system: A critical review. *British Food Journal*, *123*(8), 2907–2935.

Ananno, A. A., Masud, M. H., Chowdhury, S. A., Dabnichki, P., Ahmed, N., & Arefin, A. M. E. (2021). Sustainable food waste management model for Bangladesh. *Sustainable Production and Consumption*, *27*, 35–51.

Aramyan, L., Grainger, M., Logatcheva, K., Piras, S., Setti, M., Stewart, G., & Vittuari, M. (2021). Food waste reduction in supply chains through innovations: A review. *Measuring Business Excellence*, *25*(4), 475–492.

Arshadi, M., Attard, T. M., Lukasik, R. M., Brncic, M., da Costa Lopes, A. M., Finell, M., & Yuste-Cordoba, F. J. (2016). Pre-treatment and extraction techniques for recovery of added value compounds from wastes throughout the agri-food chain. *Green Chemistry*, *18*(23), 6160–6204.

Aschemann-Witzel, J., De Hooge, I., Amani, P., Bech-Larsen, T., & Oostindjer, M. (2015). Consumer-related food waste: Causes and potential for action. *Sustainability*, *7*(6), 6457–6477.

Aschemann-Witzel, J., Giménez, A., & Ares, G. (2018). Convenience or price orientation? Consumer characteristics influencing food waste behaviour in the context of an emerging country and the impact on future sustainability of the global food sector. *Global Environmental Change*, *49*, 85–94.

Attiq, S., Chau, K. Y., Bashir, S., Habib, M. D., Azam, R. I., & Wong, W. K. (2021). Sustainability of household food waste reduction: A fresh insight on youth's emotional and cognitive behaviors. *International Journal of Environmental Research and Public Health*, *18*(13), 7013.

Awasthi, M. K., Chen, H., Awasthi, S. K., Liu, T., Wang, M., Duan, Y., & Li, J. (2019). Greenhouse gas emissions through biological processing of solid waste and their global warming potential. *Biological Processing of Solid Waste*, 111–127.

Aydin, A. E., & Yildirim, P. (2021). Understanding food waste behavior: The role of morals, habits and knowledge. *Journal of Cleaner Production, 280*, 124250.

Bangar, S. P., Chaudhary, V., Kajla, P., Balakrishnan, G., & Phimolsiripol, Y. (2023). Strategies for upcycling food waste in the food production and supply chain. *Trends in Food Science & Technology*, *143*, 104314.

Banožić, M., Banjari, I., Jakovljević, M., Šubarić, D., Tomas, S., Babić, J., & Jokić, S. (2019). Optimization of ultrasound-assisted extraction of some bioactive compounds from tobacco waste. *Molecules*, *24*(8), 1611.

Battista, F., Fino, D., & Mancini, G. (2016). Optimization of biogas production from coffee production waste. *Bioresource Technology*, *200*, 884–890.

Ben Hassen, T., & El Bilali, H. (2022). Impacts of the Russia-Ukraine war on global food security: towards more sustainable and resilient food systems?. *Foods*, *11*(15), 2301.

Bilal, M., & Iqbal, H. M. (2019). Sustainable bioconversion of food waste into high-value products by immobilized enzymes to meet bio-economy challenges and opportunities–A review. *Food Research International*, *123*, 226–240.

Block, L., Vallen, B., & Austin, M. P. (2022). Food waste (mis) takes: The role of (mis) perception and (mis) estimation. *Current Opinion in Psychology*, *46*, 101327.

Bravi, L., Francioni, B., Murmura, F., & Savelli, E. (2020). Factors affecting household food waste among young consumers and actions to prevent it. A comparison among UK, Spain and Italy. *Resources, Conservation and Recycling*, *153*, 104586.

Bravi, L., Murmura, F., Savelli, E., & Viganò, E. (2019). Motivations and actions to prevent food waste among young Italian consumers. *Sustainability*, *11*(4), 1110.

Bux, C., Varese, E., Amicarelli, V., & Lombardi, M. (2022). Halal food sustainability between certification and clockchain: A review. *Sustainability, 14*(4), 2152.

Caldeira, C., De Laurentiis, V., Corrado, S., van Holsteijn, F., & Sala, S. (2019). Quantification of food waste per product group along the food supply chain in the European Union: A mass flow analysis. *Resources, Conservation and Recycling, 149*, 479–488.

Chaudhary, P., Garg, S., George, T., Shabin, M., Saha, S., Subodh, S., & Sinha, B. (2021). Underreporting and open burning–the two largest challenges for sustainable waste management in India. *Resources, Conservation and Recycling, 175*, 105865.

Chen, H., Zhang, H., Tian, J., Shi, J., Linhardt, R. J., Ye, T. D. X., & Chen, S. (2019). Recovery of high value-added nutrients from fruit and vegetable industrial wastewater. *Comprehensive Reviews in Food Science and Food Safety, 18*(5), 1388–1402.

Chuyen, H. V., Nguyen, M. H., Roach, P. D., Golding, J. B., & Parks, S. E. (2018). Microwave-assisted extraction and ultrasound-assisted extraction for recovering carotenoids from Gac peel and their effects on antioxidant capacity of the extracts. *Food Science & Nutrition, 6*(1), 189–196.

Ciccullo, F., Cagliano, R., Bartezzaghi, G., & Perego, A. (2021). Implementing the circular economy paradigm in the agri-food supply chain: The role of food waste prevention technologies. *Resources, Conservation and Recycling, 164*, 105114.

Corrado, S., & Sala, S. (2018). Bio-economy contribution to circular economy. In E. Benetto & K. Gericke (Eds.), *Designing sustainable technologies, products and policies: From science to innovation* (pp. 49–59). Springer International Publishing.

Dabas, S., Sharma, S., & Manaktola, K. (2021). Adoption of digital marketing tools in independent businesses: Experiences of restaurant entrepreneurs in India and United Kingdom. *Worldwide Hospitality and Tourism Themes, 13*(2), 214–235.

da Costa, T. P., Gillespie, J., Cama-Moncunill, X., Ward, S., Condell, J., Ramanathan, R., & Murphy, F. (2022). A systematic review of real-time monitoring technologies and its potential application to reduce food loss and waste: Key elements of food supply chains and IoT technologies. *Sustainability, 15*(1), 614.

Dahiya, S., Kumar, A. N., Sravan, J. S., Chatterjee, S., Sarkar, O., & Mohan, S. V. (2018). Food waste biorefinery: Sustainable strategy for circular bioeconomy. *Bioresource technology, 248*, 2–12.

de Los Mozos, E. A., Badurdeen, F., & Dossou, P. E. (2020). Sustainable consumption by reducing food waste: A review of the current state and directions for future research. *Procedia Manufacturing, 51*, 1791–1798.

Despoudi, S., Bucatariu, C., Otles, S., & Kartal, C. (2021). Food waste management, valorization, and sustainability in the food industry. In C. M. Galanakis (Ed.), *Food waste recovery* (pp. 3–19). Academic Press.

Dhiman, S., & Mukherjee, G. (2021). Present scenario and future scope of food waste to biofuel production. *Journal of Food Process Engineering, 44*(2), e13594.

Dung, T. N. B., Sen, B., Chen, C. C., Kumar, G., & Lin, C. Y. (2014). Food waste to bioenergy via anaerobic processes. *Energy Procedia, 61*, 307–312.

Esteban, J., & Ladero, M. (2018). Food waste as a source of value-added chemicals and materials: A biorefinery perspective. *International Journal of Food Science & Technology, 53*(5), 1095–1108.

European Commission. (2014). *Food waste and its impacts: European week for waste reduction*. European Commission.

European Commission. (2020). Farm to fork strategy: For a fair, healthy and environmentally-friendly food system. *Communication from the Commission to the European Parliament, the Council, the European Economic and Social Committee and the Committee of the Regions, 381*, 1–9.

FAO, F. (2014). *Food wastage footprint full-cost accounting*. Food Wastage Footprint.

FAO, I. (2019). *The state of food and agriculture 2019: Moving forward on food loss and waste reduction* (pp. 2–13). FAO.

Fassio, F., & Chirilli, C. (2023). The circular economy and the food system: A review of principal measuring tools. *Sustainability, 15*(13), 10179.

Fattibene, D., Recanati, F., Dembska, K., & Antonelli, M. (2020). Urban food waste: A framework to analyse policies and initiatives. *Resources, 9*(9), 99.

Ferdes, M., Zăbavă, B. Ș., Paraschiv, G., Ionescu, M., Dincă, M. N., & Moiceanu, G. (2022). Food waste management for biogas production in the context of sustainable development. *Energies, 15*(17), 6268.

Fesenfeld, L., Rudolph, L., & Bernauer, T. (2022). Policy framing, design and feedback can increase public support for costly food waste regulation. *Nature Food, 3*(3), 227–235.

Galli, F., Cavicchi, A., & Brunori, G. (2019). Food waste reduction and food poverty alleviation: A system dynamics conceptual model. *Agriculture and Human Values, 36*, 289–300.

Garcia-Garcia, G., Woolley, E., & Rahimifard, S. (2015). A framework for a more efficient approach to food waste management. *International Journal of Food Engineering, 1*(1), 65–72.

Garcia-Garcia, G., Woolley, E., Rahimifard, S., Colwill, J., White, R., & Needham, L. (2017). A methodology for sustainable management of food waste. *Waste and Biomass Valorization*, *8*(6), 2209–2227.

Garrido, T., Gizdavic-Nikolaidis, M., Leceta, I., Urdanpilleta, M., Guerrero, P., de la Caba, K., & Kilmartin, P. A. (2019). Optimizing the extraction process of natural antioxidants from chardonnay grape marc using microwave-assisted extraction. *Waste Management*, *88*, 110–117.

Geueke, B., Groh, K., & Muncke, J. (2018). Food packaging in the circular economy: Overview of chemical safety aspects for commonly used materials. *Journal of Cleaner Production*, *193*, 491–505.

Ghisellini, P., Santagata, R., Zucaro, A., & Ulgiat, S. (2019). Circular patterns of waste prevention and recovery. In *E3S web of conferences* (Vol. 119, p. 00003). EDP Sciences.

Girotto, F., Alibardi, L., & Cossu, R. (2015). Food waste generation and industrial uses: A review. *Waste management*, *45*, 32–41.

Gokarn, S., & Kuthambalayan, T. S. (2017). Analysis of challenges inhibiting the reduction of waste in food supply chain. *Journal of Cleaner Production*, *168*, 595–604.

Goossens, Y., Wegner, A., & Schmidt, T. (2019). Sustainability assessment of food waste prevention measures: Review of existing evaluation practices. *Frontiers in Sustainable Food Systems*, *3*, 476725.

Gowman, A. C., Picard, M. C., Lim, L. T., Misra, M., & Mohanty, A. K. (2019). Fruit waste valorization for biodegradable biocomposite applications: A review. *BioResources*, *14*(4), 10047–10092.

Grassino, A. N., Brnčić, M., Vikić-Topić, D., Roca, S., Dent, M., & Brnčić, S. R. (2016). Ultrasound assisted extraction and characterization of pectin from tomato waste. *Food chemistry*, *198*, 93–100.

Grosso, M., & Falasconi, L. (2018). Addressing food wastage in the framework of the UN sustainable development goals. *Waste Management & Research*, *36*(2), 97–98.

Grasso, S., & Asioli, D. (2020). Consumer preferences for upcycled ingredients: A case study with biscuits. *Food Quality and Preference*, *84*, 103951.

Gunders, D., & Bloom, J. (2017). Wasted: How America is losing up to 40 percent of its food from farm to fork to landfill. Natural Resources Defense Council. Retrieved March 7, 2019, from https://www.nrdc.org/sites/default/files/wasted-2017-report.pdf

Haldar, D., Shabbirahmed, A. M., Singhania, R. R., Chen, C. W., Dong, C. D., Ponnusamy, V. K., & Patel, A. K. (2022). Understanding the management of household food waste and its engineering for sustainable valorization-a state-of-the-art review. *Bioresource Technology*, *358*, 127390.

Harvey, J., Smith, A., Goulding, J., & Illodo, I. B. (2019). Food sharing, redistribution, and waste reduction via mobile applications: A social network analysis. *Industrial Marketing Management*, *88*, 437–448.

Hashempour-Baltork, F., Farshi, P., & Khosravi-Darani, K. (2022). Vegetable and fruit wastes as substrate for production of single-cell protein and aquafeed meal. In R. C. Ray (Ed.), *Fruits and vegetable wastes: Valorization to bioproducts and platform chemicals* (pp. 169–187). Springer Nature Singapore.

HLPE, F. (2014). Food losses and waste in the context of sustainable food systems. *A Report by the High Level Panel of Experts on Food Security and Nutrition of the Committee on World Food Security.* Retrieved October 2, 2017, from http://www.fao.org/3/a-i3901e.pdf

IPCC. (2019). IPCC special report on climate change, desertification, land degradation, sustainable land management, food security, and greenhouse gas fluxes in terrestrial ecosystems. In P. R. Shukla, J. Skea, E. Calvo Buendia, V. Masson-Delmotte, H.-O. Portner, D. C. Roberts, P. Zhai, R. Slade, S. Connors, R. van Diemen, M. Ferrat, E. Haughey, S. Luz, S. Neogi, M. Pathak, J. Petzold, J. Portugal Pereira, P. Vyas, E. Huntley, K. Kissick, M. Belkacemi, & J. Malley (Eds.), *IPCC summary for policymalers* (pp. 1–472). In press.

Irani, Z., Sharif, A. M., Lee, H., Aktas, E., Topaloğlu, Z., van't Wout, T., & Huda, S. (2018). Managing food security through food waste and loss: Small data to big data. *Computers & Operations Research*, *98*, 367–383.

Ishangulyyev, R., Kim, S., & Lee, S. H. (2019). Understanding food loss and waste – Why are we losing and wasting food?. *Foods*, *8*(8), 297.

Issock Issock, P. B., Roberts-Lombard, M., & Mpinganjira, M. (2020). Understanding household waste separation in South Africa: An empirical study based on an extended theory of interpersonal behaviour. *Management of Environmental Quality: An International Journal*, *31*(3), 530–547.

Jagtap, S., & Rahimifard, S. (2019). The digitisation of food manufacturing to reduce waste–Case study of a ready meal factory. *Waste Management*, *87*, 387–397.

Jaouhari, Y., Travaglia, F., Giovannelli, L., Picco, A., Oz, E., Oz, F., & Bordiga, M. (2023). From industrial food waste to bioactive ingredients: A review on the sustainable management and transformation of plant-derived food waste. *Foods*, *12*(11), 2183.

Joshi, P., & Visvanathan, C. (2019). Sustainable management practices of food waste in Asia: Technological and policy drivers. *Journal of Environmental Management*, *247*, 538–550.

Kamble, S. S., Gunasekaran, A., Parekh, H., & Joshi, S. (2019). Modeling the internet of things adoption barriers in food retail supply chains. *Journal of Retailing and Consumer Services*, *48*, 154–168.

Kaur, M., Singh, A. K., & Singh, A. (2023). Bioconversion of food industry waste to value added products: Current technological trends and prospects. *Food Bioscience*, 102935. https://doi.org/10.1016/j.fbio.2023.102935

Kazancoglu, Y., Ozkan-Ozen, Y. D., & Ozbiltekin, M. (2018). Minimizing losses in milk supply chain with sustainability: An example from an emerging economy. *Resources, Conservation and Recycling*, *139*, 270–279.

Kharola, S., Ram, M., Mangla, S. K., Goyal, N., Nautiyal, O. P., Pant, D., & Kazancoglu, Y. (2022). Exploring the green waste management problem in food supply chains: A circular economy context. *Journal of Cleaner Production*, *351*, 131355.

Kiran, E. U., Trzcinski, A. P., Ng, W. J., & Liu, Y. (2014). Bioconversion of food waste to energy: A review. *Fuel*, *134*, 389–399.

Kwan, T. H., Ong, K. L., Haque, M. A., Kwan, W. H., Kulkarni, S., & Lin, C. S. K. (2018). Valorisation of food and beverage waste via saccharification for sugars recovery. *Bioresource Technology*, *255*, 67–75.

Lawrence, M., & Friel, S. (Eds.). (2019). *Healthy and sustainable food systems*. Routledge.

Leverenz, D., Schneider, F., Schmidt, T., Hafner, G., Nevárez, Z., & Kranert, M. (2021). Food waste generation in Germany in the scope of European legal requirements for monitoring and reporting. *Sustainability*, *13*(12), 6616.

Lin, C. S. K., Pfaltzgraff, L. A., Herrero-Davila, L., Mubofu, E. B., Abderrahim, S., Clark, J. H., & Luque, R. (2013). Food waste as a valuable resource for the production of chemicals, materials and fuels. Current situation and global perspective. *Energy & Environmental Science*, *6*(2), 426–464.

Lipinski, B., Hanson, C., Waite, R., Searchinger, T., & Lomax, J. (2013). *Reducing food loss and waste*. World Resources Institute.

Liu, C., Hotta, Y., Santo, A., Hengesbaugh, M., Watabe, A., Totoki, Y., & Bengtsson, M. (2016). Food waste in Japan: Trends, current practices and key challenges. *Journal of Cleaner Production*, *133*, 557–564.

Magarini, A., de Cunto, A., & Porreca, E. (2018). *Food losses and waste in European cities*. Comune di Milano.

Maina, S., Kachrimanidou, V., & Koutinas, A. (2017). A roadmap towards a circular and sustainable bioeconomy through waste valorization. *Current Opinion in Green and Sustainable Chemistry*, *8*, 18–23.

Mak, T. M., Xiong, X., Tsang, D. C., Iris, K. M., & Poon, C. S. (2020). Sustainable food waste management towards circular bioeconomy: Policy review, limitations and opportunities. *Bioresource technology*, *297*, 122497.

Martin-Rios, C., Hofmann, A., & Mackenzie, N. (2020). Sustainability-oriented innovations in food waste management technology. *Sustainability*, *13*(1), 210.

Martínez-Abad, A., Ramos, M., Hamzaoui, M., Kohnen, S., Jiménez, A., & Garrigós, M. C. (2020). Optimisation of sequential microwave-assisted extraction of essential oil and pigment from lemon peels waste. *Foods*, *9*(10), 1493.

Maurya, D. K., Kumar, A., Chaurasiya, U., Hussain, T., & Singh, S. K. (2021). Modern era of microbial biotechnology: opportunities and future prospects. In M. K. Solanki, P. L. Kashyap, R. A. Ansari, & B. Kumari (Eds.), *Microbiomes and plant health* (pp. 317–343). Academic Press.

McCarthy, B., & Liu, H. B. (2017). Food waste and the 'green'consumer. *Australasian Marketing Journal*, *25*(2), 126–132.

Mirabella, N., Castellani, V., & Sala, S. (2014). Current options for the valorization of food manufacturing waste: A review. *Journal of Cleaner Production*, *65*, 28–41.

Mishra, B., Varjani, S., Pradhan, I., Ekambaram, N., Teixeira, J. A., Ngo, H. H., & Guo, W. (2020). Insights into interdisciplinary approaches for bioremediation of organic pollutants: Innovations, challenges and perspectives. *Proceedings of the National Academy of Sciences, India Section B: Biological Sciences*, *90*, 951–958.

Moshtaghian, H., Bolton, K., & Rousta, K. (2021). Challenges for upcycled foods: Definition, inclusion in the food waste management hierarchy and public acceptability. *Foods*, *10*(11), 2874.

Mu'azu, N. D., Blaisi, N. I., Naji, A. A., Abdel-Magid, I. M., & AlQahtany, A. (2019). Food waste management current practices and sustainable future approaches: A Saudi Arabian perspectives. *Journal of Material Cycles and Waste Management*, *21*, 678–690.

Mustafa, F., & Andreescu, S. (2018). Chemical and biological sensors for food-quality monitoring and smart packaging. *Foods*, *7*(10), 168.

Nagarajan, M., Saha, R., Kumar, R., & Sathasivam, D. (2022). Impact of peer influence and environmental knowledge on green consumption: Moderated by price premium. *International Journal of Social Ecology and Sustainable Development (IJSESD), 13*(6), 1–16.

Narasimmalu, A., & Ramasamy, R. (2020, November). Food processing industry waste and circular economy. In *IOP Conference Series: Materials science and engineering* (Vol. 955, Issue 1, p. 012089). IOP Publishing.

Negi, S., & Anand, N. (2015). Cold chain: A weak link in the fruits and vegetables supply chain in India. *IUP Journal of Supply Chain Management, 12*(1), 48.

Negi, S., Hu, A., & Kumar, S. (2021). Circular bioeconomy: Countries' case studies. In A. Pandey, R. D. Tyagi, & S. Varjani (Eds.), *Biomass, biofuels, biochemicals* (pp. 721–748). Elsevier.

Nguyen, D. D., Yeop, J. S., Choi, J., Kim, S., Chang, S. W., Jeon, B. H., & Ngo, H. H. (2017). A new approach for concurrently improving performance of South Korean food waste valorization and renewable energy recovery via dry anaerobic digestion under mesophilic and thermophilic conditions. *Waste Management, 66*, 161–168.

Nguyen, X. C., Tran, T. P. Q., Nguyen, T. T. H., La, D. D., Nguyen, V. K., Nguyen, T. P., & Nguyen, D. D. (2020). Call for planning policy and biotechnology solutions for food waste management and valorization in Vietnam. *Biotechnology Reports, 28*, e00529.

Nguyen Hao, H. T., Karthikeyan, O. P., & Heimann, K. (2015). Bio-refining of carbohydrate-rich food waste for biofuels. *Energies, 8*(7), 6350–6364.

Nicewicz, R., & Bilska, B. (2021). Analysis of changes in shopping habits and causes of food waste among consumers before and during the COVID-19 pandemic in Poland. *Environmental Protection and Natural Resources, 32*(3), 8–19.

Ogunmoroti, A., Liu, M., Li, M., & Liu, W. (2022). Unraveling the environmental impact of current and future food waste and its management in Chinese provinces. *Resources, Environment and Sustainability, 9*, 100064.

Ong, K. L., Kaur, G., Pensupa, N., Uisan, K., & Lin, C. S. K. (2018). Trends in food waste valorization for the production of chemicals, materials and fuels: Case study South and Southeast Asia. *Bioresource Technology, 248*, 100–112.

Otles, S., Despoudi, S., Bucatariu, C., & Kartal, C. (2015). Food waste management, valorization, and sustainability in the food industry. In C. M. Galanakis (Ed.), *Food waste recovery* (pp. 3–23). Academic Press.

Pal, M. S., & Bhatia, M. (2022). Current status, topographical constraints, and implementation strategy of municipal solid waste in India: a review. *Arabian Journal of Geosciences, 15*(12), 1176.

Parfitt, J., Barthel, M., & Macnaughton, S. (2010). Food waste within food supply chains: quantification and potential for change to 2050. *Philosophical Transactions of the Royal Society B: Biological Sciences, 365*(1554), 3065–3081.

Pearson, D., Minehan, M., & Wakefield-Rann, R. (2013). Food waste in Australian households: Why does it occur. *The Australasian-Pacific Journal of Regional Food Studies, 3*(1), 118–132.

Pellegrini, G., Sillani, S., Gregori, M., & Spada, A. (2019). Household food waste reduction: Italian consumers' analysis for improving food management. *British Food Journal, 121*(6), 1382–1397.

Pham, T. P. T., Kaushik, R., Parshetti, G. K., Mahmood, R., & Balasubramanian, R. (2015). Food waste-to-energy conversion technologies: Current status and future directions. *Waste Management, 38*, 399–408.

Pino, M. S., Rodríguez-Jasso, R. M., Michelin, M., Flores-Gallegos, A. C., Morales-Rodriguez, R., Teixeira, J. A., & Ruiz, H. A. (2018). Bioreactor design for enzymatic hydrolysis of biomass under the biorefinery concept. *Chemical Engineering Journal, 347*, 119–136.

Pleissner, D., Qi, Q., Gao, C., Rivero, C. P., Webb, C., Lin, C. S. K., & Venus, J. (2016). Valorization of organic residues for the production of added value chemicals: A contribution to the bio-based economy. *Biochemical Engineering Journal, 116*, 3–16.

Porpino, G., Wansink, B., & Parente, J. (2016). Wasted positive intentions: The role of affection and abundance on household food waste. *Journal of Food Products Marketing, 22*(7), 733–751.

Principato, L., & Principato, L. (2018). Food policies to tackle food waste: A classification. *Food Waste at Consumer Level: A Comprehensive Literature Review*, 35–42. http://dx.doi.org/10.1007/978-3-319-78887-6_3

Ramanathan, R., Duan, Y., Ajmal, T., Pelc, K., Gillespie, J., Ahmadzadeh, S., & Ramanathan, U. (2023). Motivations and challenges for food companies in using IoT sensors for reducing food waste: some insights and a road map for the future. *Sustainability, 15*(2), 1665.

Rose, K., Eldridge, S., & Chapin, L. (2015). The internet of things: An overview. *The Internet Society (ISOC), 80*, 1–50.

Sahoo, A., Dwivedi, A., Madheshiya, P., Kumar, U., Sharma, R. K., & Tiwari, S. (2023). Insights into the management of food waste in developing countries: With special reference to India. *Environmental Science and Pollution Research, 31*, 17887–17913.

Saijo, T. (2020). Future design: Bequeathing sustainable natural environments and sustainable societies to future generations. *Sustainability, 12*(16), 6467.

Sarker, A., Ahmmed, R., Ahsan, S. M., Rana, J., Ghosh, M. K., & Nandi, R. (2024). *A comprehensive review of food waste valorization for the sustainable management of global food waste*. Sustainable Food Technology.

Scacchi, A., Catozzi, D., Boietti, E., Bert, F., & Siliquini, R. (2021). COVID-19 lockdown and self-perceived changes of food choice, waste, impulse buying and their determinants in Italy: QuarantEat, a cross-sectional study. *Foods, 10*(2), 306.

Schanes, K., Dobernig, K., & Gözet, B. (2018). Food waste matters-A systematic review of household food waste practices and their policy implications. *Journal of Cleaner Production, 182*, 978–991.

Schott, A. B. S., & Andersson, T. (2015). Food waste minimization from a life-cycle perspective. *Journal of Environmental Management, 147*, 219–226.

Septianto, F., Kemper, J. A., & Northey, G. (2020). Thanks, but no thanks: The influence of gratitude on consumer awareness of food waste. *Journal of Cleaner Production, 258*, 120591.

Sharma, P., Gaur, V. K., Sirohi, R., Varjani, S., Kim, S. H., & Wong, J. W. (2021). Sustainable processing of food waste for production of bio-based products for circular bioeconomy. *Bioresource Technology, 325*, 124684.

Sheldon, M. (2020). *South Korea recycles food waste in effort to become zero-waste society*. Hunter College New York City Food Policy Center.

Šimanskienė, L., Paužuolienė, J., Fiore, M., & Župerkienė, E. (2022). Responsible Consumption: Society Habits in Time of Crisis. In: Vrontis, D., Thrassou, A., Weber, Y., Shams, S. M. R., Tsoukatos, E., Efthymiou, L. (Eds.), *Business under crisis, Volume III*. Palgrave Studies in Cross-disciplinary Business Research, In Association with EuroMed Academy of Business. Palgrave Macmillan, Cham. https://doi.org/10.1007/978-3-030-76583-5_12 –

Singhania, R. R., Patel, A. K., Soccol, C. R., & Pandey, A. (2009). Recent advances in solid-state fermentation. *Biochemical Engineering Journal, 44*(1), 13–18.

Sinha, S., & Tripathi, P. (2021). Trends and challenges in valorisation of food waste in developing economies: A case study of India. *Case Studies in Chemical and Environmental Engineering, 4*, 100162.

Sodhi, A. S., Sharma, N., Bhatia, S., Verma, A., Soni, S., & Batra, N. (2022). Insights on sustainable approaches for production and applications of value added products. *Chemosphere, 286*, 131623.

Stenmarck, Â., Jensen, C., Quested, T., Moates, G., Buksti, M., Cseh, B., & Östergren, K. (2016). *Estimates of European food waste levels*. IVL Swedish Environmental Research Institute.

Tamasiga, P., Miri, T., Onyeaka, H., & Hart, A. (2022). Food waste and circular economy: Challenges and opportunities. *Sustainability, 14*(16), 9896.

Tchonkouang, R. D., Onyeaka, H., & Miri, T. (2023). From waste to plate: Exploring the impact of food waste valorisation on achieving zero hunger. *Sustainability, 15*(13), 10571.

Teigiserova, D. A., Hamelin, L., & Thomsen, M. (2020). Towards transparent valorization of food surplus, waste and loss: Clarifying definitions, food waste hierarchy, and role in the circular economy. *Science of the Total Environment, 706*, 136033.

Thapa, B., Patidar, S. K., Khatiwada, N. R., Kc, A. K., & Ghimire, A. (2019). Production of ethanol from municipal solid waste of India and Nepal. In *Waste valorisation and recycling: 7th IconSWM—ISWMAW 2017, Volume 2* (pp. 47–58). Springer Singapore.

Thi, N. B., Kumar, G., & Lin, C. Y. (2015). An overview of food waste management in developing countries: Current status and future perspective. *Journal of Environmental Management, 157*, 220–229. doi: 10.1016/j.jenvman.2015.04.022. Epub 2015 Apr 21. PMID: 25910976.

Thyberg, K. L., & Tonjes, D. J. (2016). Drivers of food waste and their implications for sustainable policy development. *Resources, Conservation and Recycling, 106*, 110–123.

Torres-León, C., Chávez-González, M. L., Hernández-Almanza, A., Martínez-Medina, G. A., Ramírez-Guzmán, N., Londoño-Hernández, L., & Aguilar, C. N. (2021). Recent advances on the microbiological and enzymatic processing for conversion of food wastes to valuable bioproducts. *Current Opinion in Food Science, 38*, 40–45.

Tropea, A. (2022). Food waste valorization. *Fermentation, 8*(4), 168.

Tsang, Y. F., Kumar, V., Samadar, P., Yang, Y., Lee, J., Ok, Y. S., & Jeon, Y. J. (2019). Production of bioplastic through food waste valorization. *Environment International, 127*, 625–644.

Tsegaye, B., Jaiswal, S., & Jaiswal, A. K. (2021). Food waste biorefinery: Pathway towards circular bioeconomy. *Foods*, *10*(6), 1174.

UN report (2022 on global hunger. https://www.fao.org/newsroom/detail/un-report-global-hunger-SOFI-2022-FAO/en accessed on 20.01.2024.

UNEP food waste index report (2021). https://www.unep.org/resources/report/unep-food-waste-index-report-2021 accessed on 20.01.2024.

Urbinati, A., Chiaroni, D., & Chiesa, V. (2017). Towards a new taxonomy of circular economy business models. *Journal of Cleaner Production*, *168*, 487–498.

Venghaus, S., Henseleit, M., & Belka, M. (2022). The impact of climate change awareness on behavioral changes in Germany: Changing minds or changing behavior? *Energy, Sustainability and Society*, *12*(1), 8.

Wang, Y., Yuan, Z., & Tang, Y. (2021). Enhancing food security and environmental sustainability: A critical review of food loss and waste management. *Resources, Environment and Sustainability*, *4*, 100023.

Xiong, X., Iris, K. M., Tsang, D. C., Bolan, N. S., Ok, Y. S., Igalavithana, A. D., & Vikrant, K. (2019). Value-added chemicals from food supply chain wastes: State-of-the-art review and future prospects. *Chemical Engineering Journal*, *375*, 121983.

Xue, L., Liu, G., Parfitt, J., Liu, X., Van Herpen, E., Stenmarck, Å., & Cheng, S. (2017). Missing food, missing data? A critical review of global food losses and food waste data. *Environmental Science & Technology*, *51*(12), 6618 6633.

Yadav, V. S., Singh, A. R., Raut, R. D., Mangla, S. K., Luthra, S., & Kumar, A. (2022). Exploring the application of Industry 4.0 technologies in the agricultural food supply chain: A systematic literature review. *Computers & Industrial Engineering*, *169*, 108304.

Zamri, G. B., Azizal, N. K. A., Nakamura, S., Okada, K., Nordin, N. H., Othman, N., & Hara, H. (2020). Delivery, impact and approach of household food waste reduction campaigns. *Journal of Cleaner Production*, *246*, 118969.

Zhong, Y., Pu, M., Liu, M., Niu, K., & Zhang, L. (2020). The impact of COVID-19 on China's food security and suggestions for ensuring stable grain production. *Issues in Agricultural Economy*, *4*, 13–12.

2 Methods for Food Waste Characterization

Yogesh Kumar, Swarnima Dey, and Kunal Balhara

2.1 INTRODUCTION

A considerable surge in waste generation occurred around the sixteenth century with the migration of people from rural areas to cities, pushed by the industrial revolution. The population boom that resulted from this urban migration also led to a significant rise in the amount and variety of waste produced within city boundaries (Amasuomo & Baird 2016). Waste is the term used to describe the dumped byproducts of human activity that contain materials that were present in the original, beneficial product but were judged to be no longer desirable. Waste can take on a multitude of forms, and there are several criteria that can be used to categorize it. Physical states, physical qualities, potential for reuse, biodegradability, source of production, and level of environmental effect are common factors used in waste classification. Based on their physical properties, waste can be broadly divided into three types: gaseous, solid, and liquid. Although several classifications are used in different nations, the most often used ones are outlined as household/domestic waste; industrial waste; food waste; agricultural waste; commercial waste; demolition and construction waste; and mining waste (Amasuomo & Baird 2016).

The term food waste (FW) refers to leftovers or waste or byproducts generated in households, restaurants, lodging establishments, catering establishments, and other food-related sectors. FW is the remaining material whose economic worth is less than the expense of collection and recovery, which forces it to be disposed of (Kavitha et al., 2020). It includes anything that may be recycled or disposed of that is abandoned from the food supply chain (FSC), whether it is edible or inedible. This covers the management of food waste using a variety of techniques, including open dumping, landfilling, incineration for the generation of bioenergy, drainage disposal, landfilling, and disposal into aquatic bodies. Food losses can happen before, during, or after food preparation as well as during the production, packaging, handling, distribution, and marketing phases. Food leftovers include partially unwanted and inedible materials such as leaves, stalks, and skins. They also comprise leftovers from meals prepared in hotels, cafés, restaurants, and other food enterprises without social utilization policies. Byproducts are useful and commercial goods or services that come from a stage of production other than the primary one (Amasuomo & Baird, 2016). Typically, edible byproducts of food processing and preparation get diverted from the human FSC and are used as animal feed. Animal organs and body parts fall into the category of food derivatives originating from animals. Waste that can be avoided includes food waste. However, food that has been thrown out contains value and is perfectly acceptable for use. Food items that have been dropped, deteriorated, smashed, or damaged are called FW (Kavitha et al., 2020). This includes sealed or whole packets as well as individual products that are not eaten. According to Foresight (2010), FW can arise in the FSC at any point because of the subpar performance of the creator of the FW in the food chain.

2.2 ORIGIN OF FOOD WASTE

As outlined by the European Commission (2014), FW can be divided into three categories: (1) food items lost during the production stage known as food loss; (2) food items lost during the consumption stage called undeniable food waste (e.g., banana peels, fruit cores); and (3) unconsumed

DOI: 10.1201/9781032706030-2

food items lost during the consumption stage called superfluous food waste. Food waste production has been categorized at different FSC phases (Gustavsson et al., 2011). These phases comprise the production of food waste from agriculture, postharvest management, storage, distribution, and consumption. Food loss (FW) is related to seller and consumer behavior and occurs during the marketing and final consumption phase of the FSC (Parfitt et al., 2010). FW was described by Lipinski et al., (2013) as food items that can be deemed suitable for human consumption and have a higher value, but are not used because they are thrown away before or after they deteriorate. Food supply chain waste (FSCW) is defined as an organic-rich product produced for human consumption that is discarded, lost, or breaks down, mostly during the production and marketing stages. It includes food waste resulting from spoilage or insect damage. All phases of the FSC involve the generation of FW, which is most noticeable during the merchandising and customer phases. According to reports from the FAO, around 51% of food produced is wasted or thrown away both before and after it is consumed by the consumer (Parfitt et al., 2010). Worldwide, the production of food is close to 1300 million tonnes per year. This aptly shows a major ecological and social problem (Gustavsson et al., 2011).

Following the FAO's revelation that around one-third of all food produced worldwide is wasted, the issue has gained more prominence and is now a particular sub-goal of Agenda 2030's SDG 12. According to the most current UNEP survey, 61% of the approximately 931 million tonnes of food waste produced worldwide in 2019 originated from homes (Giordano & Franco, 2021). Food waste, regardless of where it is produced, is either dumped in landfills or treated chemically or biologically to minimize its impact on the environment. Brazil, the USA, China, and India are the top four countries that contribute to food waste worldwide (Sinha & Tripathi, 2021). In underdeveloped nations, conventional techniques such as composting and landfilling are frequently used for the management and disposal of food waste. However, problems including the discharge of harmful gases, offensive odors, and related environmental contamination make these approaches unfeasible (Cabello et al., 2020). Food waste happens at every stage of the food supply chain, starting with agricultural production. This waste might be the consequence of factors like subpar produce or co-products with little to no economic worth. Companies are shifting towards more sustainable practices to cut expenses and improve process efficiency, which will enable them to develop creative yet financially viable FW reutilization techniques (Kavitha et al., 2020). A range of approaches can be investigated to develop cutting-edge techniques for the by-products and residues of food waste that are valuable. These include large amounts of biomolecules that are plentiful, easily obtainable, renewable, and reusable, such as proteins, polysaccharides, triglycerides, lipids, and phenolic chemicals. Products with additional value can be collected, processed, and repurposed as healthy meals, oils, and flavoring compounds from different pools of food waste (Sinha & Tripathi, 2021). Improving value estimation techniques might lead to improvements in food waste management and more sustainable food supply chains by addressing major issues in the food sector. Considering the serious problems with agro-based food waste, these techniques can potentially solve problems with food waste's origin and treatment. These issues include reducing the amount of trash that ends up in landfills, cutting greenhouse gas emissions, reducing the amount of inorganic material leakage that contaminates water supplies, and enhancing the effectiveness of traditional food waste treatment techniques like composting and burning. Leading examples of these systems are the developments of closed-loop supply chain models. These models clarify that all types of food waste may be recycled along the food supply chain; packaged food waste, for example, can be repurposed (Kavitha et al., 2020). Food waste is mostly produced during the final consumer stage of the food supply chain, such as domestic garbage; however, it is not the only instance. Conversely, food loss occurs during a number of phases, such as the processing, distribution, marketing, final utilization, and post-consumption stages. Food loss during the processing step is typically caused by spillage or mechanical damage that occurs during harvesting tasks like crop classification, fruitlet collection, or separation. Food loss is mostly caused by environmental factors, such as differences in climate and climate change, as well as economic issues, such as laws, policies, and private and public standards pertaining to appearance and quality (Kader, 2009). Food loss happens in several ways, including distribution, handling, storage, postharvest, and management. Food loss in the postharvest phase is ascribed to problems including leaks and decomposition, insufficient storage spaces, and difficulties in transporting the food from the farm (Kader, 2009). Food spoils significantly in storage, mostly because of insect and microbial activity. On the other hand, food waste during processing – for example, during juice preparation and canning – may arise from leakage or decomposition. Food waste can also happen during the processing step when things are being cleaned, shredded, sliced, and steam-cooked (Kavitha et al., 2020). Food wastage is ascribed to time restrictions, improper transportation, poor packing, problems with dealer/customer interactions, and organizational inefficiencies throughout the distribution stage. Food waste is the term used to describe food loss at the store level, which frequently results from a conscious choice to throw food away. The Department of Environment Food & Rural Affairs states that this food is trash even if it is declared non-toxic and healthy for human consumption (Defra, 2009). At the retailer level, food waste results from miscalculated demand, improper record-keeping, temperature swings, unfavorable weather during transit, unsold item disposal, improper packing, and a lack of clarity or advice on food standards.

Large amounts of food bearing dates like "best before," "sell by," or "use by" are thrown out by retailers (Cabello et al., 2020). Some businesses work with redistribution groups and charities, including food banks, to give away unsold food or advise consumers on how to cut back on food waste. Food waste occurs at the consumption stage for a variety of reasons, including personal buying preferences, ignorance, poor understanding of effective food utilization, social influences, improper shopping estimation, packaging problems, and other related difficulties (Defra, 2009).

2.3 COMPOSITION OF FOOD WASTE

The waste obtained from plant and animal sources during the production, processing, storage, transportation, and consumption is known as food waste (Zhu et al., 2023). On the basis of food groups, the Food and Agriculture Organization (FAO) classified food waste into different groups, which includes 30% wastage of cereals, pulses, and oilseeds, 35% wastage of meat, poultry, and fish, 20% wastage of dairy and dairy products, and 45–50% wastage produced from fruits and vegetables (Zhu et al., 2023). According to the report of FAO (2023), in a year, a total amount of 570 million tons of food is wasted from households, and 74 kg/year of food is wasted by each of us globally. It has been identified that industrialized and developed countries are responsible for food waste worth $680 billion, which is higher compared to developing countries ($310 billion) (FAO, 2011). The reduction of food waste and the utilization of food waste to develop value-added products can meet the goal of "Change of Our World: Sustainable Development 2030," which provides solutions to global development issues including poverty eradication, social protection, climate change, and employment promotion (Zhu et al., 2023). To attain the value addition of food waste, food waste should first be categorized into various group and the essential components identified using different techniques (Table 2.1).

The fruit and vegetables group consists of peels, seeds, pomace, pulp, skin, and shell as waste. When produce undergoes changes in colour, biochemical properties, microbial infestation, freeze burn, thermal treatment, burnishing of fruits and vegetables, and overripeness levels that are unpalatable to customers, they are sometimes deemed waste. From farm to fork, waste can happen at any stage during the food supply system. In the course of harvesting, shipping, selling, and processing, around 30% of the fruits and vegetables are lost (Wang et al., 2022). The details are given in Table 2.2, which includes the waste from fruit and vegetables with their essential components that have a therapeutic activity. By-products of grain waste are produced through several technical stages. Wastes from the cleaning process of grains include straw and cob; bran is created when the cereal is used to extract certain chemicals, including rice and corn oil, or during the polishing of rice during processing (Viganó et al., 2015). Wheat bran, wheat germ, rice bran, rice germ, corn germ, corn bran, corn germ, barley bran, and brewery spent grain are the cereal-based wastes that may be used

TABLE 2.1
Essential Components of Food Waste Sources

Essential Components	Food Waste Sources	References
Vitamin E, phytates, insoluble dietary fiber, phenols, γ-Oryzanol, tocopherols, carbohydrates, lipid-soluble vitamins, folic acid, phytosterols, amino acids, oligosaccharides, peptides, activated carbon, proteins, lipids, fatty acids, minerals	Cereal, pulses, and oilseeds	Papageorgiou and Skendi (2018); Wanyo et al., (2014); Balandrain-Quintana et al., (2015); Kiran et al., (2014); Ben-Othman et al., (2020)
Prebiotics, biodiesel, ethanol, whey protein, lactose, galactooligosaccharides, baker's yeast, minerals, ethanol, single cell protein, animal feed, whey protein, lactose	Dairy and dairy products	Hua et al., (2010); Lappa et al., (2019)
Feather meal, lactic acid, animal feed, blood meal, meat and bone meal, probiotics, chitin, calcium carbonate, protein, astaxanthin, chitinase	Meat, poultry, eggs, and seafood	Yaakob et al., (2019); Marques et al., (2016); Ashayerizadeh et al., (2017)

for the extraction of bioactive components (Cassano et al., 2018). Stems, leaves, husks from rice, maize, millets, and barley used as fodder, fertilizer, and can be used to extract phytates, phenolics, and insoluble dietary fiber (Liu et al., 2023).

Refining processes include protocols like bleaching, neutralization, degumming, and deodorization, which marks the production of waste at each step of oil processing. Edible oils degrade and turn into waste mostly due to oxidative or hydrolytic rancidity. The causes of this rancidity include the effects of light, heat, oxygen, water, and bacteria, as well as the gradual hydrolytic process or oxidation. Further production of lubricious acid and glycerol occurs due to breakdown of neutral grease or the formation of peroxide during the cleavage of unsaturated chains within the fatty acids, which break down into various materials, including ketones, lower-grade fatty acids, and aldehydes. The extraction of squalene, tocopherols, and sterols – waste products produced through the edible oil industry – leads to these substances being utilized as crude ingredients in numerous sectors, including processes for medicinal compositions and the synthesis of food-grade single-cell oil or protein (Diwan et al., 2018; Sherazi et al., 2016).

Each year, nearly 128 million tons of milk are wasted worldwide. The dairy sector produces the majority of its waste as an outcome of processing, microbial deterioration, and inappropriate handling. The complex organic makeup of dairy products consists of carbohydrates, lipids, proteins, and a few dietary additives. Along with this, by dissociating milk, proteins consisting of nucleic acids, urea, proteins, nitrites, nitrates, and ammonium ions may be utilized to obtain value-added products (ethanol, enzyme synthesis) (Kushwaha et al., 2011; Parashar et al., 2016). The percentages of slaughterhouse waste, animal by-products, and wastewater generated through the meat, poultry, and egg industries comprises 49% from cattle, 47% from lamb and sheep waste, 44% from pigs, and 37% of chicken waste. Common waste produced by industries includes feathers, hair, skin, horns, hooves, soft tissue, deboning remnants, and bones. The nutrient-rich source of abattoir waste may be processed into a range of valuable products, such as blood used for both culinary and industrial applications, methane, biomass, and biogas. Among these products are fish feed, binders, fertilizers, puddings, clots, blood cakes, and a sustainable energy replacement made from wastewater digested anaerobically. Pork fat waste has been converted to biodiesel by fermenting it with *Staphylococcus xylosus*, producing biochar from chicken dung using pseudo-catalytic transesterification processes, and using eggshells (Adhikari et al., 2018; Marques et al., 2016). Every year, 50–70% of raw seafood is tossed away, including fish scales, crab shells, prawn shells, prawn waste, and endoskeleton shells from crustaceans. Crustacean waste holds astaxanthin, a pigment akin to lutein, also recognized

TABLE 2.2
Fruits and Vegetable Waste with Essential Component and Therapeutic Activity

Essential Component	Waste	Therapeutic Activity	References
Caffeic acid	Shells of coffee	Antibacterial, antioxidant	Marinova et al., (2009)
Polyphenols	Peels of apple, artichoke, date, cauliflower, banana, mango, orange, carrot, and onion	Antimicrobial, antioxidant	Zhu et al., (2023)
Carotenoids	Pomace of grape, pomegranate, carrot, tomato seeds and skin, persimmon and lemon peel	Antioxidant	Andrés et al., (2017); Bellur et al., (2015); Hwang et al., (2011)
Sulforaphane	Broccoli seeds	Anticarcinogenic, antibacterial	Xing et al., (2018)
Flavonoids	Peels of satsuma, orange, lemon, banana, grape seeds and skin	Antibacterial, antioxidant	Zhu et al., (2023)
Dietary fiber	Peel of grape, orange, lemon, potato, carrot pomace, tamarind seeds, watermelon rinds	Antiflatulence	Zhu et al., (2023)
Pectins	Peel of citrus, pumpkin, pistachio green hull, tomato wastes, sugar beet	Help in bowel movements	Zhu et al., (2023)
Unsaturated fatty acids	Tomato seeds, pistachio pomace	Antioxidant	Elbadrawy et al., (2016); Abdolshahi et al., (2015)
Amino acids and proteins	Peels of pineapple, papaya, kinnow mandarin waste	Increase bone density	El-Safy et al., (2012)
Saponins	Sapota seeds	Antibacterial	Kothari et al., (2010)
Glycosides	Banana stem, apple peels	Anticancer, antioxidant	Milner and Romagnolo (2010)

as 3,30-dihydroxy-carotene-4,40-dione. This compound may be retrieved by the oxidation process of β-carotene or zeaxanthin found in microalgae consumed as feed. The waste generated by the prawn processing industry can be utilized to make this kind of substance. Studies have demonstrated the considerable antibacterial properties of chitosan, a biopolymer derived from the waste of crustaceans, against a range of bacterial species, including *Enterococcus faecalis*, *Escherichia coli*, *Staphylococcus aureus*, and *Candida albicans*. Glycosaminoglycans extracted from marine animals were shown to be of higher quality when compared to those from terrestrial creatures in a comparative analysis (Prameela et al., 2017; Valcarcel et al., 2017).

2.4 CHARACTERIZATION OF FOOD WASTE

Food waste characterization presents many challenges given its high complexity and heterogeneous collection of organic elements with diverse composition. For extraction, isolation, or concentration of certain components as demanded by some characterization techniques, analytical procedures frequently call for thorough sample preparation. Further, the interpretation of data in the form of complicated chromatograms or spectra comprising numerous peaks dedicated to a vast number of compounds within the food waste sample calls for experience. It is, therefore, difficult for a single approach to offer a thorough or standalone characterization, and sometimes it is essential to combine various procedures, which may be expensive and time-consuming. Variations in the characterization of food waste arising from the absence of reference materials and standardized methodologies also pose a major challenge. Despite these challenges, several studies have attempted to characterize food waste through advanced techniques (Figure 2.1), some of which are discussed in this section.

2.4.1 Nuclear Magnetic Resonance (NMR)

NMR is a technique consisting of specific nuclei that, under external magnetic fields, may exist in particular kinds of nuclear spin states which are responsible for specific transitions due to changes in their chemical or molecular structure, which can be characterized by NMR. One of the limits of NMR spectroscopy is that it cannot detect nuclei (if the spin is equal to zero). The NMR technique is useful in characterizing molecular structure in various fields of research, such as biochemistry, chemistry, physics, and cell structure (Diehl, 2008). In a study, mixed food waste was allowed to torrefy at 300°C at a heating rate of 10°C per minute for generating bio-oil. NMR was utilized as both a quantitative and qualitative method, and the components discovered using GC-MS were cross checked via NMR for the bio-oil (Modak et al., 2023). In another investigation, anaerobic digestion was used at two hydraulic retention times to break down the organic materials found in food waste in mesophilic and thermophilic environments. ^{1}H nuclear magnetic resonance monitored the evolution of the digestive process. Compared to thermophilic stabilization, NMR techniques indicated that longer retention durations would be needed for food waste stabilization under mesophilic settings. NMR techniques revealed that the stabilization procedure involved two stages. The first involved the formation of complex organic compounds, which were then digested to provide adequate hydraulic retention time. In mesophilic environments, longer hydraulic retention durations were needed to stabilize food waste (Gómez et al., 2010). Organic solid wastes can serve as viable feedstocks for the synthesis of liquid biofuels, which could serve as acceptable substitutes for fossil fuels in the transportation and heating sectors, as well as for industrial applications. The wet biomass is partially converted into a water-immiscible, oil-like organic substance known as bio-oil by hydrothermal liquefaction. For the purpose of characterizing the hydrothermal liquefaction of bio-oil at the molecular level, an integrated NMR spectroscopy/mass spectrometry technique was devised. The aromatic carbon content of the combination was determined, and functional groups were identified using ^{1}H and ^{13}C NMR spectroscopy (Leonardis et al., 2013).

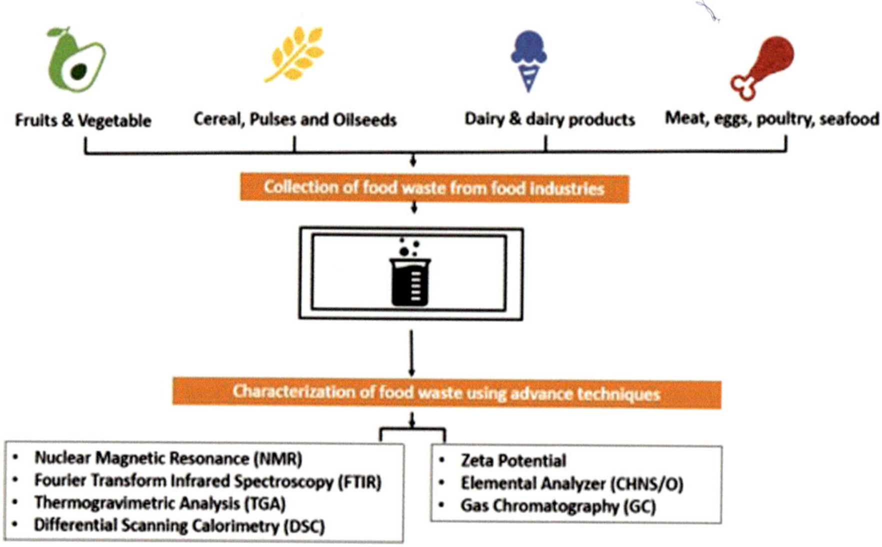

FIGURE 2.1 Characterization of food waste using various techniques.

2.4.2 Fourier Transform Infrared (FTIR) Spectroscopy

A tried-and-true method for identifying and analyzing the structure of chemical substances is infrared spectroscopy. The numerous chemical bonds and functional groups that are present in the molecules are linked to the peaks in the infrared spectrum of a sample, which indicate the excitation of the vibrational modes of the molecules in the sample. Consequently, one of the compound's most distinctive physical characteristics is its infrared spectrum, which is sometimes referred to as its "fingerprint." Given that the concentration of a molecule is directly correlated with its infrared energy absorption, infrared spectroscopy is an effective method for quantitative examination as well. IR spectroscopy can be used in the quantitative and qualitative investigation of food systems (Mohamed et al., 2017). The infrared spectral pattern of waste products reflects the stages of organic matter degradation. As the waste material is biologically treated, indicator bands shift. Process control can be based on how the band intensities change. Divergent compost properties can be linked to different spectral patterns of yard-waste composting processes. Municipal solid waste can be treated mechanically and biologically to produce a final product that can be dumped in a landfill. The quality of the finished product is shown by a few missing bands that indicate metabolic processes. Comparing unknown materials to pure compounds or landfill components can help identify them. Based on a spectral pattern corresponding to a particular stage of decomposition, waste organic matter in abandoned landfills is evaluated (Smidt & Schwanninger, 2005). Food waste thermal degradation has garnered a lot of attention lately. Acidification and evaporation were used to create nutrient-dense solid digestates from raw liquid anaerobic digestate (milk and brewery wastes and expired foods from grocery stores). Ammonium was preserved in the digestate by acidification, which also contributed the anion to liberate ammonium and create an ammonium salt. Sulfuric, nitric, and phosphoric acids were added to the digestate to produce ammonium sulfate, ammonium nitrate, and ammonium phosphate. Further ammonium salts were characterized using FTIR between acidified and unacidified digestible samples (O'Connor et al., 2024).

2.4.3 Thermogravimetric Analysis (TGA)

The mass change (gain or loss) of the substances with respect to temperature and time is measured by TGA. The main purposes of these measurements are to ascertain a material's composition and forecast its thermal stability. Materials exhibiting weight gain or loss as a result of oxidation, reduction, breakdown, or sorption/desorption of volatiles can be described using this approach (Prime et al., 2009). Food waste was converted to biochar at high temperatures of 200–350 °C. TGA was used to characterize the hydrochar samples in order to get a comprehensive understanding of their physicochemical and structural features. In direct proportion to the heating temperature, food waste's carbon content and greater heating value rose significantly, rising from 39 to 73% and 15 to 31 MJ/kg, respectively. The thermal behavior of the coal and hydrochar blends made in three distinct ratios (5%, 10%, and 15%) varied. The co-combustion study's overall findings demonstrated that as temperature increased, the activation energy of the hydrochar samples dropped from 56.78 KJ/mol to 29.80 KJ/mol. With a coal blending ratio of 10%, hydrochar made at 300 °C had the lowest activation energy of 19.45 KJ/mol. Furthermore, thermal gravimetric examination of the samples demonstrated that hydrochar's combustion capabilities may be enhanced by high-temperature carbonization (Saqib et al., 2018). Using TGA, the three food waste products that were chosen – carrots, cucumbers, and tomatoes – were examined. One food component (carrot), one binary combination (carrot + cucumber), and one ternary blend (carrot, cucumber, and tomato) were all included in the biochar analysis. Kinetic modeling investigations with the Arrhenius and Coats-Redfern models were conducted utilizing two heating rates. The gas analysis's findings show that mixing the food wastes increased the production of hydrogen (Elkhalifa et al., 2022). Using TGA coupled with an FTIR and a pyrolyzer coupled with a gas chromatograph and mass spectrometer, the thermal decomposition behavior, volatile release characteristics, pyrolytic product composition, and distribution of two typical food waste components (pork and rice) were studied. The major pyrolysis reaction and the volatilization of moisture were the two distinct pyrolysis processes that both rice and pork underwent, according to the pyrolysis studies (Ming et al., 2020).

2.4.4 Differential Scanning Calorimetry (DSC)

DSC is a technique that evaluates the changes in the physical state of substances with changes in temperature and time. It may also it may evaluate the heat released and heat absorbed during the change in the physical state of substance which can be further measured by enthalpy. Thermal analysis in DSC can be carried out using two common methods, i.e., heat flux and power compensation. In first method, exothermic or endothermic reactions take place during the thermal analysis, which can be measured by the temperature difference of the sample unit and the reference unit. In power consumption, temperature of the sample units and reference units is changed to a predetermined value, and the temperature difference between the two units can be measured. DSC has been used to analyze the degradation properties of food wastes. For instance, four food wastes were studied for their slow pyrolysis properties: acorn cups and pericarp, shells, almonds, and nuts. The thermogravimetric (TG) study was combined with DSC and mass spectrometry (MS). The water vaporization, devolatilization, and formation of char corresponded to the three phases of thermal deterioration of these samples, and changes in the thermal behavior of the samples were associated with compositional variations in the samples (Malika et al., 2016). DSC has also been used to characterize three distinct streams of organic municipal wastes, which are usually mixed with food waste: composted sewage sludge and municipal waste, and dried sewage sludge. The organic matter in the CS was the most stable, while the organic matter in the TS was the least stable, according to the total quantities of CO_2 respired (Fernández et al., 2012). Municipal solid wastes are used to obtain compost/vermi-compost, which can be characterized for their maturity indices. In comparison to compost, DSC analysis indicated that vermicompost showed a significant breakdown of complex and bigger macromolecules into simpler compounds (Soobhany et al., 2017).

2.4.5 Zeta Potential

Zeta potential is a technique used to describe the electrokinetic potential of colloidal dispersions (liquid-liquid and solid-liquid). In zeta potential, the particles in colloidal dispersions become charged when they are introduced into the zeta cell due to particle movement towards the electrode in the presence of an electric field. This phenomenon is also known as electrophoresis, electroosmosis, and sedimentation potential. LASER Doppler velocimetry is used to quantify their velocity, or electrophoretic mobility, under the influence of an electric field (Prakash et al., 2014). Henry's equation is then applied to determine the zeta potential:

$$U_E = \frac{2eZ}{3n} f(Ka)$$

U_E: electrophoretic mobility
e: dielectric constant
Z: zeta potential
n: viscosity
$f(Ka)$: Henry's function

The hydrothermal dewatering process of digestate from food waste (DFW) at various temperatures was examined in this work. Divalent cations raised the zeta potential, which aided in particle settling, while membrane rupture and solid size reduction caused water to be released between 120 and 180°C. Temperatures above 180°C resulted in the polymerization and aromatization of organic materials, which decreased bound water and formed hydrophobic humic compounds. By combining with cationic metals, these compounds improved DFW dewaterability and enhanced nanoparticle agglomeration while decreasing zeta potential. It is essential to understand these procedures in order to manage and value DFW effectively (Zhang et al., 2022).

2.4.6 Elemental Analyzer (CHNS/O)

The elemental analyzer is a tool for the quantitative analysis of organic materials that contain the elements carbon, hydrogen, nitrogen, sulfur, and oxygen. The device's primary method of operation involves burning the sample at a temperature of at least 1100°C. An inert gas (helium, He) is employed as the carrier to send the gaseous sample to the chromatography column. To complete the quantitative combustion process, the gaseous sample is burned with oxygen (O_2) gas and moved through the oxidation zone before being oxidized with the appropriate catalysts. The elements C, H, N, and S are transformed into the gases CO_2, H_2O, N_2, and SO_2, and they are then transported to a thermal conductivity detector, which records an electrical signal proportional to the amount of each gas. Next, based on the curve areas found in the spectrum, this electrical signal provides, for instance, the percentage of elemental composition. A combination of hydrogen and helium (H_2-He) gases pyrolyzes the sample to determine the element oxygen. The catalytic system transforms all oxygen-containing combustion byproducts into carbon monoxide, and elemental determination is also performed (Fadeeva et al., 2008).

The food waste samples (tofu, carrot, corn, chicken, biscuit, rice, and beef) were analyzed for the energy using a CHNS/O analyzer (Ahmad et al., 2019). The absence of sulfur in all food waste samples suggests the advantages of using food waste as a renewable energy source. With 53% of all elements detected in beef, carbon has the largest percentage dispersion ever measured. Carrots' low carbon content suggests that they have limited promise as an energy fuel. However, given that carrots have a high ash level, this suggests that they might be added to organic waste for composting (Ahmad et al., 2019).

2.4.7 Gas Chromatography

Gas chromatography is a technique in which a specific analyte in the sample is moved by the mobile phase (commonly helium gas due to its low molecular weight) and further separated by the stationary phase present in column, which segregates the analytes according to their molecular weight. When analyzing volatile molecules that contribute to food's flavor and fragrance, gas chromatography (GC) is an essential tool for evaluating food quality, authenticity, and sensory attributes. The volatiles developed (i.e., CO_2, CH_4, C_2H_4, and C_2H_6) during the pyrolysis of food wastes (raw and digested) were characterized independently using gas chromatography (Opatokun et al., 2016). Despite the different heating rates, phenols, esters, saturated and unsaturated hydrocarbons, and their derivatives were detected; nevertheless, different quantities of these compounds were formed. In general, the oils generated from the RFW have greater chemical concentrations than the digestate (Opatokun et al., 2016).

2.5 CONCLUSIONS AND PERSPECTIVES

Characterizing food waste is essential to comprehending its makeup, its uses, and its effects on the environment. To do this, a variety of analytical methods are used, each having benefits and drawbacks. NMR spectroscopy offers qualitative information on the constituents of food waste. It provides insights into molecular architecture and chemical surroundings by looking at the nuclear spin states of certain nuclei. NMR, however, might not be appropriate for zero-spin nuclei. NMR provides extensive information; however, for cross-validation, it needs supporting methods such as GC-MS. Food waste may be easily identified by its chemical structures using FTIR spectroscopy. It offers a distinct "fingerprint" of the material by using its functional groups and chemical linkages. FTIR can be quantitative or qualitative, which helps track the different phases of organic matter breakdown. But deciphering spectral patterns necessitates knowledge and contrast with recognized substances. The composition and thermal stability of a material may be ascertained by measuring the weight change and rate of change in response to temperature using TGA. Although TGA is useful for analyzing the carbon content and investigating the thermal deterioration of food waste, it cannot offer comprehensive molecular information. DSC monitors how a sample's physical properties vary as a function of temperature. While it might not provide comprehensive compositional data, it is helpful for researching material transitions. The electrokinetic potential of colloidal dispersions is measured by the zeta potential, which indicates the stability and charge of the particles. Understanding complicated electrokinetic processes is necessary for interpreting results, as it aids in comprehending food waste dewaterability and interactions with cations. This instrument measures the amount of carbon, hydrogen, nitrogen, sulfur, and oxygen in organic materials quantitatively. It provides accurate elemental composition data, but neither molecule structures nor functional groups are covered. Food waste may be evaluated for quality and sensory qualities by using GC, which separates and detects volatile components in the trash. Although extremely sensitive, it could need complicated sample preparation and calibration.

REFERENCES

Abdolshahi, A.; Majd, M.H.; Rad, J.S.; Taheri, M.; Shabani, A.; Da Silva, J.A.T. Choice of solvent extraction technique affects fatty acid composition of pistachio (Pistacia vera L.) oil. *J. Food Sci. Technol.* 2015, 52, 2422–2427.

Adhikari, B.B.; Chae, M.; Bressler, D.C. Utilization of slaughterhouse waste in value-added applications: Recent advances in the development of wood adhesives. *Polymers* 2018, 10, 176.

Ahmad, N.; Sahrin, N.; Talib, N.; Ghani, F.A. Characterization of energy content in food waste by using thermogravimetric analyser (TGA) and elemental analyser (CHNS-O). In *Journal of Physics: Conference Series 2019* (Vol. 1349, No. 1, p. 012140). IOP Publishing.

Amasuomo, E.; Baird, J. The concept of waste and waste management. *J. Mgmt. & Sustainability* 2016, 6, 88.

Andrés, A.; Petrón, M.; Adámez, J.; López, M.; Timón, M. Food by-products as potential antioxidant and antimicrobial additives in chill stored raw lamb patties. *Meat Sci.* 2017, 129, 62–70.

Ashayerizadeh, O.; Dastar, B.; Samadi, F.; Khomeiri, M.; Yamchi, A.; Zerehdaran, S. Study on the chemical and microbial composition and probiotic characteristics of dominant lactic acid bacteria in fermented poultry slaughterhouse waste. *Waste Manag.* 2017, 65, 178–185.

Balandrán-Quintana, R.R.; Mercado-Ruiz, J.N.; Mendoza-Wilson, A.M. Wheat bran proteins: A review of their uses and potential. *Food Rev. Int.* 2015, 31, 279–293.

Barp, L.; Višnjevec, A. M.; Moret, S. Pressurized liquid extraction: A powerful tool to implement extraction and purification of food contaminants. *Foods* 2023, 12(10), 2017.

Bellur Nagarajaiah, S.; Prakash, J. Nutritional composition, acceptability, and shelf stability of carrot pomace-incorporated cookies with special reference to total and _-carotene retention. *Cogent Food Agric.* 2015, 1, 1039886.

Ben-Othman, S.; Jõudu, I.; Bhat, R. Bioactives from agri-food wastes: Present insights and future challenges. *Molecules* 2020, 25, 510.

Boateng, I.D. Recent processing of fruits and vegetables using emerging thermal and non-thermal technologies. A critical review of their potentialities and limitations on bioactives, structure, and drying performance. *Crit. Rev. Food Sci. Nutr.* 2022. https://doi.org/10.1080/10408398.2022.2140121

Cassano, A.; Galanakis, C.M. Membrane technologies for the fractionation of compounds recovered from cereal processing by-products. In *Sustainable Recovery and Reutilization of Cereal Processing By-Products*; Galanakis, C.M., Ed.; Woodhead Publishing: Sawston, UK, 2018; pp. 159–187.

Defra, 2009. UK food security assessment: our approach. http://archive.defra.gov.uk/foodfarm/food/pdf/foodassess-approach-0908.pdf (Accessed 25.01.12).

Diehl, B. Principles in NMR spectroscopy. In *NMR Spectroscopy in Pharmaceutical Analysis*. Elsevier, 2008, pp. 1–41.

Diwan, B.; Parkhey, P.; Gupta, P. From agro-industrial wastes to single cell oils: A step towards prospective biorefinery. *Folia Microbiol.* 2018, 63, 547–568.

Elbadrawy, E.; Sello, A. Evaluation of nutritional value and antioxidant activity of tomato peel extracts. *Arab. J. Chem.* 2016, 9, S1010–S1018.

Elkhalifa, Samar.; Prakash, Parthasarathy.; Hamish, R. Mackey.; Tareq, Al-Ansari.; Omar, Elhassan.; Said, Mansour.; Gordon, McKay. Biochar development from thermal TGA studies of individual food waste vegetables and their blended systems. *Biomass Convers. Bior.* 2022, 1–18.https://doi.org/10.1007/s13399-022-02441-0

El-Safy, F.S.; Salem, R.H.; Abd El-Ghany, M. Chemical and nutritional evaluation of different seed flours as novel sources of protein. *World J. Dairy Food Sci.* 2012, 7, 59–65.

European Commission. *Food waste and its impacts: European week for waste reduction.* 2014.

European Parliament Council. Council directive of 15 July 1975 on waste 75/442/EEC. *Off. J. L.* 1975, 194, 3941.

Fadeeva, V.P.; Tikhova, V.D.; Nikulicheva, O.N. Elemental analysis of organic compounds with the use of automated CHNS analyzers. *Journal of analytical chemistry.* 2008 Nov;63:1094–106.

FAO. *State of food security and nutrition in the world report.* 2023.

Fernández, José M.; César, Plaza.; Alfredo, Polo.; Alain F. Plante. Use of thermal analysis techniques (TG–DSC) for the characterization of diverse organic municipal waste streams to predict biological stability prior to land application. *Waste Manag.* 2012, 32(1), 158–164.

Foresight. 2010. *How can waste reduction help to healthily and sustainably feed a future global population of nine billion people?* London. http://www.bis.gov.uk/assets/foresight/docs/food-and-farming/workshops/11-608-w4-expert-forum-reduction-of-foodwast

Giordano, C.; Franco, S. Household food waste from an international perspective. *Sustainability* 2021, 13(9), 5122.

Gómez, X.; Cuetos, M.J.; Tartakovsky, B.; Martínez-Núñez, M. F.; Moran, A. A comparison of analytical techniques for evaluating food waste degradation by anaerobic digestion. *Bioprocess Biosyst. Eng.* 2010, 33, 427–438.

Grassino, A. N.; Ostojić, J.; Miletić, V.; Djaković, S.; Bosiljkov, T.; Zorić, Z.; Brnčić, M. Application of high hydrostatic pressure and ultrasound-assisted extractions as a novel approach for pectin and polyphenols recovery from tomato peel waste. *Innov. Food Sci. Emerg. Technol.* 2020-, 64, 102424.

Gustavsson, J.; Cederberg, C.; Sonesson, U.; van Otterdijk, R.; Meybeck, A. *Global food losses and food waste: Extent, causes and prevention.* Food and Agriculture Organisation of the United Nations (FAO): Rome, Italy, 2011.

Hua, X.; Yang, R.; Zhang, W.; Fei, Y.; Jin, Z.; Jiang, B. Dual-enzymatic synthesis of lactulose in organic-aqueous two-phase media. *Food Res. Int.* 2010, 43, 716–722.

Hwang, I.-W.; Jeong, M.-C.; Chung, S.-K. The physicochemical properties and the antioxidant activities of persimmon peel powders with different particle sizes. *J. Korean Soc. Appl. Biol. Chem.* 2011, 54, 442–446.

Kader, A.A. Handling of horticultural perishables in developing vs. developed countries. *VI International Post Harvest Symposium* 2009, 877, 121126.

Kavitha, S.; Kannah, R. Y.; Kumar, G.; Gunasekaran, M.; Banu, J. R. Introduction: Sources and characterization of food waste and food industry wastes. In *Food waste to valuable resources* (pp. 1–13). Academic Press, 2020.

Kiran, E.U.; Trzcinski, A.P.; Ng, W.J.; Liu, Y. Bioconversion of food waste to energy: A review. *Fuel* 2014, 134, 389–399.

Kothari, V.; Seshadri, S. In vitro antibacterial activity in seed extracts of Manilkara zapota, Anona squamosa, and Tamarindus indica. *Biol. Res.* 2010, 43, 165–168.

Kushwaha, J.P.; Srivastava, V.C.; Mall, I.D. An overview of various technologies for the treatment of dairy wastewaters. *Crit. Rev. Food Sci. Nutr.* 2011, 51, 442–452.

Lappa, I.K.; Papadaki, A.; Kachrimanidou, V.; Terpou, A.; Koulougliotis, D.; Eriotou, E.; Kopsahelis, N. Cheese whey processing: Integrated biorefinery concepts and emerging food applications. *Foods* 2019, 8, 347.

Leonardis, Irene; Stefano, Chiaberge; Tiziana, Fiorani; Silvia, Spera; Ezio, Battistel; Aldo, Bosetti; Pietro, Cesti; Samantha, Reale; Francesco, De Angelis. Characterization of bio-oil from hydrothermal liquefaction of organic waste by NMR spectroscopy and FTICR mass spectrometry. *Chem. Sus. Chem.* 2013, 6(1), 160–167.

Lipinski, B.; Hanson, C.; Lomax, J.; Kitinoja, L.; Waite, R.; Searchinger, T. *Reducing food loss and waste.* World Resource Institute, Washington, DC, Working Pap. 140, 2013..

Liu, Z.; de Souza, T. S.; Holland, B.; Dunshea, F.; Barrow, C.; Suleria, H. A. Valorization of food waste to produce value-added products based on its bioactive compounds. *Processes* 2023, 11(3), 840.

Malika, A.; Jacques, N.; Jaafar, E.F.; Fatima, B.; Mohammed, A. Pyrolysis investigation of food wastes by TG-MS-DSC technique. *Biomass Convers. Biorefin.* 2016, 6, 161–172.

Marinova, E.M.; Toneva, A.; Yanishlieva, N. Comparison of the antioxidative properties of caffeic and chlorogenic acids. *Food Chemistry.* 2009 Jun 15; 114(4):1498–502.

Marques, R.V.; Paz, M.F.d.; Duval, E.H.; Corrêa, L.B.; Corrêa, É.K. Staphylococcus xylosus fermentation of pork fatty waste: Raw material for biodiesel production. *Braz. J. Microbiol.* 2016, 47, 675–679.

Milner, J.A.; Romagnolo, D.F. Cancer biology and nutrigenomics. In *Bioactive compounds and cancer.* Springer: Berlin/Heidelberg.

Ming, Xue; Fanfan, Xu; Yuan, Jiang; Peijie, Zong; Bo, Wang; Jun, Li; Yingyun, Qiao; Yuanyu, Tian. Thermal degradation of food waste by TG-FTIR and Py-GC/MS: Pyrolysis behaviors, products, kinetic and thermodynamic analysis. *J Clean. Prod.* 2020, 244, 118713.

Modak, S.; Jain, S.; Katiyar, P.; Yadav, S. Compositional study of bio-oil obtained from torrefaction of mixed food waste using H-NMR, 13C-NMR, DEPT 90 & 135, and GC-MS. *Biofuels* 2023, 14(2), 183–189.

Mohamed, M. A.; Jaafar, J.; Ismail, A.F.; Othman, M.H.D.; Rahman, M.A. Fourier transform infrared (FTIR) spectroscopy. In *Membrane characterization.* 2017, pp. 3–29. Elsevier.

Niksiar, A.; Nasernejad, B. Activated carbon preparation from pistachio shell pyrolysis and gasification in a spouted bed reactor. *Biomass Bioenerg.* 2017, 106, 43–50.

Opatokun, S.A.; Kan, T.; Al Shoaibi, A.; Srinivasakannan, C.; Strezov, V. Characterization of food waste and its digestate as feedstock for thermochemical processing. *Energy & Fuels* 2016, 30(3), 1589–1597.

Papageorgiou, M.; Skendi, A. Introduction to cereal processing and by-products. In *Sustainable recovery and reutilization of cereal processing by-products.* Elsevier: Amsterdam, The Netherlands, 2018; pp. 1–25.

Parashar, A.; Jin, Y.; Mason, B.; Chae, M.; Bressler, D.C. Incorporation of whey permeate, a dairy effluent, in ethanol fermentation,to provide a zero waste solution for the dairy industry. *J. Dairy Sci.* 2016, 99, 1859–1867.

Parfitt, J.; Barthel, M.; Macnaughton, S. 2010. Food waste within food supply chains: Quantification and potential for change to 2050. *Philos. Trans. R. Soc. B Biol. Sci.* 365, 30653081.

Prakash, S.; Mishra, R.; Malviya, R.; Sharma, P.K. Measurement techniques and pharmaceutical applications of zeta potential: a review. *J. Chronother. Drug*, 2014, 5(2), 33–40.

Prameela, K.; Venkatesh, K.; Immandi, S.B.; Kasturi, A.P.K.; Krishna, C.R.; Mohan, C.M. Next generation nutraceutical from shrimp waste: The convergence of applications with extraction methods. *Food Chem.* 2017, 237, 121–132.

Prime, R.B.; Bair, H.E.; Vyazovkin, S.; Gallagher, P.K.; Riga, A. Thermogravimetric analysis (TGA). In J. D. Menczel and R. B. Prime (Eds.), *Thermal analysis of polymers: Fundamentals and applications.* John Wiley & Sons, New York, 2009, pp. 241–317.

Ren, Y.; Yu, M.; Wu, C.; Wang, Q.; Gao, M.; Huang, Q.; Liu, Y. A comprehensive review on food waste anaerobic digestion: Research updates and tendencies. *Bioresour. Technol.* 2018, 247, 1069–1076.

Saqib; Najam, Ul.; Saeid, Baroutian.; Ajit, K. Sarmah. Physicochemical, structural and combustion characterization of food waste hydrochar obtained by hydrothermal carbonization. *Bioresource Technology* 2018, 266, 357–363.

Sherazi, S.T.H.; Mahesar, S.A. Vegetable oil deodorizer distillate: A rich source of the natural bioactive components. *J. Oleo Sci.* 2016, 65, 957–966.

Sinha, S.; Tripathi, P. Trends and challenges in valorisation of food waste in developing economies: A case study of India. *CSCEE* 2021, 4, 100162.

Smidt, E.; M. Schwanninger. Characterization of waste materials using FTIR spectroscopy: process monitoring and quality assessment. *Spectroscopy Letters* 2005, 38(3), 247–270.

Soobhany, N.; Gunasee, S.; Rago, Y.P.; Joyram, H.; Raghoo, P.; Mohee, R.; Garg, V.K. Spectroscopic, thermogravimetric and structural characterization analyses for comparing Municipal Solid Waste composts and vermicomposts stability and maturity. *Bioresour Technol.* 2017, 236, 11–19.

Valcarcel, J.; Novoa-Carballal, R.; Pérez-Martín, R.I.; Reis, R.L.; Vázquez, J.A. Glycosaminoglycans from marine sources as therapeutic agents. *Biotechnol. Adv.* 2017, 35, 711–725.

Viganó, J.; Machado, A.P.D.F.; Martínez, J. Sub- and supercritical fluid technology applied to food waste processing. *J. Supercrit. Fluids* 2015, 96, 272–286.

Wang, Y.; Jiang, L.; Dai, L.; Yu, Z.; Liu, Y.; Ruan, R.; Fu, G.; Zhou, Y.; Fan, L.; Duan, D. Microwave-assisted catalytic co-pyrolysis of soybean straw and soapstock for bio-oil production using SiC ceramic foam catalyst. *J. Anal. Appl. Pyrolysis* 2018, 133, 76–81.

Wang, Y.; Pan, S.; Yin, J.; Feng, H.;Wang, M.; Chen, T. Resource potential and global warming potential of fruit and vegetable waste in China based on different treatment strategies. *Waste Manag.* 2022, 140, 225–232.

Xing, J.-J.; Cheng, Y.-L.; Chen, P.; Shan, L.; Ruan, R.; Li, D.; Wang, L.-J. Effect of high-pressure homogenization on the extraction of sulforaphane from broccoli (Brassica oleracea) seeds. *Powder Technol.* 2018, 358, 103–109.

Yaakob, M.A.; Mohamed, R.M.S.R.; Al-Gheethi, A.; Tiey, A.; Kassim, A.H.M. Optimising of Scenedesmus sp. biomass production in chicken slaughterhouse wastewater using response surface methodology and potential utilisation as fish feeds. *Environ. Sci. Pollut. Res.* 2019, 26, 12089–12108.

Zhang, C.; Shao, M.; Wu, H.; Wang, N.; Wang, X.; Wang, Q.; Xu, Q. Mechanism insights into hydrothermal dewatering of food waste digestate for products valorization. *Sci. Total Environ.* 2022, 804, 150145.

Zhu, Y.; Luan, Y.; Zhao, Y.; Liu, J.; Duan, Z.; Ruan, R. Current technologies and uses for fruit and vegetable wastes in a sustainable system: *A review. Foods.* 2023 May 11;12(10):1949.

3 Thermo-Chemical Processes for Food Waste Treatment

Ayon Tarafdar, Ranjna Sirohi and Uadal Singh

3.1 INTRODUCTION

Food waste management is a critical global challenge, with significant environmental, economic, and social implications. Thermo-chemical processes offer promising solutions for the treatment of food waste, enabling its conversion into valuable products such as energy, biofuels, and chemicals. There are several processes that are used for the conversion of food waste to these valuable products. Some of the most salient processes include pyrolysis, gasification, hydrothermal conversion, torrefaction, and incineration.

Pyrolysis is a thermo-chemical conversion process that involves the decomposition of organic materials in the absence of oxygen at elevated temperatures. During pyrolysis, food waste is heated to temperatures typically ranging from 300 °C to 800 °C, leading to the release of volatile compounds, char, and gases (Xiong et al., 2018). These products can be further utilized for energy generation or as feedstocks for various industrial processes. The pyrolysis process comprises several stages, including drying, devolatilization, and char formation. Initially, moisture present in the food waste is removed during the drying stage. Subsequently, organic compounds undergo thermal decomposition, resulting in the release of volatile gases such as methane, carbon dioxide, and hydrogen. The remaining solid residue, known as biochar, contains carbonaceous material and can serve as a soil amendment or carbon sequestration agent. Recent advancements in pyrolysis technology include the development of integrated systems for on-site food waste treatment in commercial kitchens and urban areas. For instance, companies such as Agri-Tech Producers LLC have introduced compact pyrolysis units capable of processing food waste into biochar and bio-oil, reducing the environmental footprint associated with waste disposal.

Gasification is another thermochemical process that converts organic materials into a combustible gas mixture known as syngas, consisting primarily of carbon monoxide, hydrogen, and methane. Unlike pyrolysis, gasification involves the partial oxidation of food waste using a controlled amount of oxygen or steam, operating at temperatures typically exceeding 700 °C (Murugesan et al., 2022). During gasification, food waste is subjected to high temperatures and a controlled environment, promoting the breakdown of organic matter into gaseous components. The reactions occurring in the gasification process include combustion, pyrolysis, and reforming, leading to the production of syngas. This syngas can be utilized for heat and power generation, as well as for the synthesis of valuable chemicals such as methanol and ammonia.

Advanced gasification technologies, such as plasma gasification, have gained attention for their efficiency in treating food waste and producing clean syngas. Companies such as Sierra Energy have developed innovative plasma gasification systems capable of converting various waste streams, including food waste, into syngas with minimal emissions and environmental impact.

Hydrothermal processing involves the treatment of organic materials in hot, pressurized water, typically at temperatures above 200 °C. This thermo-chemical process facilitates the hydrolysis, solubilization, and conversion of food waste into biofuels, organic acids, and other valuable products. In hydrothermal processing, food waste is subjected to high temperatures and pressures, leading to the breakdown of complex organic compounds into simpler molecules. The presence of water

DOI: 10.1201/9781032706030-3

acts as a solvent, enhancing the dissolution and conversion of organic matter. Various reactions such as hydrolysis, dehydration, and decarboxylation occur, resulting in the formation of liquid, gas, and solid products.

Recent research efforts have focused on optimizing hydrothermal processing parameters and exploring novel catalysts to improve the efficiency and selectivity of product formation. For instance, studies have investigated the use of heterogeneous catalysts derived from waste materials such as zeolites, activated carbons, and metal oxides to enhance the conversion of food waste into biofuels and value-added chemicals.

This chapter explores various thermochemical processes utilized for food waste treatment, discussing their mechanisms, recent advancements, and potential applications.

3.2 TREATMENT OF FOOD WASTE BY PYROLYSIS

Depending on the settings and characteristics of the operation, there are many types of pyrolysis processes. Among these processes, slow and fast pyrolysis are the most often utilized methods. The process selection determines which kind of reactor is best to use in pyrolysis units (Roy & Dias, 2017). Rotary kilns, drums, and screw/auger reactors are the reactor types most frequently employed in slow pyrolysis. Conversely, reactors such as fluidized beds, spinning cones, entrained flow, vacuum, and ablative reactors are useful for rapid pyrolysis (Marshall, 2013). Reactors of the auger type can be used for both slow and fast pyrolysis. Catalytic microwave processes for in situ production of bioproducts from biomass and other waste reserves have also emerged (Irfan et al., 2023). This technology can rapidly (2–4 min) convert biomass with minimal residues. Where the segregation system for food waste is not very prominent, food wastes can often be found mixed with food plastic containers. Microwave-assisted pyrolyzers can be suitable in such situations for the conversion of plastic waste to carbon nanotubes and combustible fuels. Some of the most salient pyrolyzers have been shown in Figure 3.1.

Fast pyrolysis refers to heating the biomass medium for a brief period of time (1 to 10 sec) between 800 and 1300 °C at a rate of heating that varies from 10 to 200 °C. Since the oil product yield is higher than the char and gas product yields, fast pyrolysis is mostly utilized to produce bio-oil. Usually, 60–75% of the biofuel liquid products, 10–15% of the noncondensable gaseous products, and 15–20% of the solid biochars are produced during the rapid pyrolysis process (Heidari et al., 2019). The primary goal of the fast pyrolysis process is to minimize the exposure period, which encourages the creation of chars, while exposing the biomass feedstock to a temperature high enough to induce thermal cracking. In order to transform the feed biomass into liquid biofuel products, high heating rates are also applied during the rapid pyrolysis process to prevent the creation of unwanted char.

The normal form of pyrolysis, known as slow pyrolysis, comprises extended residence times and modest heating rates. Biomass is heated to a temperature of around 400–500 °C by pyrolysis, while maintaining a heating rate of 0.1–1 °C/s for a suitable duration ranging from 5 to 30 minutes. Although the generation of solid carbonaceous char is aided by the slow pyrolysis process, relatively modest amounts of liquid fuel and gas products are also formed (Demirbas & Arin, 2002).

Gasification and pyrolysis are the two methods available for producing biochars; however, pyrolysis is preferred over gasification since it produces greater yields of biochar and emits less hazardous gases, including SOX and NOX. The ability to handle a variety of feedstock types and operating circumstances with flexibility makes pyrolysis advantageous, as it can produce end products with desired properties. The products' yields were split across the liquid, solid, and gas residuals. For both wastes, the char yield was around 25%, and in both cases, the gas yield was between 15% and 25%. Conversely, the liquid residue yielded the most, averaging around 46% for the leftover cereal and 62% for the peanut crisps. The high oxygen content of the two types of feedstocks that were employed may be the cause of this notable volume of liquid. The yields of charcoal and gas were clearly impacted by temperature. In contrast to higher temperatures, lower

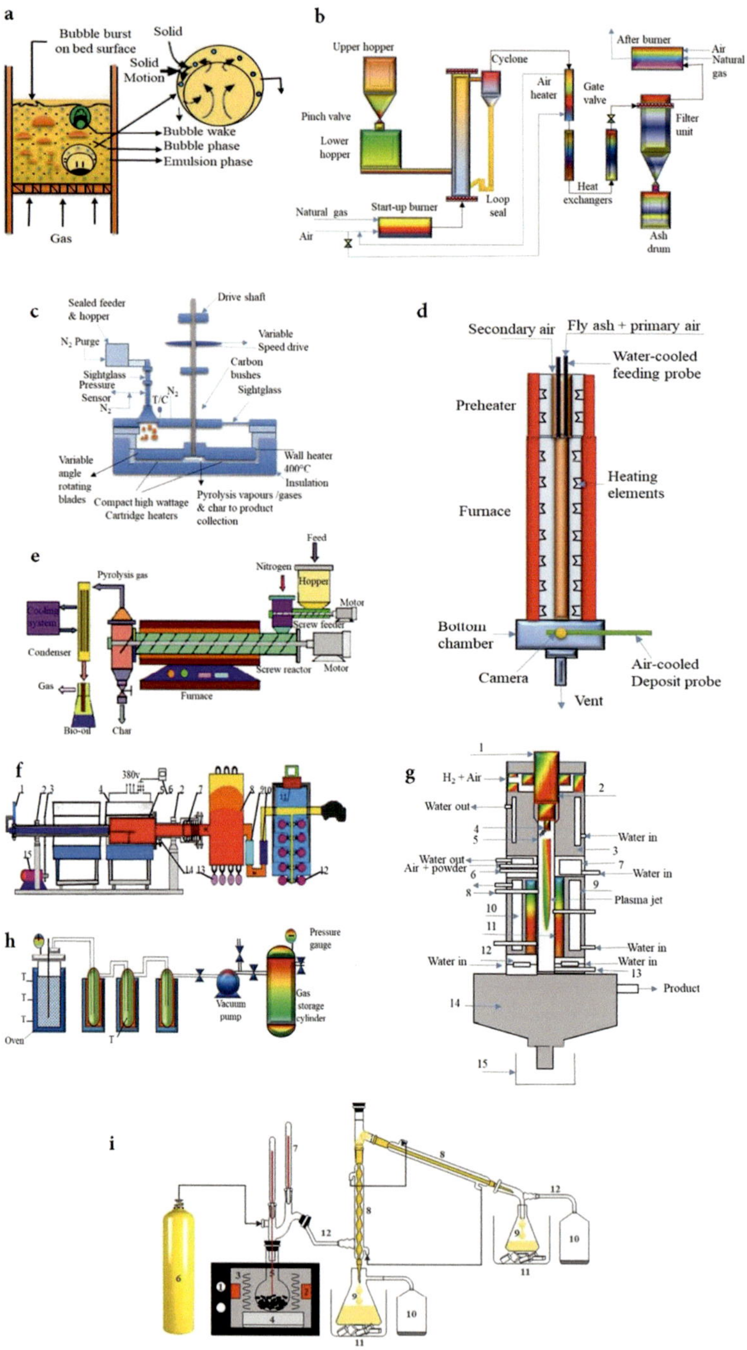

FIGURE 3.1 Types of pyrolysis reactors: a. bubbling fluidized bed reactor; b. circulating fluidized bed reactor; c. ablative pyrolysis reactor; d. entrained flow reactor; e. auger pyrolysis reactor; f. rotary kiln pyrolyzer with (1) thermometer, (2) bearing, (3) gear transmission, (4) electrical furnace, (5) rotary kiln, (6) temperature controller, (7) seal, (8) tube type condenser, (9) filter, (10) total flow meter, (11) computer, (12) gas sampling device, (13) tar reservoir, (14) feed and discharge opening, and (15) feed and discharge; g. plasma pyrolysis reactor with 1) cathode, (2) insulation, (3) anode, (4) arc, (5, 11) graphite linings, (6, 8, 13) sampling holes, (7) injector, (9) reactor, (10) cooling water, (12) water spray, (14) separator, (15) water filter; h. vacuum pyrolysis reactor; i. microwave pyrolysis reactor. (Adapted from Raza et al., (2021) and Irfan et al., (2023)).

temperatures promoted a higher production of charcoal, while raising the pyrolysis temperature sped up the evolution of hydrogen.

3.3 TREATMENT OF FOOD WASTE BY GASIFICATION

Food waste can be converted to biochar, bio-oil, and hydrogen through gasification (Figure 3.2). It is also an effective way to recover energy from food waste matrices, as it produces synthetic gas or syngas consisting of carbon monoxide and hydrogen. Gasification of food waste takes longer time than pyrolysis; however, the inorganic material in food char can act as a catalyst to improve the rate of gasification (Ahmed and Gupta, 2010). Studies show that during gasification, an increase in process temperature and time of operation, along with lowering of feeding rate, favors the production of hydrogen (Elgazar et al., 2024). Many advances have been made in the gasification process, among which the process of supercritical water gasification (SCWG) stands out, converting wet biomass into hydrogen-rich syngas. Unlike a traditional gasification process, it occurs at high pressure and low temperature. Furthermore, this technique uses very little energy for feedstock drying and pretreatment. A crucial component of the entire water-gas shift process is water. Figure 3.3 shows the supercritical water gasification system for the conversion of food waste to energy.

One of the biggest challenges of biomass gasification is overcoming the high initial and ongoing expenditures. The lower plant sizes (500 tonnes/day), which are determined by the intrinsic biomass availability and the cost of transportation in the scattered resource, are the cause of this issue. The presence of alkaline materials, such as potassium and sodium, which result in slag and fouling concerns, is another challenge. Reduced temperatures also cause unintended tar production, which causes major issues with functioning. Although a number of catalysts and process configurations have been developed to address these challenges, the problem of tar still exists. To break down the tar into smaller molecules, a catalytic tar cracker has been developed and added to the gasifier output. Despite these challenges, gasification is still a quick way to turn grain processing waste into power and heat.

3.4 TREATMENT OF FOOD WASTE BY HYDROTHERMAL CARBONIZATION

Hydrothermal carbonization, or HTC, transforms wet biomass or feedstock into solid (hydrochar), liquid (oils after extraction), or gaseous products without the need for a pretreatment step. This process utilizes minimal energy and can operate at temperatures well below 350 °C (Funke et al.,

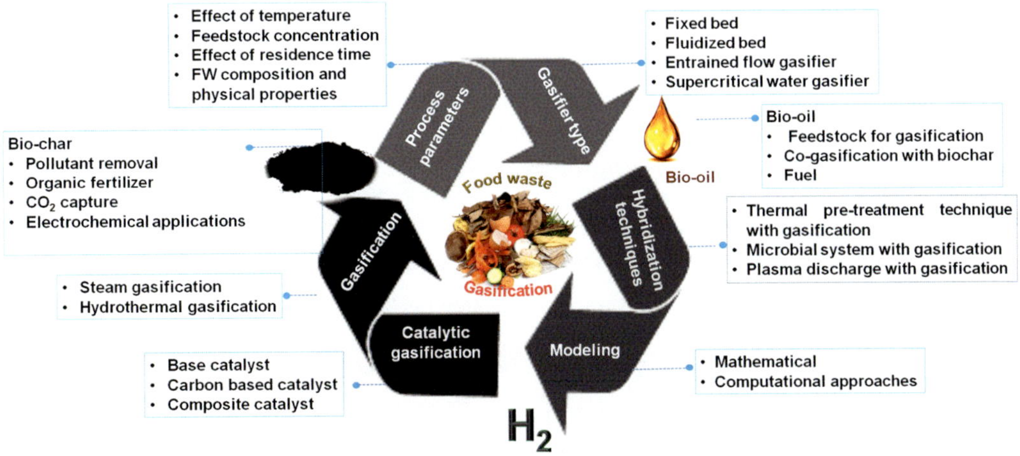

FIGURE 3.2 Overview of gasification process in food waste management (Adapted from Murugesan et al., (2022) with permission).

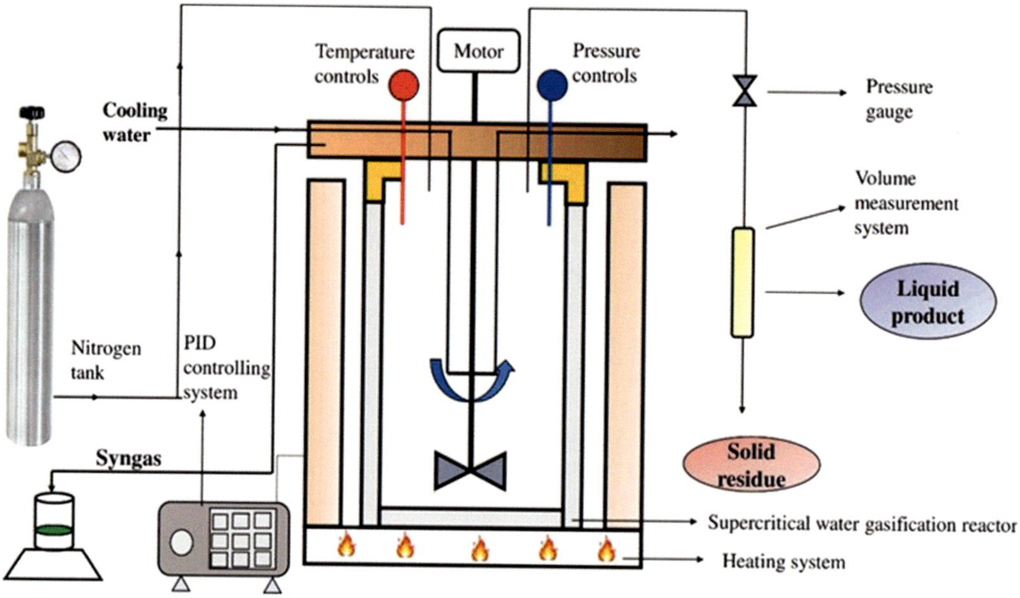

FIGURE 3.3 Schematic of a supercritical water gasifier for food waste conversion (Adapted from Sridhar et al., (2021) with permission).

2012). The moisture available in the food waste can be utilized as a solvent in the HTC process. A representative HTC process has been depicted in Figure 3.4a. The environmental advantages of HTC on various food waste, food residues, and nutrient recovery have been well established. For instance, in an investigation, the effect of residence time and temperature during the treatment of wet agro-industrial waste using hydrothermal carbonization was explored. The findings demonstrated not only a volume drop of less than 30% but also energy savings of more than 50% compared to thermal treatments (Benavente et al., 2015). Furthermore, the qualities of the hydrochar differed based on the kind of waste that was processed. In order to ascertain how process conditions impact the HTC process, food waste and packaging materials were the subject of another investigation. The study concluded that the inclusion of packaging materials – such as paper, plastic, cups, and plastic utensils – had a significant impact on the recovered solids' energy content. In another study, a net positive energy balance was obtained with a 5% (by weight) solids concentration of plastic materials at 275 °C. On the other hand, it was shown that the retention rate dropped as the amount of plastic increased (Li et al., 2013). Similar studies have been conducted to understand the heat recovery rate, moisture absorption capacity, and self-ignition properties of biomass. Their findings indicate that the treatment procedure was energy-rich, with a moisture absorption rate during pre-treatments of less than 3.8% and a cumulative heat output of 887 kJ/kg at 300 °C. Additionally, HTC has been assessed for utilizing a variety of waste residues, including pasta, noodles, chicken meat, tomato sauce, and green vegetables. The results showed a rise in MJ/kg from 25.1 to 33.1 when compared to traditional treatment.

3.5 TREATMENT OF FOOD WASTE BY INCINERATION

One of the methods for treating food waste is through its combustion using incineration, which produces energy and heat (Mayer et al., 2019). In processing industries, this recovered energy may be used to run turbines or heat exchangers. In addition, this method can cut the amount of solid waste that is delivered for disposal by 80–85% (Sridhar et al., 2021). Figure 3.4b shows a representative

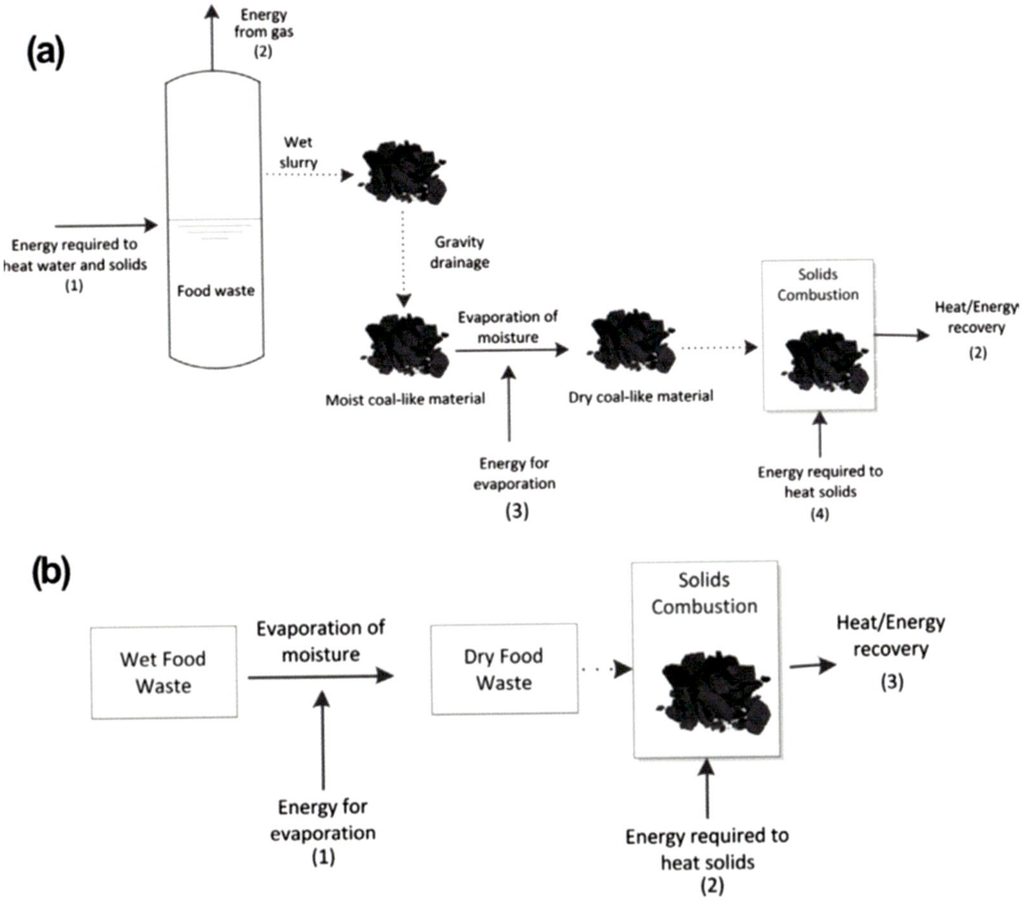

FIGURE 3.4 Food waste to energy conversion process through: a. hydrothermal carbonization and b. incineration (Adapted from Li et al., (2013) with permission).

food waste incineration process. Incineration is one of the safest disposal methods, but because of significant emissions and air pollution from heavy metals, it is not a practical approach. Ash, another consequence of incineration, must also be disposed of properly since it includes inorganic waste. A Korean investigation on energy recoveries stated that the food and municipal waste processed at temperatures of 850 to 1100 °C produced roughly 37.7 kJ of heat with a carbon credit of –315 kg of CO_2 per gram of food waste. The results are encouraging; however, it should be highlighted that if food waste is processed alone, it would result in increased energy loss because of organic debris resulting from fruit and vegetable peels, manure, etc. (Kim et al., 2013). Even after segregation, food waste accounted for 20–25% of municipal garbage, according to another Swedish research (Svensson et al., 2014) which could affect the fuel quality, operational costs, and the economy of food waste conversion.

3.6 CONVERSION OF FOOD WASTE BY HYDROTHERMAL LIQUEFACTION

Hydrothermal liquefaction, or HTL, transforms a variety of biomass sources into energy-dense bio-crude oil and its byproducts, including char, gases, and an aqueous phase. HTL breaks down complicated macromolecules into lower molecular weight products that polymerize into bigger compounds to create bio-crude oil by using sub-critical water (200–370 °C; 10–30 MPa). Compared to products from previous thermochemical conversions (pyrolysis, gasification), HTL bio-crude oil has a lower oxygen content and a greater energy density. It may be converted into conventional hydrocarbon fuels or utilized directly as a source of bio-phenols and bio-polyols (Li et al., 2016).

Numerous researchers have examined the HTL of food waste. For meats, cheeses, and fruits, the highest bio-crude oil recovered from HTL of food residues at 300 °C was 90, 75, and 10 wt%, respectively (Kostyukevich et al., 2018, Motavaf & Savage, 2021). Biocrude oil output rose from 25 to 43 wt% when an acid catalyst was added to the hydrothermal liquid of food waste high in carbohydrates (Posmanik et al., 2018). The production of biocrude oil can also be enhanced with an accelerated thermal reaction rate by adding affordable red mud and red clay catalysts to the heat-treated liquor of food waste at 300 °C for 1 h. It was noted that the molecular families in the biocrude oil with and without catalysts were identical (Chen et al., 2020). When fruit peels were heated to 300–350 °C for 30–120 minutes, the temperature during HTL had a significant impact on the yield and characteristics of the bio-crude oil; the yield of bio-crude oil produced at 350 °C was comparatively low. HTL of food waste at moderate temperatures (240–295 °C) with reaction periods ranging from 0–60 min, and 15% wt% solids input has been used in a batch reactor to produce bio-crude oil. Under such conditions, the greatest bio-crude oil energy content (40.2 MJ/kg) was recorded at 295 °C, while the maximum HTL bio-crude oil production (27.5 wt%) and energy recovery (49%) were obtained at 240 °C and 30 min (Bayat et al., 2021). It is important to note that food waste may sometimes contain certain other materials such as plastic waste which can alter the course of HTL. A study on the co-HTL of pistachio hulls and polypropylene (PP) demonstrated that, under normal circumstances (<350 °C), HTL might be a viable approach for the valorization of food waste and related plastic materials with the inclusion of hydrogen donor solvents and catalysts (Hongthong et al., 2020). This implies that catalytic HTL can be used for industrial processes for the conversion of food waste to bio-crude.

In order to determine the commercialization potential of HTL technology, researchers have recently concentrated on its continuous and up-scale processing. Few studies have compared bio-crude oil produced from bench and pilot size reactors due to the complexity and difficulty of building a continuous HTL system. The historical evolution of continuous HTL was recently the subject of a critical assessment, which also described the effects of feedstock selection and dry matter concentration on the output of biocrude oil. A 1 L tubular catalytic reactor was created by Pacific Northwest National Laboratory, and it produced 38–62.6%, 58.8%, and 50–56% of the product with microalgae, macroalgae, and grape pomace, respectively, at 350 °C and 20 MPa (Elliot et al., 2014). While the majority of the earlier continuous HTL reactors were between 0.68 and 1.5 L in capacity, there have been reports of several bigger reactors as well. A 5 L PFR created by the University of Sydney was used to convert macroalgae with organic co-solvents, yielding a 25% yield (He et al., 2016). A bigger pilot size reactor to study the continuous food waste valorization is necessary to expand the use of continuous HTL technology as food waste output worldwide rises (Aierzhati et al., 2021). The techno-economic viability and possible financial benefits from upscale HTL technology also provide obstacles for the commercialization of biocrude oil production technology, even though technical difficulties still stand in the way of HTL technology's eventual upscaling and commercialization.

3.7 CONCLUSIONS AND PERSPECTIVES

Thermo-chemical processes offer versatile and sustainable approaches for the treatment of food waste, enabling its conversion into valuable resources such as energy, biofuels, and chemicals. Recent advancements in pyrolysis, gasification, and hydrothermal processing have demonstrated the feasibility and potential of these technologies for mitigating food waste disposal challenges and promoting resource recovery. However, further research is needed to address key challenges such as process optimization, scale-up, and economic viability. Integrating thermo-chemical processes with other waste management strategies such as anaerobic digestion and composting could enhance overall efficiency and resource recovery. Additionally, regulatory support, public awareness, and industry collaborations are essential for accelerating the adoption of thermo-chemical technologies for food waste treatment on a global scale. In general, thermo-chemical processes present promising opportunities to transform food waste into valuable products, contributing to environmental sustainability, resource conservation, and circular economy initiatives. With continued innovation and collaboration, thermo-chemical-based solutions can play a significant role in addressing the complex challenges associated with food waste management in the years to come.

REFERENCES

Ahmed, I. I., & Gupta, A. K. (2010). Pyrolysis and gasification of food waste: Syngas characteristics and char gasification kinetics. *Applied Energy*, 87(1), 101–108.

Aierzhati, A., Watson, J., Si, B., Stablein, M., Wang, T., & Zhang, Y. (2021). Development of a mobile, pilot scale hydrothermal liquefaction reactor: Food waste conversion product analysis and techno-economic assessment. *Energy Conversion and Management*, 10, 100076.

Bayat, H., Dehghanizadeh, M., Jarvis, J. M., Brewer, C. E., & Jena, U. (2021). Hydrothermal liquefaction of food waste: Effect of process parameters on product yields and chemistry. *Frontiers in Sustainable Food Systems*, 5, 658592.

Benavente, V., Calabuig, E., & Fullana, A. (2015). Upgrading of moist agro-industrial wastes by hydrothermal carbonization. *Journal of Analytical and Applied Pyrolysis*, 113, 89–98.

Cheng, F., Tompsett, G. A., Murphy, C. M., Maag, A. R., Carabillo, N., Bailey, M., Hemingway, J. J., Romo, C. I., Paulson, A. D., Yelvington, P. E., & Timko, M. T. (2020). Synergistic effects of inexpensive mixed metal oxides for catalytic hydrothermal liquefaction of food wastes. *ACS Sustainable Chemistry & Engineering*, 8(17), 6877–6886.

Demirbas, A., & Arin, G. (2002). An overview of biomass pyrolysis. *Energy -Sources*, 24(5), 471–482.

Elgazar, Y. G., Khalifeh, H. A., Alkhedher, M., & Ramadan, M. (2024). A review of hydrogen production from food waste through gasification process. *International Journal of Hydrogen Energy*. https://doi.org/10.1016/j.ijhydene.2024.03.070

Elliott, D. C., Hart, T. R., Neuenschwander, G. G., Rotness, L. J., Roesijadi, G., Zacher, A. H., & Magnuson, J. K. (2014). Hydrothermal processing of macroalgal feedstocks in continuous-flow reactors. *ACS Sustainable Chemistry & Engineering*, 2(2), 207–215.

Funke, A., & Ziegler, F. (2012). Perspective: Jatropha cultivation in southern India: Assessing farmers' experiences. *Biofuels, Bioprod Biorefining,* 6, 1–18.

He, Y., Liang, X., Jazrawi, C., Montoya, A., Yuen, A., Cole, A. J., Neveux, N., Paul, N. A., de Nys, R., Maschmeyer, T., & Haynes, B. S. (2016). Continuous hydrothermal liquefaction of macroalgae in the presence of organic co-solvents. *Algal Research,* 17, 185–195.

Heidari, A., Khaki, E., Younesi, H., & Lu, H. R. (2019). Evaluation of fast and slow pyrolysis methods for bio-oil and activated carbon production from eucalyptus wastes using a life cycle assessment approach. *Journal of Cleaner Production*, 241, 118394.

Hongthong, S., Leese, H. S., & Chuck, C. J. (2020). Valorizing plastic-contaminated waste streams through the catalytic hydrothermal processing of polypropylene with lignocellulose. *ACS Omega*, 5(32), 20586–20598.

Irfan, M., Saleem, R., Shoukat, B., Hussain, H., Shukrullah, S., Naz, M. Y., & Jakubowski, T. (2023). Production of combustible fuels and carbon nanotubes from plastic wastes using an in-situ catalytic microwave pyrolysis process. *Scientific Reports*, 13(1), 9057.

Kim, M. H., Song, H. B., Song, Y., Jeong, I. T., & Kim, J. W. (2013). Evaluation of food waste disposal options in terms of global warming and energy recovery: Korea. *International Journal of Energy and Environmental Engineering*, 4, 1–12.

Kostyukevich, Y., Vlaskin, M., Borisova, L., Zherebker, A., Perminova, I., Kononikhin, A., Popov, I., & Nikolaev, E. (2018). Investigation of bio-oil produced by hydrothermal liquefaction of food waste using ultrahigh resolution Fourier transform ion cyclotron resonance mass spectrometry. *European Journal of Mass Spectrometry*, 24(1), 116–123.

Li, B., Feng, S. H., Niasar, H. S., Zhang, Y. S., Yuan, Z. S., Schmidt, J., & Xu, C. (2016). Preparation and characterization of bark-derived phenol formaldehyde foams. *RSC Advances,* 6(47), 40975–40981.

Li, L., Diederick, R., Flora, J. R. V., & Berge, N. D. (2013). Hydrothermal carbonization of food waste and associated packaging materials for energy source generation. *Waste Management,* 33(11), 2478–2492.

Marshall, A. J. (2013). *Commercial application of pyrolysis technology in agriculture*. Ontario Federation of Agriculture (OFA).

Mayer, F., Bhandari, R., & Gäth, S. (2019). Critical review on life cycle assessment of conventional and innovative waste-to-energy technologies. *Science of the Total Environment*, 672, 708–721.

Motavaf, B., & Savage, P. E. (2021). Effect of process variables on food waste valorization via hydrothermal liquefaction. *ACS ES&T Engeering,* 1(3), 363–374.

Murugesan, P., Raja, V., Dutta, S., Moses, J. A., & Anandharamakrishnan, C. (2022). Food waste valorisation via gasification–A review on emerging concepts, prospects and challenges. *Science of The Total Environment*, 851, 157955.

Posmanik, R., Martinez, C. M., Cantero-Tubilla, B., Cantero, D. A., Sills, D. L., Cocero, M. J., & Tester, J. W. (2018). Acid and alkali catalyzed hydrothermal liquefaction of dairy manure digestate and food waste. *ACS Sustainable Chemistry & Engineering*, 6(2), 2724–2732.

Raza, M., Inayat, A., Ahmed, A., Jamil, F., Ghenai, C., Naqvi, S. R., & Park, Y. K. (2021). Progress of the pyrolyzer reactors and advanced technologies for biomass pyrolysis processing. *Sustainability*, 13(19), 11061.

Roy, P., Dias, G., 2017. Prospects for pyrolysis technologies in the bioenergy sector: A review. *Renewable and Sustainable Energy Reviews*. https://doi.org/10.1016/j.rser.2017.03.136

Sridhar, A., Kapoor, A., Kumar, P. S., Ponnuchamy, M., Balasubramanian, S., & Prabhakar, S. (2021). Conversion of food waste to energy: A focus on sustainability and life cycle assessment. *Fuel*, 302, 121069.

Svensson Myrin, E., Persson, P. E., & Jansson, S. (2014). The influence of food waste on dioxin formation during incineration of refuse-derived fuels. *Fuel,* 132, 165–169.

Xiong, Z., Wang, Y., Syed-Hassan, S. S. A., Hu, X., Han, H., Su, S., & Xiang, J. (2018). Effects of heating rate on the evolution of bio-oil during its pyrolysis. *Energy Conversion and Management*, 163, 420–427.

4 Biological Processes for Food Waste Treatment

Prachi Gaur, Komal Kumari, Poonam Sharma, Vartika Gupta, and Vivek Kumar Gaur

4.1 INTRODUCTION

Globally, about 1.3 billion tons of food, equating to one-third of the world's total food production, is wasted across the entire food supply chain, causing various social, environmental, and economic issues (FAO, 2020). Waste emerges across the entirety of the food chain, starting from primary production (agricultural residues), moving through secondary production (industrial processing residues), and culminating in household and catering consumption (food surpluses). An expanding body of research endeavors to quantify the total loss of food across the supply chain, consistently underscoring its adverse effects on the food system. This loss incurs unnecessary burdens on land, water, and energy resources, consequently amplifying greenhouse gas (GHG) emissions. The economic ramifications of food wastage from households, retail, catering, and manufacturing sectors are profound, amounting to approximately US $680 billion in industrialized nations and US $310 billion in developing ones (Morone et al. (2019). Forecasts predict a significant escalation in these figures over the next three decades due to the projected global population surge. Projections indicate that the global populace will burgeon by nearly one billion individuals in the coming decade and reach 9.6 billion by 2050 (FAO, 2013). This will drive heightened demand for food and its by-products, thereby exacerbating pressure on food supply chains globally. The issue of global food security and supply is a pressing concern. Several countries are actively implementing mitigation and prevention measures to reduce food waste (FW). Along these approaches, there exists a dire need for different consumption models and novel technological approaches for valorizing the FW.

Traditionally, FW has been disposed of through dumping or incineration within municipal solid waste systems. However, these methods pose significant environmental and health risks (Sharma et al. 2021). Incineration, due to FW's high moisture content, can lead to the production of dioxins, while open dumping contributes to environmental degradation. FW contributes substantially to greenhouse gas emissions, estimated at 3.3 billion tonnes of CO2 per year (Sharma et al. 2020). Consequently, there is growing global interest in exploring alternatives to enhance environmental sustainability and address socio-economic concerns. Valorizing FW by converting it into value-added products has emerged as a promising solution, garnering attention from researchers worldwide. Therefore, this chapter aims to highlight the biological methods, including traditional composting and anaerobic digestion, as well as advanced ones such as enzymatic hydrolysis and microbial fuel cells (MFC) for the treatment of FW.

4.2 BIOLOGICAL TREATMENT OF FW

In the global arena, food waste generation has surged due to intensified industrial and human activities, contributing significantly to environmental degradation and resource scarcity. With annual disposal reaching 1.3 million tons, comprising 44% of the world's organic solid waste, and human activities accounting for 33%, totaling 1.6 billion metric tons per year, the inefficient use of resources

DOI: 10.1201/9781032706030-4

exacerbates these challenges (Zan et al. 2022). Additionally, food waste disposal in landfills releases substantial greenhouse gases (GHGs), notably methane (CH4) and carbon dioxide (CO2), contributing to global warming and emphasizing the critical need for effective waste management strategies (Slorach et al. 2019).

Food waste originates from various sectors such as food processing industries and households, amounting to nearly 1.3 billion tons lost along the food supply chain . Projections indicate a continued rise in food waste over the next 25 years, particularly in Asian countries, driven by economic and population growth, posing significant environmental and resource utilization challenges (Singh et al. 2022). Approximately, 1.4 billion hectares of fertile land, representing 28% of the world's agricultural area, are annually dedicated to producing food that ultimately goes to waste. Coupled with its substantial carbon footprint, this emphasizes the critical need for sustainable management practices (Bhatia et al. 2023).

Presently, developing nations predominantly utilize five primary methods for food waste recovery, viz., animal feeding, landfills, incineration, composting, and anaerobic digestion (AD). While anaerobic fermentation offers the potential to generate biogas from food waste, it is hindered by challenges such as high costs, strict startup conditions, and long fermentation periods (Gladchenko et al. 2017). Traditional landfilling requires extensive land area and results in significant GHG emissions (Bin Hu et al. 2024). Moreover, the incineration of food waste poses environmental risks and impedes nutrient recovery and valorization (Mohan and Joseph, 2021). Composting emerges as a prevalent method for the treatment of food waste, facilitating organic waste decomposition and producing valuable fertilizer and biogas (Ferdeș et al. 2022). Anaerobic digestion (AD) presents an attractive solution for bolstering global energy security by converting food waste into biogas, addressing waste management, and promoting nutrient recycling (Fei et al. 2020). This technology, focusing on biogas production, has gained prominence for its potential in food waste treatment and renewable energy acquisition. Generally, the biological treatment of food waste offers an environmentally sustainable solution for managing organic waste while also producing valuable resources such as compost, biogas, and organic acids. It helps divert waste from landfills, reduces GHG emissions, and contributes to the circular economy by recycling nutrients back into the soil. Additionally, the concept of sustainable consumption and production is gaining traction, emphasizing efficient resource utilization to ensure economic, environmental, and social sustainability.

4.3 COMPOSTING AND VERMICOMPOSTING

4.3.1 Composting

Composting is an effective technology to deal with FW, which can transform this waste into fertilizer (Zhang et al. 2022). There are various characteristics of FW, like high moisture content, low carbon to nitrogen ratio, and high acidity. The pH, moisture, and carbon to nitrogen (C:N) ratio of FW can be maintained by adding bulking agents like wood chips, sawdust, straw, and rice husk. Adding these bulking agents will increase the efficiency of the composting process. Various studies have used other alternatives to bulking agents to increase the efficiency of the composting process. Wang and colleagues employed mature compost to expedite the composting process, conducting experiments with two distinct groups: the control group (CO) and the C20 group (Wang et al. (2022). In the CO group, a mixture of 160 kg of FW and 40 kg of mature compost was utilized, while the C20 group employed 20 kg of mature compost, 160 kg of FW, and 20 kg of sawdust. This experiment was investigated for 35 days, and results indicated that mature compost significantly accelerated the breakdown of FW, resulting in a shorter composting period. The FW composting process can be improved by using multistage inoculums wherein microbial strains viz. *Lactobacillus amylophilus*, *Geobacillus thermoleovorans*, and *Bacillus subtilis* can be used. In a study, multistage inoculums were used to increase the efficiency of the composting process of FW. For multistage inoculums, three different bacterial strains, *Lactobacillus amylophilus*, *Geobacillus thermoleovorans*, and *Bacillus subtilis*, were used. The experiment was done in three groups, named T1, T2, and CK. In group T1, the three bacterial strains were added on 0, 2, and 14 days of composting. In group T2, these three microbes were added on 0 days of composting. In group CK, the same amount of sterilized media was added on 0, 2, and 14 days of composting. The composting feedstock contained FW, biogas residue, and sawdust. Their findings show that multiple inoculums extend high-temperature periods, improve the quality of compost, and enrich the bacteria that are helpful in degrading organic matter (Xu et al. (2022).

4.3.2 Vermicomposting

Vermicomposting has emerged as a more efficient technology compared to other composting methods due to its cost-effectiveness and eco-friendly nature in managing food and solid waste. This process entails overcoming the problems and disadvantages associated with utilizing synthetic fertilizers (Figure 4.1). Combining organic waste with other types of waste can further enhance the effectiveness of vermicomposting. In 2022, Cai and colleagues conducted a study demonstrating the effectiveness of combining soybean meal (SM) and sugarcane bagasse (SCB) in vermicomposting. This combination is advantageous because SM is highly biodegradable and serves as an excellent food source for earthworms, while SCB helps maintain earthworm populations. The study evaluated different ratios of SM to SCB (0:100, 25:75, 50:50, 75:25, and 100:0) and found that a 50:50 ratio produced the highest quality vermicompost. This balanced blend resulted in optimal enzyme activity, pH levels, microbial populations, reduced phytotoxicity, enhanced nutrient retention capacity, and promoted the growth and reproduction of earthworms (Eisenia fetida) (Cai et al. 2022). Mangrove fungi were found to be effective in enhancing vermicomposting of FW. In a study, three mangrove fungi – *Acrophialophora jodhpurensis*, *Talaromyces diversus*, and *Acremonium potronii* –along with amendments of biochar and cow dung were used. FW vermicomposting was studied for 90 days. This study revealed that the use of mangrove fungi, biochar, and cow dung during the vermicomposting of FW was most effective in improving the compost quality (Ameen and Al-Homaidan, 2022). Vermicomposting of waste biomass from potato crops using *Eisenia fetida* in two separate pot experiments was investigated. In the first pot, potato plant biomass was utilized as

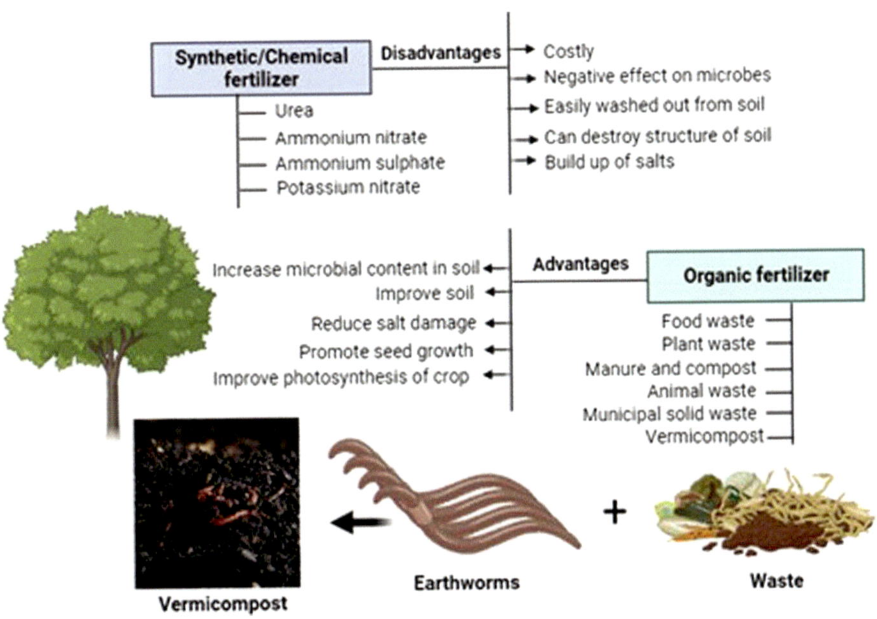

FIGURE 4.1 Advantages and disadvantages of synthetic versus organic fertilizers.

raw material, while in the second pot, potato plant biomass was combined with cow dung in a 5:1 ratio. The findings indicated a decrease in the C:N ratio, and improvements in nutrient composition and nitrogen fixation. The inclusion of cow dung resulted in an overall enhancement of both the quality and quantity of the final vermicompost product.

4.4 ANAEROBIC DIGESTION OF FW

Anaerobic digestion (AD) has emerged as a cost-effective technology for renewable energy production and waste treatment due to the detrimental environmental impacts linked to landfilling, incineration, or composting of FW. AD stands as a well-established technology capable of converting various biodegradable residues, particularly FW, into biogas (a methane-rich gas mixture) through anaerobic microbes (Figure 4.2). This technology is highly adaptable across all stages of the FW chain and is applicable to diverse biodegradable waste types, including the organic fraction of municipal solid waste, agricultural waste, industrial waste, and more. FW emerged as a promising substrate for biogas generation, owing to its composition and moisture content (Fernandes et al. 2021; Abbasi et al. 2024) (Table 4.1).

4.4.1 Anaerobic Digestion Process

The biogas production process involves several distinct stages, each driven by specific microbial activities. Initially, hydrolysis takes place, during which water and extracellular enzymes, such as amylase, cellulase, lipase, protease, and pectinase, play a pivotal role in breaking down the intricate

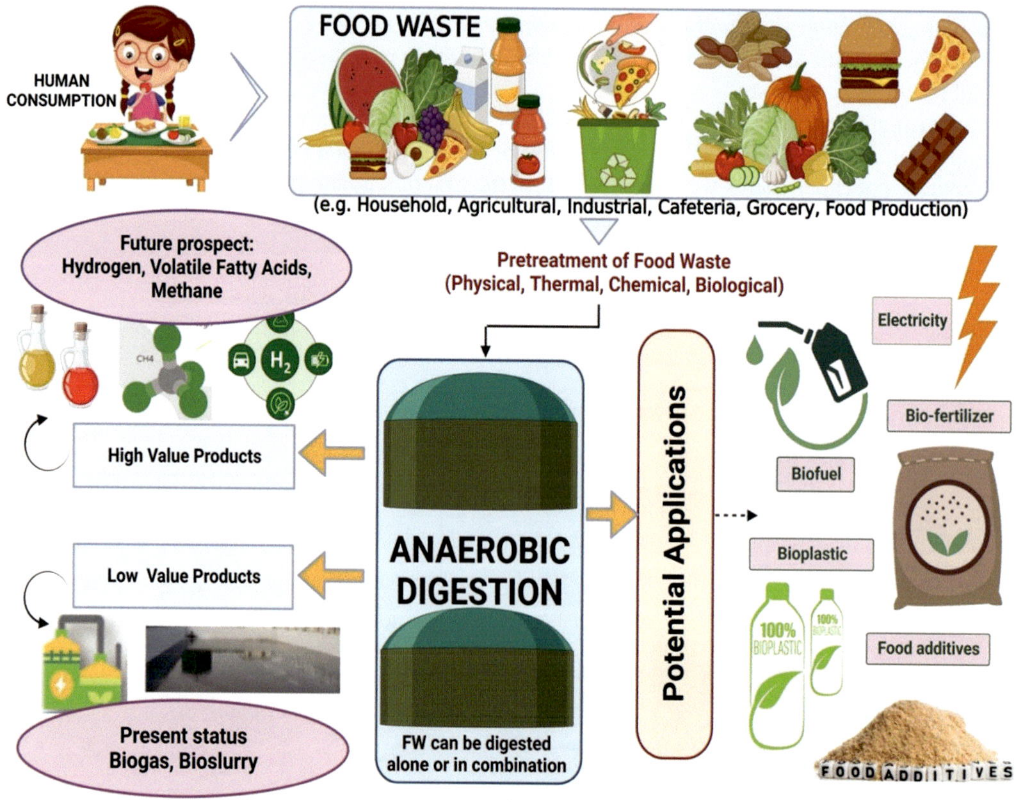

FIGURE 4.2 The conceptualization of anaerobic digestion for food waste, aimed at generating value-added products and exploring their potential applications.

TABLE 4.1
The Biodegradability of FW Mentioned by Several Researchers

Operational Scale	Substrate Type 1	Substrate Type 2	Ratio	CH_4 Yield ($mlCH_4/gVS_{deg}$)	Enhancement: Supplemental Effect	Enhancement: Synergistic Effect	Enhancement: Global Effect	References
Batch	Fruit Waste	Cattle Manure	20:80	380	Nutrient and C/N ratio balance. High Buffer Capacity	CH_4 yield increases while addition of FW increases		Callaghan et al. (2002)
	Fruit Waste	Cattle Manure	30:70	340				
	Fruit Waste	Cattle Manure	40:60	380				
	Fruit Waste	Cattle Manure	50:50	450				
Lab scale	Sewage Sludge	Food waste	50:50	215	Enhanced OLR	24.27	85.3	Kim et al. (2006)
	Sewage Sludge	Food waste	20:80	157	Enhanced buffering capacity from ammonia	54.6	35.3	
Continuous pilot scale	Sewage Sludge	Food waste	80:20	326	Enhanced buffering capacity from ammonia	10.8	21	la Cour Jansen et al. (2004)
A semi-continuously fed and mixed reactor	Waste Activated Sludge	Food waste	90:10	186	Nutrient balance. Enhanced buffering capacity from ammonia	CH_4 yield raises while with the addition of FW raises		Heo et al. (2004)
	Waste Activated Sludge	Food waste	50:50	321				
	Waste Activated Sludge	Food waste	10:90	346				
----	Sewage Sludge	Food waste	75:25	439		47.81	----	Sosnowski et al. (2008)
Batch	Cattle Manure	Food waste	50:50	298	Elevated buffering capacity of ammonia	----	44	Li et al. (2009)
Lab scale	Food waste	Slaughterhouse Waste	83:17	300	Trace element supplement. Ammonia accumulation	----	-61.9	Cuetos et al. (2010)
Continuous Lab scale	Food waste	Piggery waste water	93:7	358	Trace elements supplements	----	----	Zhang et al. (2011)
	Food waste	Piggery waste water	83:17	388				
Continuous	Cattle Manure	Food waste	67:33	388	Elevated buffering capacity and trace element supplement	-----	41.1	Zhang et al. (2013)

polymeric structures of cellulose, starch, and proteins. This hydrolytic action transforms these complex structures into simpler units like glucose, fatty acids, and amino acids. Subsequently, acidogenesis ensues, during which the hydrolytic products are primarily converted into short-chain volatile fatty acids (VFA), including acetic acid, propionic acid, formic acid, and lactic acid, as well as alcohols (ethanol, methanol) and ketones (glycerol and acetone). By-products such as CO_2, H_2, NH_3, alcohol, and trace substances are also generated.

Moving to the acetogenesis stage, acetogenic bacteria play a crucial role in converting the products from acidogenesis, along with some long-chain fatty acids from the hydrolysis stage, into acetate, CO_2, and H_2. Finally, in the methanogenesis phase, CH_3COOH (acetate) and H_2 undergo conversion into CO_2 and CH_4 (Ni et al. (2022). This transformation is orchestrated by two distinct groups of bacteria: acetophilic bacteria, responsible for converting acetate into CH_4 and CO_2, and hydrogenophilic bacteria, which convert H_2 and CO_2 into CH_4. The sequential occurrence of these stages highlights the intricate biochemical processes underlying the efficient production of biogas through anaerobic digestion.

Several critical parameters, including temperature, pH, VFAs concentration, carbon-to-nitrogen ratio, organic loading rate, and hydraulic retention time, influence the efficiency of AD in the context of biogas production, emphasizing the significance of controlling specific environmental conditions.

***Temperature*:** This plays a crucial role, with distinctions made between psychrophilic (below 20°C), mesophilic (25–40°C), and thermophilic (45–60°C) conditions (Rodríguez-Jiménez et al. 2022). Psychrophilic conditions, although less favored in FW AD research due to prolonged digestion times and lower CH_4 and CO_2 yields, exhibit a biochemical methane potential (BMP) ranging from 250 to 460 mL/g VS Lou et al. (2012).

pH: This is another pivotal factor, with an optimal range of pH 7, ensuring fermented stability. Bacteria involved in AD can function within a pH range of 5.5–8.5, with an optimum range of 6.5–8.0 Pilarska et al. (2023).

Volatile fatty acid (VFA) concentrations: The products of the acidogenesis stage significantly impact biogas composition, particularly methane. The optimal concentration falls within the range of 2 g/L to 3 g/L (Pilarska et al. 2023).

Carbon-to-nitrogen (C:N) ratio: This is crucial, with a range of 15–30:1 considered optimal, preventing nutrient imbalance and enhancing productivity. A high C:N ratio may indicate nutrient imbalance, leading to rapid nitrogen degradation (Morales-Polo et al. 2018). Thus, achieving an optimal balance between C and N is imperative. Numerous researchers have identified a C:N ratio within the range of 20 to 30 as the most favorable (Zhang et al. 2014; Beniche et al. 2021). It was observed that the maximum methane production occurred at a C:N ratio of 27 (Wang et al. 2012; Chidambarampadmavathy et al. 2017). Organic fraction municipal solid waste (OFMSW) encompasses organic materials, including FW, yard waste, and other residues. However, each of these organic fractions exhibits distinct C:N ratios. For instance, food and yard waste typically possess <20, while mixed paper may have C:N ratios of >100 (Tyagi et al. 2018). However, an ideal C:N ratio is contingent upon both the substrate and the inoculums used in the process.

Organic loading rate (OLR): This is a vital parameter, where a high value can result in the accumulation of inhibitory substances like VFA and ethanol, while a lower-than-optimal OLR inhibits biogas generation. Ammonia content, directly related to the C:N ratio, influences AD by enhancing buffer capacity but acting as an inhibitor for bacterial growth, especially acetoclastic methanogens (Hao et al. 2022).

Hydraulic retention time (HRT): This is temperature-dependent, with mesophilic conditions typically requiring 14–40 d and thermophilic conditions 14–20 d for optimal bioreactor operation (Zhang et al. 2020b)

4.4.2 Methods to Enhance Anaerobic Digestion of FW

4.4.2.1 Pretreatment

Physical and mechanical pretreatments are essential for enhancing the physical properties of FW, such as surface area. Methods like milling, chopping, grinding, screw press, lysis-centrifugation, liquid shear/collision, and high-pressure homogenization can be alternately employed to improve FW's physical and mechanical characteristics. Thermal pre-treatment methods, classified as wet-type and dry-type, are commonly applied to FW before AD. Operating under elevated temperatures and pressure, wet-type thermal pre-treatment aids in solubilizing sugars through improved hydrolysis, resulting in a more homogeneous pulp for AD. However, this process can lead to the formation of recalcitrant compounds, such as melanoidins, which may impact product recovery at temperatures over 150°C with short retention times or below 100°C with longer retention times.

Chemical pre-treatment relies on strong to mild chemical agents to modify the physico-chemical and biological properties of FW. Acid pre-treatment efficiently solubilizes carbohydrates, while alkali pre-treatment is effective in solubilizing proteins, lignin, and lipid saponification. Acid pre-treatment involves corrosive chemicals (e.g., H_2SO_4, HCl) and requires non-corrosive coatings for equipment. The FW to acid/alkali ratio needs optimization based on total solid content and the strength of the acid/alkali used (Karthikeyan et al. 2018).

Biological pretreatment involves separating various phases in two different reactors (hydrolysis-acidogenesis followed by acetogenesis-methanogenesis), achieving optimal conditions in both processes. The pH is lowered to acidic conditions (4 to 6) in one reactor for hydrolysis-acidogenesis, breaking down inhibiting chemicals for the second-stage microorganisms. The other reactor maintains a pH value from 6.5 to 8 for proper growth of the methanogenic population, with fewer VFAs, as they are easily converted into methane, thus avoiding VFA accumulation. This two-stage digestion approach results in increased biogas production, up to 20%, compared to single-stage anaerobic digestion (De Gioannis et al. 2017). Despite its effectiveness, two-stage digestion is less commonly used in full-scale plants due to the simplicity of operation in single-stage reactors.

4.4.2.2 Co-Digestion Process

Anaerobic co-digestion (CoAD) involves mixing substrates, such as FW and sewage sludge, FW and manure, or even straw and slurry, to achieve nutrient balance in the digester, creating a more stable environment for anaerobic bacteria (Li et al. 2009; Yong et al. 2015). The primary advantage of CoAD lies in the synergy between substrates, resulting in enhanced biogas yield, altered biogas composition, and enriched methane content. Co-digestion of FW with sewage sludge from wastewater treatment plants stands out as an excellent solution, offering benefits in terms of synergy in biogas production, methane yield, and process stability.

4.5 FERMENTATION

Fermentation, a biological process, works in various applications, seamlessly blending tradition with innovation. Two distinct approaches, Bokashi and yeast fermentation, define the versatility of microbial processes in addressing contemporary challenges while promoting sustainability, such as FW treatment.

4.5.1 Bokashi Fermentation

Anaerobic fermentation, characterized by its capacity to accommodate diverse substrates, is well suited for the treatment of FW (Luo et al. 2019). An effective approach for FW treatment is Bokashi fermentation, derived from a traditional Japanese method involving the fermentation of organic matter utilizing specific microorganisms (Formowitz et al. 2007). This innovative and straightforward process aims to recycle and valorize various FW or biowaste types into a nutrient-rich product, serving as an organic fertilizer (Olle, 2021). The initial step involves Bokashi substrate preparation, utilizing almost any biodegradable material in shredded form, inoculated with a blend of beneficial microorganisms known as effective microorganisms (EM) (Lew et al. 2021). Flourishing in anaerobic and acidic environments, these microbes, including purple non-sulfur bacteria (PNSB), lactobacilli (LAB), yeasts, and actinomycetes, efficiently suppress pathogen viability and enhance environmental benefits (Budihardjo et al. 2023).

However, EM, applicable as a liquid or mixed with nutrient-rich organic matter in the form of fermented compost, synergistically interacts with organic matter. This interaction is based on EM's ability to ferment organic material, releasing nutrients and nutrient-rich organic acids beneficial for plant utilization (Boechat et al. 2013). Unlike anaerobic fermentation for hydrogen production, Bokashi fermentation eliminates the need for expensive equipment. The inoculant preparation involves immersing the host material in a carbohydrate-rich brew within closed bioreactors (Olle, 2021). Following a brief fermentation period, the inoculated material can be stored long-term as a Bokashi starter, yielding two distinct fractions: a solid component and a liquid known as "Bokashi tea," both serving as valuable fertilizers (Boechat et al. 2013). Therefore, the optimal fermentation conditions require moisture regulation and temperatures exceeding 18–20°C. Industrial fertilizer product standardization involves substrate mixing to adjust the C:N ratio and balance macronutrients (Aguinaga et al. 2023). The final products can undergo extended storage without altering properties when stored in dark, airtight conditions at lower temperatures. Studies indicate that the application of EM, such as LAB, PNSB, and yeast, alongside organic matter, enhances environmental benefits (Lew et al. 2021). EM can be applied as a liquid or mixed with nutrient-rich organic matter in the form of fermented compost, known as "Bokashi." The synergy between EM and organic matter lies in EM's capacity to ferment organic material, releasing nutrients and nutrient-rich organic acids beneficial for plant utilization (Boechat et al. 2013) (Table 4.2).

4.5.2 Yeast Fermentation

For millennia, yeasts, notably *Saccharomyces cerevisiae*, have played a pivotal role in alcohol production, particularly in the realms of brewing and winemaking. *S. cerevisiae* is preferred in industrial ethanol production owing to its capacity to endure a broad pH range, enhancing the resilience of the process against potential contamination (Walker and Stewart, 2016). However, the efficacy of ethanol production is based upon the characteristics of the utilized feedstocks. Generally, ethanol production entails three key phases: (a) obtaining a solution rich in fermentable sugars, (b) converting sugars to ethanol through fermentation, and (c) subsequently separating and purifying the ethanol (Behera et al. 2010). Feedstocks undergo pretreatment to reduce particle size and facilitate subsequent processes. Following this, hemicellulose and cellulose are hydrolyzed into fermentable sugars (Wyman, 2018). Yeasts are entrusted with fermenting these sugars to produce ethanol. Various separation technologies are deployed to recover ethanol, rendering it suitable for use as a fuel source (Wyman, 2018). Assessments of fermentation performances under different cultivation conditions highlight that the selection of a yeast strain depends not only on ethanol tolerance but also on the preferential utilization of carbon resources from biological residues (Azhar et al. 2017).

However, yeast faces common challenges during fermentation, including elevated temperatures (35–45 °C) and high ethanol concentrations (over 20%). While yeast growth and metabolism generally increase with temperature until reaching an optimum value, higher temperatures can pose challenges. Similarly, increased ethanol concentration during fermentation can inhibit microorganism growth and viability (Chatterjee et al. 2006). The inability of *S. cerevisiae* to grow in media containing high levels of alcohol leads to the inhibition of ethanol production (Fiedurek et al. 2011). Additionally, *S. cerevisiae* can only ferment hexoses, not pentoses, posing a challenge in bioethanol production, as certain feedstocks contain both types of sugars (Kumar et al. 2009). Therefore, to enhance ethanol production, yeast strains that are tolerant to inhibitors should be used (Tofighi et al. (2014). Ethanol-tolerant and thermotolerant yeast strains, resistant to stresses, can be isolated from

TABLE 4.2
An Overview of Bokashi Fermentation Characteristics, with Their Advantages and Limitations

S. No.	Feature	Advantage	Limitations
1	Biowaste fractions	Diverse biowaste types (household, animal, bakery, brewery)	–
2	Chemical composition	Less critical than traditional composting	–
3	Process efficiency	Faster, scalable, and adaptable	Initial setup costs, industrial scaling challenges, potential for incomplete decomposition
4	Speed of process	Faster, scalable, ambient	Time-consuming process
5	Pest issues	No pest issues	-
6	Odor concerns	Odor-free process	Potential for odors during fermentation
7	Greenhouse gas emission	Minimal compared to composting	–
8	Nutrient retention	No loss of nutrients, resulting in higher nutritional values	–
9	Beneficial microorganisms	Fertilizer rich in beneficial microorganisms, especially effective microorganisms (EM)	Reliance on specific microorganisms may limit flexibility
10	pH antimicrobial properties	Low pH and antimicrobial compounds inactivate pathogens	
11	Soil health improvement	Enhances soil health, structure, and carbon sequestration	Lack of mechanical aeration, limited carbon to nitrogen ratio adjustability
12	Fertilizer quality enhancement	Improved fertilizer quality and uptake by plants	pH management, limited carbon-to-nitrogen ratio adjustable
13	Heavy metal and pesticide treatment	Addresses heavy metals and pesticide residues in soils	-
14	Health and food quality	Enhances plant and animal health, improving food quality	-
15	Other aspects		Temperature sensitivity, moisture regulation, space requirements, transportation and handling challenges, pathogen inactivation time, quality variation

natural resources like soil, water, plants, and animals. Ethanol fermentation at high temperatures is advantageous, as it selects for thermo-tolerant microorganisms and eliminates the need for cooling costs and cellulose (Fonseca et al. 2008).

Both Bokashi and yeast fermentation underscore the significance of harnessing microbial processes for sustainable bioprocessing applications. These approaches showcase the potential for environmentally friendly and resource-efficient solutions in waste management and biofuel production. As we navigate the complexities of modern challenges, incorporating such bioprocessing methods not only aligns with sustainable practices but also paves the way for the development of a more resilient and eco-friendly industrial landscape (Table 4.3).

TABLE 4.3
The Distinctions between Bokashi Fermentation and Yeast Fermentation, Outlining the Key Differences and Comparative Aspects of These Two Fermentation Processes

S. No.	Feature	Bokashi Fermentation	Yeast Fermentation
1	Primary microorganism	Anaerobic bacteria (lactic acid bacteria, yeasts, etc.)	Saccharomyces ceravisiae (common yeast) and other species
2	Oxygen requirement	Anaerobic process, minimal oxygen involvement	Aerobic process, requires oxygen for yeast metabolism
3	Key substrates	Organic kitchen waste	Sugars and starches
4	End produts	Acidic fermented material (Bokashi pre-compost)	Ethanol and carbon dioxide
5	Nutrient retention	Preserves more nutrients due to anaerobic conditions	May result in some nutrient loss due to aerobic breakdown
6	Time for completion	2 weeks to several weeks	Shorter time frame
7	Microbial diversity	Diverse microbial community	Predominantly yeast population
8	Container type	Typically uses an airtight container for anaerobic processes	Can be carried out in open or closed containers

4.6 ENZYMATIC HYDROLYSIS OF FW

Enzymatic hydrolysis stands as a pivotal technique in the paradigm of sustainable FW management, offering a beacon of hope within the framework of the circular economy. As the global population reaches the count of 8 billion, the demand for food and agro products is increasing, thereby generating significant unmanaged amounts of FW. In such instances, enzymatic hydrolysis becomes a key technique for extracting valuable elements from organic waste. FW is a complex treasure of organic nutrients (carbohydrates, lipids, protein) and bioactive compounds, so in this process, complex organic compounds are transformed into simpler forms, facilitating the recovery of valuable materials. Through incepting enzymatic hydrolysis, the circular economy aims towards zero-waste and zero-emission by minimizing the dependence upon non-renewable resources while emphasizing resource and energy recovery from FW. This is a sustainable practice that aligns with environmental goals by mitigating greenhouse gas emissions and raising a more sustainable approach to waste management (Szopa et al. 2023).

Enzymatic hydrolysis is an ideal biological pretreatment process producing simpler sugars, such as fructose, glucose, xylose, galactose, and ribose from FW. The allure of enzymatic hydrolysis lies in its inherent advantages over chemical and physical alternatives. With its simple bio-reaction mechanism, enzymatic hydrolysis boasts higher specificity and lower energy requirements. It operates under mild reaction conditions, ensuring high product purity and precision. Moreover, the integration of in situ-produced enzymes derived from FW presents a cost-effective alternative to expensive commercial enzymes, effectively dismantling economic barriers and enhancing accessibility to sustainable waste management solutions (Zou et al. 2020).

Enzymatic hydrolysis of FW can exhibit a high level of efficiency for acquiring nutrient-rich hydrolysates for downstream implementation (Piercy et al. 2023). Enzymatic hydrolysis of proteins in FW has the capability to generate smaller chains of amino acids, resulting in the production of well-defined peptides, which can improve the nutritional and functional characteristics of FW (Tavano, 2013). The hydrolysis process can increase the protein content, improve the amino acid score, increase the biological value, and increase the in vitro protein digestibility of food (Piercy et al. 2023). Various proteases lead to distinct degrees of hydrolysis, liberating explicit outcomes pertaining to amino acid composition, molecular weight distribution, and sequence variation, thereby influencing the nutritional and functional properties of the resulting hydrolysate (Schlegel et al. 2019). Enzymatic hydrolysis additionally facilitates the liberation of free polypeptides and amino acids from FW (Avramidou, 2020). Enzymatic hydrolysis presents promising applications in the food industry, offering a range of potential nutritive and functional products for various uses in the food industry, including animal feed formulation (Table 4.4).

Integrative approaches combining enzyme hydrolysis with various pre-treatment techniques (hydrothermolysis, sonolysis, electrochemical oxidation, sonoelectrolysis, and solid-liquid extraction) offer enhanced efficiency in FW processing (Sahu et al. 2023). Significant improvements in solid degradation, and starch and cellulose solubilization can be noticed using such approaches. Enzymatic hydrolysis in conjunction with sonolysis yields higher solid, starch degradation, and cellulose solubilization, while electrocatalysis significantly enhances cellulose solubilization. Integrating enzymatic hydrolysis with ultrasound treatment saves time and boosts glucose yield by 10% compared to conventional methods. Integrative approaches aid in saccharification of lignocellulosic biomass through disrupting crystalline structure of cellulose and increasing the accessible surface area, thereby separating lignin and hemicellulose from cellulose. These alterations make the cellulose more amenable to enzymes during hydrolysis (Singhvi and Gokhale, 2019; Zou et al. 2020).

Physical pretreatment methods including ultrasound and milling have also demonstrated the ability to decrease the particle size, crystallinity index, and polymerization degree of biomass (Khan et al. 2021). On the other hand, chemical pretreatment methods such as acid pretreatment have proven to be effective in disrupting the structure of the biomass and enhancing the accessibility of cellulose to enzymes during hydrolysis (J. Zhang et al. 2020a). The amalgamation of physical and chemical methods can reduce pretreatment time, enhance enzymatic efficiency, and augment the potential for industrial applications (Periyasamy et al. 2022). Pretreatment methods encompassing thermal, sono-electrochemical, and solid-liquid extraction techniques have been investigated to enhance the efficiency of enzymatic hydrolysis of FW (Ramasamy and Subramanian, 2022). The combination of enzymatic hydrolysis and ultrasound treatment has also been observed to save time and boost glucose yield by 10% compared to conventional methods. Ultrasound treatment enhances the enzymatic hydrolysis process by expediting the breakdown of cell wall structure and promoting the release of glucose, ultimately resulting in an elevated glucose yield (Córdova et al. 2022; Wang et al. 2017). These findings underscore the efficacy of integrated approaches, highlighting their potential in sustainable waste management and resource recovery.

4.7 MICROBIAL FUEL CELLS

A microbial fuel cell (MFC) is an electrochemical approach that harnesses the power of microorganisms and organic matter to anaerobically utilize carbon sources and generate electricity via bacterial metabolism from a wide range of inorganic and organic substrates. The effectiveness of MFCs in converting chemical energy into electrical energy through bio-catalysis primarily relies on the biodegradation efficiency of various waste including FW, agriculture, and household waste using microorganisms. Therefore, it presents a practical solution to address the challenge of FW treatment by environmentally friendly degradation and green energy generation, serving as a sustainable source of electricity (Kumar et al. 2022). Sambavi et al. (2021) demonstrated the production of bioelectricity using homogenized FW as a substrate in a series of three microbial fuel cells. With an initial substrate concentration of 19,800 mg COD/L, the system achieved a potential of 2250 mV and a current of 22.50 mA with an external resistance of 100Ω in 65 h. This demonstration not only addresses the need for energy but also has a significant role in enhancing the efficient environmental management of waste. Another study explored the valorization of cheese whey by using a single-chamber MFC with an air cathode involving the filter sterilization and pre-treatment acidification of the cheese whey to maximize high energy output (Antonopoulou et al. 2021). The findings revealed that utilizing MFCs for cheese whey produced a maximum power density of 3.26 W/m^3

TABLE 4.4
Table Showing the Potential of Enzymes in Valorizing Various FW, Associated Challenges, and Applications in Food Industry

FW Source	Enzyme Used	Extracted Compound	Challenges	Potential Applications	References
Meat	Protease (alcalase, trypsin, pepsin, and pancreatin), lipase (resinase)	Protein hydrolysate, collagen, and fatty acids	Heterogeneity of the waste produced, selection of the appropriate enzymes, optimization of the enzyme-to-substrate ratio, control of the reaction conditions such as pH and temperature.	Production of nutritional foods, animal feed, biodiesel/biogas, and dietetic products	Angulo and Márquez (2023); Rivera et al. (2022)
Fish	Alcalase, Flavourzyme, Neutrase, Protamex, and PTN	Protein hydrolysates, bioactive compounds such as phenolics and flavonoids, and oils	Enzyme recovery and purification, optimization of hydrolysis conditions	Production of protein hydrolysate	Szucs et al. (2021); Angulo and Márquez (2023)
Grains (rice and wheat)	α-amylase and glucoamylase	Phenolic compounds, monosaccharides, and protein hydrolysate	Selection of appropriate enzymes, optimization of extraction condition, cost of enzyme production, enzyme stability and activity, recovery of target compounds	Production of nutritional foods, bioplastics, bioenergy, animal feed, and the production of valuable chemicals	Sirohi et al. (2021)
Corn	Cellulose, hemicellulose, lactase, manganese peroxidase, and lignin peroxidase	---	Optimize the enzymatic process to improve ethanol production	Production of bioethanol, extraction of bioactive compounds	Wang et al. (2021); Wang et al. (2020)
Bread	BAN 240 L (α-amylase from Bacillus amyloliquefaciens) and AMG 300 L (glucoamylase from Aspergillus niger)	Glucose, organic acids, enzymes/biocatalysts, and bioactive compounds such as phenolics and flavonoids	---	Generation of high-glucose slurries	Szymczak et al. (2021); Mihajlovski et al. (2020); Rosa-Sibakov et al. (2022)

(Continued)

TABLE 4.4 (CONTINUED)
Table Showing the Potential of Enzymes in Valorizing Various FW, Associated Challenges, and Applications in Food Industry

FW Source	Enzyme Used	Extracted Compound	Challenges	Potential Applications	References
Pizza	α-amylase, glucoamylase	Fermentable sugars, such as glucose	Composition of waste		Torres-León et al. (2021)
Hamburgers	α-amylase, glucoamylase	Fermentable sugars, such as glucose	Need for effective pre-treatment to break down the complex components	Generation of fermentable sugars, bioethanol production	Han et al. (2020); Torres-León et al. (2021)
Pomegranate	Pectinase, protease, and cellulase	polyphenols, ellagitannins, and other phenolic substances	Optimize the enzymatic process, different enzyme strategies, optimization of hydrolysis conditions	Extraction of bioactive lipids	Saleem et al. (2020); Chaudhary et al. (2021)
Orange	Cellulose, pectinase, α-amylase, and glucoamylase	Pectic oligosaccharides, fermentable sugars, such as glucose and fructose,	Inhibitory effect of limonene, optimization of hydrolysis conditions	Production of food additives, encapsulants, prebiotics, and essential oils	de la Torre et al. (2020); Ayala et al. (2021); Utekar et al. (2021)
Potato	α-amylase and glucoamylase	Glyceryl tribehenate, ethyl iso, glycoalkaloids, and bioactive compounds such as phenolics and flavonoids		Food preservation, pharmaceutical ingredients, renewable energy, animal feed, and the production of bio plastics and biodegradable substitutes	Suresh et al. (2020); Ma et al. (2022)
Banana	Pectinase and β-glucosidase	Fermentable sugars, cellulose, nanocellulose, and phenolic compounds	The cost and complications in enzyme usage	Development of valuable products and contribution to waste reduction	Baruah et al. (2022); Sawarkar et al. (2022); Phirom-on and Apiraksakorn (2021)
Apple	Cellulose, pectinase	Polyphenols, ellagitannins, and other phenolic substances	Non-productive adsorption of enzymes on lignin	Production of value-added products	de Oliveira et al. (2022); Procentese et al. (2020)
Cassava	α-amylase, β-amylase, and glucoamylase	Fermentable sugars, such as glucose and fructose	--	Production of biofuels and bio refinery chemicals	Aruwajoye et al. (2020); Dorleku et al. (2022)

(equivalent to 0.33 MJ/kg COD) when the filter-sterilized whey was diluted to an initial concentration of 0.8 g COD/L. A cassette-electrode microbial fuel cell (CE-MFC) was developed using sweet potato shochu waste as a substrate and a microbial consortium comprising *Bacteroides*, *Bacillus*, and *Methanobacterium* (Iigatani et al. 2019). The CE-MFC exhibited a maximum power density of 1.2 W/m^3, and 67.4±1.8% chemical oxygen demand chromium (CODCr/L) removal efficiency was achieved when utilizing 10 g CODCr/L shochu waste.

The bioenergy production from Nigerian corn starch wastewater as a substrate was studied using a dual-chamber MFC setup with iron electrodes for electron transfer (Nwaokocha et al. 2021). The maximum values of current and power densities calculated were 8.10 mA/cm^2 and 7.70 mW/cm^2, respectively, within 9 days. This application of MFC technology in corn starch wastewater treatment not only proves effective but also presents a promising avenue for an economically viable energy source for off-grid power solutions during the Covid-19 and post-Covid-19 eras. The efficient utilization of household organic waste as a resource was suggested as a key contributor enabling energy generation using MFC by many bacterial species including *Pseudomonas aeruginosa*, *Pseudomonas nitroreducens*, *Enterobacter sp.*, and *Acinetobacter schindleri* (Yaakop et al. 2023). Their study revealed the generation of 110 mV voltage over a 12-day period with an external resistance of 500 Ω. The MFC, featuring an internal resistance of 117 Ω, achieved a maximum power and current density of 0.1047mW/m^2 and 21.84mA/m^2, respectively. The sustainable production of energy from effluent mud, mixed waste, and FW using the cost-effective MFC electrodes was also conducted (Sathyamoorthy and Sushmitha, 2022). The maximum open circuit voltage obtained for effluent mud, mixed waste, and FW were 1.160 mV, 0.989 mV, and 0.5 mV with COD removal efficiencies of 95.68%, 82.56%, and 69.78%, respectively. The power density and current density obtained for the above three wastes were 14.01, 21.36, 16.44 W/m^3 and 4.676, 16.96, 9.55 A/m^3, respectively. Thus, it was found that FW mixed with effluent mud could be used as an effective substrate for sustainable power generation compared to sludge and FW separately.

4.8 CONCLUSIONS AND PERSPECTIVES

FW is an opportunistic venture for microbial cultivation. The biological management of this resource offers promising future prospects in addressing environmental, economic, and social challenges. The inception of modern biological processes in the recycling of FW may lead to the generation of several streams of functional ingredients, bioactive compounds, biogas, biofuel, and dietary supplements, thereby contributing to the circular bioeconomy and a significant reduction in unattended waste generation.

Biological processes such as enzymatic hydrolysis are garnering significant interest because of their higher purity and causing the least harm to the environment. Due to its high sugar content, FW is responsive to acetone–ethanol-butanol fermentation. Enzyme- or microbe-mediated transformation of FW into bioalcohols (bioethanol and biobutanol) at a commercial scale is the demand of the hour, and it is capable of replacing conventional petroleum fuels. Utilization of FW for bioalcohol production can significantly reduce the cost of raw materials and processing.

Biological processes are helpful in converting FW into a sustainable replacement for conventional plastic in the form of bioplastics (PHA, PHB) with comparable physicochemical properties. These bioplastics also offer biodegradable and biocompatible features at low cost. However, large-scale production and commercial applications of bioplastics still face limitations. To enhance production and applications, genetically engineered microbes or microalgae as biorefineries can be a viable solution to the existing challenges associated with synthetic plastic, which degrades the environment and harms flora and fauna.

Thus, it can be concluded that biological processes provide effective and sustainable solutions for food waste treatment, offering multiple benefits including waste reduction, resource recovery, and energy generation. Techniques such as composting, anaerobic digestion, fermentation, enzymatic hydrolysis, and microbial fuel cells demonstrate diverse applications in waste management.

Composting methods, both traditional and vermicomposting, facilitate organic waste decomposition and nutrient recycling. Anaerobic digestion proves instrumental in converting organic matter into valuable biogas. Fermentation processes such as Bokashi and yeast fermentation showcase potential for bioethanol production and waste valorization. Enzymatic hydrolysis offers a means to break down complex organic compounds, aiding in waste digestion. Additionally, microbial fuel cells present an innovative approach to simultaneously treat waste and generate electricity. As we envision the future, integrating these biological methods into waste management strategies is crucial for fostering sustainability and transitioning towards a circular economy model.

REFERENCES

Abbasi, Iffat Abbas et al. 2024. "Circular economy-based integrated farming system for indigenous chicken: Fostering food security and sustainability." *Journal of Cleaner Production* 436: 140368.

Aguinaga, Danny A Lizarzaburu et al. 2023. "Traditional composting, Bokashi and Takakura: Efficiency in the degradation of organic waste." *Chemical Engineering Transactions* 100: 79–84.

Ameen, Fuad and Al-Homaidan, Ali A. 2022. "Improving the efficiency of vermicomposting of polluted organic food wastes by adding biochar and mangrove fungi." *Chemosphere* 286. https://doi.org/10.1016/j.chemosphere.2021.131945

Angulo, Miguel and Márquez, Mª Carmen. 2023. "A green technology approach using enzymatic hydrolysis to valorize meat waste as a way to achieve a circular economy." *Applied Sciences* 1513: 8763.

Antonopoulou, G et al. 2021. "On the evaluation of filtered and pretreated cheese whey as an electron donor in a single chamber microbial fuel cell." *Biomass Conversion and Biorefinery* 11: 633–43.

Aruwajoye, Gabriel S; Faloye, Funmilayo D and Kana, Evariste Gueguim. 2020. "Process optimisation of enzymatic saccharification of soaking assisted and thermal pretreated cassava peels waste for bioethanol production." *Waste and Biomass Valorization* 11: 2409–20.

Avramidou, Kalliopi. 2020. "Biocatalysis for biomass valorization: Protein hydrolysates and sugar esters from agri-food wastes." https://dx.doi.org/10.13130/avramidou-kalliopi_phd2020-01-30

Ayala, José R et al. 2021. "Characterization of orange peel waste and valorization to obtain reducing sugars." *Molecules* 526: 1348.

Azhar, Siti Hajar Mohd et al. 2017. "Yeasts in sustainable bioethanol production: A review." *Biochemistry and Biophysics Reports* 10: 52–61.

Baruah, Julie et al. 2022. "Integrated pretreatment of banana agrowastes: Structural characterization and enhancement of enzymatic hydrolysis of cellulose obtained from banana peduncle." *International Journal of Biological Macromolecules* 201: 298–307.

Behera, Shuvashish et al. 2010. "Comparative study of bio-ethanol production from mahula (Madhuca latifolia L.) flowers by Saccharomyces cerevisiae cells immobilized in agar agar and Ca-alginate matrices." *Applied Energy* 187: 96–100.

Beniche, I et al. 2021. "Effects of C/N ratio on anaerobic co-digestion of cabbage, cauliflower, and restaurant food waste." *Biomass Conversion and Biorefinery* 511: 2133–45.

Bhatia, Latika et al. 2023. "Food waste utilization for reducing carbon footprints towards sustainable and cleaner environment: A review." *International Journal of Environmental Research and Public Health* 20(320): 2318.

Bin, Hu et al. 2024. "Review of waste plastics treatment and utilization: Efficient conversion and high value utilization." *Process Safety and Environmental Protection* 183: 378–98.

Boechat, Cácio Luiz; Santos, Jorge Antonio Gonzaga and Accioly, Adriana Maria de Aguiar. 2013. "Net mineralization nitrogen and soil chemical changes with application of organic wastes with 'Fermented Bokashi Compost'." *Acta Scientiarum. Agronomy* 35: 257–64.

Budihardjo, Mochamad Arief et al. 2023. "Utilization of Bokashi composting and animal feed silage for sustainable agricultural waste management and environmental impact analysis." *Global Nest Journal* 25: 1–10.

Cai, Linlin et al. 2022. "Vermicomposting with food processing waste mixtures of soybean meal and sugarcane bagasse." *Environmental Technology and Innovation* 28. https://doi.org/10.1016/j.eti.2022.102699

Callaghan, FJ et al. 2002. "Continuous co-digestion of cattle slurry with fruit and vegetable wastes and chicken manure." *Biomass and Bioenergy* 122: 71–7.

Chatterjee, Indranil et al. 2006. "Very low ethanol concentrations affect the viability and growth recovery in post-stationary-phase Staphylococcus aureus populations." *Applied and Environmental Microbiology* 472: 2627–36.

Chaudhary, Asma et al. 2021. "Pomegranate peels waste hydrolyzate optimization by response surface methodology for bioethanol production." *Saudi Journal of Biological Sciences* 928: 4867–75.

Chidambarampadmavathy, Karthigeyan; Karthikeyan, Obulisamy Parthiba and Heimann, Kirsten. 2017. "Sustainable bio-plastic production through landfill methane recycling." *Renewable and Sustainable Energy Reviews* 71: 555–62.

Córdova, Andrés et al. 2022. "Recent advances in the application of enzyme processing assisted by ultrasound in agri-foods: A review." *Catalysts* 112: 107.

Cuetos, MJ et al. 2010. "Anaerobic digestion and co-digestion of slaughterhouse waste (SHW): influence of heat and pressure pre-treatment in biogas yield." *Waste Management* 1030: 1780–9.

De Gioannis, Giorgia et al. 2017. "Energy recovery from one-and two-stage anaerobic digestion of food waste." *Waste Management* 68: 595–602.

de la Torre, Isabel; Ladero, Miguel and Santos, Victoria E. 2020. "D-lactic acid production from orange waste enzymatic hydrolysates with L. delbrueckii cells in growing and resting state." *Industrial Crops and Products* 146: 112176.

de Oliveira, Tatiane CG et al. 2022. "Sustainable valorization of apple waste in a biorefinery: A bibliometric analysis." *Biofuels, Bioproducts and Biorefining* 316: 891–919.

Dorleku, Winfred-Peck et al. 2022. "Response surface optimisation of enzymatic hydrolysis of cassava peels without chemical and hydrothermal pretreatment." *Biomass Conversion and Biorefinery* 1–14. https://doi.org/10.1007/s13399-021-02201-6

FAO, IFAD. 2013. "WFP (2013): The state of food insecurity in the world." In *Economic growth is necessary but not sufficient to accelerate reduction of hunger and malnutrition*. FAO, Rome.

Fei, Xionghui et al. 2020. "Enhancement effect of ionizing radiation pretreatment on biogas production from anaerobic fermentation of food waste." *Radiation Physics and Chemistry* 168: 108534.

Ferdeș, Mariana et al. 2022. "Food waste management for biogas production in the context of sustainable development." *Energies* 1715: 6268.

Fernandes, Rikitha S; Shetty, Nitinkumar S and Thimmappa, BHS. 2021. "A review of the current situation of municipal solid waste management in India and its potential for anaerobic digestion." *International Journal of Environment and Pollution* 1–269: 130–50.

Fiedurek, Jan; Skowronek, Marcin and Gromada, Anna. 2011. "Selection and adaptation of Saccharomyces cerevisae to increased ethanol tolerance and production." *Polish Journal of Microbiology* 160: 51–8.

Fonseca, Gustavo Graciano et al. 2008. "The yeast Kluyveromyces marxianus and its biotechnological potential." *Applied Microbiology and Biotechnology* 79: 339–54.

Formowitz, Beate et al. 2007. "The role of 'effective microorganisms' in the composting of banana (Musa ssp.) residues." *Journal of Plant Nutrition and Soil Science* 5170: 649–56.

Gladchenko, MA et al. 2017. "Methane production by anaerobic digestion of organic waste from vegetable processing facilities." *Applied Biochemistry and Microbiology* 253: 242–9.

Han, Wei et al. 2020. "Bioethanol production from waste hamburger by enzymatic hydrolysis and fermentation." *Journal of Cleaner Production* 264: 121658.

Hao, Tianwei; Xiao, Yihang and Varjani, Sunita. 2022. "Transiting from the inhibited steady-state to the steady-state through the ammonium bicarbonate mediation in the anaerobic digestion of low-C/N-ratio food wastes." *Bioresource Technology* 351: 127046.

Heo, Nam Hyo; Park, Soon Chul and Kang, Ho. 2004. "Effects of mixture ratio and hydraulic retention time on single-stage anaerobic co-digestion of food waste and waste activated sludge." *Journal of Environmental Science and Health, Part A* 739: 1739–56.

Iigatani, Reo et al. 2019. "Electricity generation from sweet potato-shochu waste using microbial fuel cells." *Journal of Bioscience and Bioengineering* 1128: 56–63.

Karthikeyan, Obulisamy Parthiba et al. 2018. "Pretreatment of food waste for methane and hydrogen recovery: A review." *Bioresource Technology* 249: 1025–39.

Khan, Muhammad Farooq Saleem et al. 2021. "A review on the role of pretreatment technologies in the hydrolysis of lignocellulosic biomass of corn stover." *Biomass and Bioenergy* 155: 106276.

Kim, Jung Kon et al. 2006. "Effects of temperature and hydraulic retention time on anaerobic digestion of food waste." *Journal of Bioscience and Bioengineering* 4102: 328–32.

Kumar, Ashish; Singh, LK and Ghosh, Sanjoy. 2009. "Bioconversion of lignocellulosic fraction of water-hyacinth (Eichhornia crassipes) hemicellulose acid hydrolysate to ethanol by Pichia stipitis." *Bioresource Technology* 13100: 3293–7.

Kumar, Shanmugam Dilip et al. 2022. "Efficiency of microbial fuel cells in the treatment and energy recovery from food wastes: Trends and applications-a review." *Chemosphere* 287: 132439.

la Cour Jansen, Jes et al. 2004. "Digestion of sludge and organic waste in the sustainability concept for Malmö, Sweden." *Water Science and Technology* 1049: 163–9.

Lew, Pei Sze et al. 2021. "Optimization of bokashi-composting process using effective microorganisms-1 in smart composting bin." *Sensors* 821: 2847.

Li, R; Chen, S and Li, X. 2009. "Anaerobic co-digestion of kitchen waste and cattle manure for methane production." *Energy Sources, Part A: Recovery, Utilization, and Environmental Effects* 2031: 1848–56.

Lou, Xian Fang; Nair, Jaya and Ho, Goen. 2012. "Field performance of small scale anaerobic digesters treating food waste." *Energy for Sustainable Development* 416: 509–14.

Luo, Jingyang et al. 2019. "Promoting the anaerobic production of short-chain fatty acids from food wastes driven by the reuse of linear alkylbenzene sulphonates-enriched laundry wastewater." *Bioresource Technology* 282: 301–9.

Ma, Qianyun et al. 2022. "The effects of enzymatic modification on the functional ingredient-Dietary fiber extracted from potato residue." *Lwt* 153: 112511.

Mihajlovski, Katarina; Rajilić-Stojanović, Mirjana and Dimitrijević-Branković, Suzana. 2020. "Enzymatic hydrolysis of waste bread by newly isolated Hymenobacter sp. CKS3: Statistical optimization and bio-ethanol production." *Renewable Energy* 152: 627–33.

Mohan, S and Joseph, Charles P. 2021. "Potential hazards due to municipal solid waste open dumping in India." *Journal of the Indian Institute of Science* 4101: 523–36.

Morales-Polo, Carlos; Cledera-Castro, María Del Mar and Moratilla Soria, B Yolanda. 2018. "Reviewing the anaerobic digestion of food waste: From waste generation and anaerobic process to its perspectives." *Applied Sciences* 108: 1804.

Morone, Piergiuseppe et al. 2019. "Food waste: Challenges and opportunities for enhancing the emerging bio-economy." *Journal of Cleaner Production* 221: 10–6.

Ni, Renjie et al. 2022. "Acetoclastic methanogenesis pathway stability despite the high microbial taxonomic variability in the transition from acidogenesis to methanogenesis during food waste anaerobic digestion." *Journal of Cleaner Production* 372: 133758.

Nwaokocha, Collins N et al. 2021. "Microbial fuel cell: Bio-energy production from Nigerian corn starch wastewater using iron electrodes." *Materials Today: Proceedings* 46: 5565–9.

Olle, Margit. 2021. "Bokashi technology as a promising technology for crop production in Europe." *The Journal of Horticultural Science and Biotechnology* 296: 145–52.

Periyasamy, Selvakumar et al. 2022. "Chemical, physical and biological methods to convert lignocellulosic waste into value-added products. A review." *Environmental Chemistry Letters* 220: 1129–52.

Phirom-on, Konlarat and Apiraksakorn, Jirawan. 2021. "Development of cellulose-based prebiotic fiber from banana peel by enzymatic hydrolysis." *Food Bioscience* 41: 101083.

Piercy, Ellen et al. 2023. "A sustainable waste-to-protein system to maximise waste resource utilisation for developing food-and feed-grade protein solutions." *Green Chemistry* 325: 808–32.

Pilarska, Agnieszka A et al. 2023. "Anaerobic digestion of food waste—A short review." *Energies* 1516: 5742.

Procentese, Alessandra et al. 2020. "Kinetic characterization of enzymatic hydrolysis of apple pomace as feedstock for a sugar-based biorefinery." *Energies* 513: 1051.

Ramasamy, Rajam and Subramanian, Parthasarathi. 2022. "Bioconversion of food waste to wealth–circular bioeconomy approach." *Biotechnology for Zero Waste: Emerging Waste Management Techniques* 421–38.

Rivera, Anna María Polania et al. 2022. "Adding value to processes, products, and by-products from the poultry industry through enzymatic technologies." In Mohammed Kuddus and Cristobal Noe Aguilar (Eds.), *Value-addition in food products and processing through enzyme technology* 235–251. Elsevier.

Rodríguez-Jiménez, Lina Mariana; Pérez-Vidal, Andrea and Torres-Lozada, Patricia. 2022. "Research trends and strategies for the improvement of anaerobic digestion of food waste in psychrophilic temperatures conditions." *Heliyon* 108: e11174.

Rosa-Sibakov, Natalia et al. 2022. "Functionality and economic feasibility of enzymatically hydrolyzed waste bread as a sugar replacer in wheat bread making." *Journal of Food Processing and Preservation* 346: e16378.

Sahu, Ram Bhajan; Pahlwani, Janki and Singh, Priyanka. 2023. "Current status of enzymatic hydrolysis of cellulosic biomass." *Production of Biobutanol from Biomass* 77–104. https://doi.org/10.1002/9781394172887.ch3

Saleem, Ayesha et al. 2020. "Acid hydrolysis optimization of pomegranate peels waste using response surface methodology for ethanol production." *Biomass Conversion and Biorefinery* 1–12. https://doi.org/10.1007/s13399-020-01117-x

Sambavi, SM et al. 2021. "Bioconversion of food waste with cosubstrate to electricity using microbial fuel cell." In *Journal of physics: Conference series* (Vol. 1979, No. 1, p. 012011). IOP Publishing.

Sathyamoorthy, GL and Sushmitha, A Sam. 2022. "Sustainable energy production from food waste using microbial fuel cell (MFC)." In *AIP Conference Proceedings*. (Vol. 2446, No. 1). AIP Publishing.

Sawarkar, Ashish N et al. 2022. "Bioethanol from various types of banana waste: A review." *Bioresource Technology Reports* 18: 101092.

Schlegel, Katharina et al. 2019. "Enzymatic hydrolysis of lupin protein isolates—changes in the molecular weight distribution, technofunctional characteristics, and sensory attributes." *Food Science & Nutrition* 87: 2747–59.

Sharma, Poonam et al. 2020. "Microbial strategies for bio-transforming food waste into resources." *Bioresource Technology* 299: 122580.

Sharma, Poonam et al. 2021. "Sustainable processing of food waste for production of bio-based products for circular bioeconomy." *Bioresource Technology* 325: 124684.

Singh, Anusuiya et al. 2022. "Production of bioethanol from food waste: Status and perspectives." *Bioresource Technology* 360: 127651.

Singhvi, Mamata S and Gokhale, Digambar V. 2019. "Lignocellulosic biomass: hurdles and challenges in its valorization." *Applied Microbiology and Biotechnology* 103: 9305–20.

Sirohi, Ranjna et al. 2021. "Two-stage enzymatic hydrolysis for fermentable sugars production from damaged wheat grain starch with sequential process optimization and reaction kinetics." *Starch-Stärke* 1–273: 2000082.

Slorach, Peter C et al. 2019. "Environmental and economic implications of recovering resources from food waste in a circular economy." *Science of The Total Environment* 693: 133516.

Sosnowski, Piotr et al. 2008. "Kinetic investigations of methane co-fermentation of sewage sludge and organic fraction of municipal solid wastes." *Bioresource Technology* 1399: 5731–7.

Suresh, T et al. 2020. "Process intensification and comparison of bioethanol production from food industry waste (potatoes) by ultrasonic assisted acid hydrolysis and enzymatic hydrolysis: Statistical modelling and optimization." *Biomass and Bioenergy* 142: 105752.

Szopa, Daniel et al. 2023. "Waste valorization towards industrial products through chemo-and enzymatic-hydrolysis." *Bioengineered* 114: 2184480.

Szucs, Michelle et al. 2021. "Meat waste valorization through protein hydrolysis using different types of proteases." *Recent Progress in Materials* 43: 1–19.

Szymczak, Tomasz et al. 2021. "Various perspectives on microbial lipase production using agri-food waste and renewable products." *Agriculture* 611: 540.

Tavano, Olga Luisa. 2013. "Protein hydrolysis using proteases: An important tool for food biotechnology." *Journal of Molecular Catalysis B: Enzymatic* 90: 1–11.

Tofighi, Azadeh et al. 2014. "Bio-ethanol production by a novel autochthonous thermo-tolerant yeast isolated from wastewater." *Journal of Environmental Health Science and Engineering* 12: 1–6.

Torres-León, Cristian et al. 2021. "Recent advances on the microbiological and enzymatic processing for conversion of food wastes to valuable bioproducts." *Current Opinion in Food Science* 38: 40–5.

Tyagi, Vinay Kumar et al. 2018. "Anaerobic co-digestion of organic fraction of municipal solid waste (OFMSW): Progress and challenges." *Renewable and Sustainable Energy Reviews* 93: 380–99.

Utekar, Pooja G; Kininge, Madhuri M and Gogate, Parag R. 2021. "Intensification of delignification and enzymatic hydrolysis of orange peel waste using ultrasound for enhanced fermentable sugar production." *Chemical Engineering and Processing-Process Intensification* 168: 108556.

Walker, Graeme M and Stewart, Graham G. 2016. "Saccharomyces cerevisiae in the production of fermented beverages." *Beverages* 42: 30.

Wang, Danli et al. 2017. "Ultrasound assisted enzymatic hydrolysis of starch catalyzed by glucoamylase: Investigation on starch properties and degradation kinetics." *Carbohydrate Polymers* 175: 47–54.

Wang, Wen et al. 2021. "Mild Urea/KOH pretreatment to enhance enzymatic hydrolysis of corn stover with liquid waste recovery for plant growth." *Journal of Cleaner Production* 284: 125392.

Wang, Xiaojiao et al. 2012. "Optimizing feeding composition and carbon–nitrogen ratios for improved methane yield during anaerobic co-digestion of dairy, chicken manure and wheat straw." *Bioresource Technology* 120: 78–83.

Wang, Xuemei et al. 2020. "Impacts of cellulase and amylase on enzymatic hydrolysis and methane production in the anaerobic digestion of corn straw." *Sustainability* 1312: 5453.

Wang, Yumei; Tang, Ya and Yuan, Zengwei. 2022. "Improving food waste composting efficiency with mature compost addition." *SSRN Electronic Journal*. https://doi.org/10.2139/ssrn.3992440

Wyman, Charles E. 2018. "Ethanol production from lignocellulosic biomass: Overview." *Handbook on Bioethanol* 1–18. http://dx.doi.org/10.1007/978-90-481-3295-9_1

Xu, Mingyue et al. 2022. "Role of multistage inoculation on the co-composting of food waste and biogas residue." *Bioresource Technology* 361. https://doi.org/10.1016/j.biortech.2022.127681

Yaakop, Amira Suriaty et al. 2023. "Domestic organic waste: A potential source to produce the energy via a single-chamber microbial fuel cell." *International Journal of Chemical Engineering* 2023. https://doi.org/10.1155/2023/2425735

Yong, Zihan et al. 2015. "Anaerobic co-digestion of food waste and straw for biogas production." *Renewable Energy* 78: 527–30.

Zan, Feixiang et al. 2022. "'Food waste-wastewater-energy/resource' nexus: Integrating food waste management with wastewater treatment towards urban sustainability." *Water Research* 211: 118089.

Zhang, Cunsheng et al. 2013. "The anaerobic co-digestion of food waste and cattle manure." *Bioresource Technology* 129: 170–6.

Zhang, Cunsheng et al. 2014. "Reviewing the anaerobic digestion of food waste for biogas production." *Renewable and Sustainable Energy Reviews* 38: 383–92.

Zhang, Jingzhi et al. 2020a. "Pretreatment of lignocellulosic biomass for efficient enzymatic saccharification of cellulose." In Abu Yousuf, Domenico Pirozzi And Filomena Sannino (Eds) *Lignocellulosic biomass to liquid biofuels* 17–65. Elsevier.

Zhang, Le et al. 2020b. "Highly efficient anaerobic co-digestion of food waste and horticultural waste using a three-stage thermophilic bioreactor: Performance evaluation, microbial community analysis, and energy balance assessment." *Energy Conversion and Management* 223: 113290.

Zhang, Lei; Lee, Yong-Woo and Jahng, Deokjin. 2011. "Anaerobic co-digestion of food waste and piggery wastewater: Focusing on the role of trace elements." *Bioresource Technology* 8102: 5048–59.

Zhang, Xuan et al. 2022. "Food waste composting based on patented compost bins: Carbon dioxide and nitrous oxide emissions and the denitrifying community analysis." *Bioresource Technology* 346: 126643.

Zou, Huijing et al. 2020. "Enhanced hydrolysis of lignocellulose in corn cob by using food waste pretreatment to improve anaerobic digestion performance." *Journal of Environmental Management* 254: 109830.

5 Production of Biohydrogen from Food Waste

Marcela Moreira Albuquerque, Walter José Martinez-Burgos, Gabriela de Bona Sartor, Carlos Ricardo Soccol, and Adriane Bianchi Pedroni Medeiros

5.1 INTRODUCTION

In 2019, around 931 million tons of food were wasted worldwide, with homes being responsible for around 61% of waste, food services by almost 26%, and retail by around 13% (UNEP, 2021). Meanwhile, in the same year, an estimated 690 million people around the world experienced hunger, and 2 billion people were affected by moderate or severe food insecurity (FAO, 2020). Most food waste is sent to landfills, where it generates compounds such as CH_4, CO_2, NH_3 and net emissions (leachate) (Regueira-Marcos et al., 2023; Singh et al., 2023), contributing to the worsening of environmental, economic, and social problems due to the waste of money and resources such as water and soil. In addition, it is responsible for around 8–10% of global greenhouse gas emissions (UNEP, 2021; Zhang et al., 2023).

For this reason, member countries of the United Nations (UN), with the adoption of sustainable development goals, have taken responsibility for halving global food waste by 2030 and reducing food losses throughout the production, retail, and consumption chains (Beretta & Hellweg, 2019). In addition to reducing food waste, another approach that has been adopted to minimize problems related to food waste is energy production (Jia et al., 2020). Food waste has a variable constitution, being rich in carbohydrates, fats, and proteins (Alavi-Borazjani et al., 2019a; Alian et al., 2021), characterizing itself as potential substrate for anaerobic digestion processes in dark fermentation systems and photofermentation, aiming at the production of biofuels such as hydrogen (Ding et al., 2022; Jung et al., 2022).

Due to the numerous benefits of using hydrogen as a fuel, such as its application in fuel cells and the generation of electrical energy with an efficiency of around 122 KJ/g, 2.75 times higher than hydrocarbon fuels, and releasing water as the only product of its combustion, it has been commonly referred to as the "fuel of the future" (Alian et al., 2021; Keskin & Hallenbeck, 2012; Niño-Navarro et al., 2020). Furthermore, the use of hydrogen can solve environmental problems related to the intensification of the greenhouse effect and other issues related to the burning of fossil fuels, depletion of this non-renewable matrix, and growing demand for energy (Bilanovic et al., 2009; Jafary et al., 2019; Rai & Singh, 2016; Rezaeitavabe et al., 2020).

The biological production of hydrogen has been considered an alternative method to its production from fossil fuels, using organic, ecological, and renewable substrates, which can be carried out under ambient temperature and pressure (Das & Veziroglu, 2008), allowing the integration of energy production and waste treatment (Boboescu et al., 2016). During dark fermentation, food residues can be oxidized by facultative or strict anaerobic microorganisms of genera such as *Clostridium*, *Enterobacter*, *Klebsiella*, *Citrobacter*, and *Bacillus* (Hawkes et al., 2007) producing hydrogen. However, during this process, the substrate is incompletely oxidized, especially in the case of the high carbohydrate/protein ratio present in these residues, resulting in the accumulation of volatile fatty acids that acidify the culture medium. This acidification harms dark fermentation microorganisms, compromising H_2 yields and maintaining high COD (chemical oxygen demand) levels (Moreno-Andrade et al., 2023; Rambabu et al., 2021a).

DOI: 10.1201/9781032706030-5

The best use of this substrate can be achieved through the use of dark fermentation effluent for the hybrid photofermentation process or sequential to dark fermentation, given the fact that purple non-sulfur bacteria (PNSB), such as those of the genera *Rhodobacter* (Pandey et al., 2015) *Rhodobium* (Anam et al., 2012), *Rhodopseudomonas* (García-Sánchez et al., 2018), and *Rhodospirillum* (Ferreira et al., 2022), which are responsible for the photofermentation process, preferentially use volatile fatty acids as substrate and light energy to produce H_2 (Das & Basak, 2021), culminating in the production of a value-added compound, and better use and treatment of the substrate (Rai & Singh, 2016).

Thus, in this chapter, some aspects related to the valorization of food waste to produce biohydrogen ($BioH_2$) are addressed, such as the composition of substrates, the processes of dark fermentation and photofermentation, the microorganisms used in each stage, as well as the processes and parameters such as pH, temperature, and pretreatments used on the substrate to obtain the best $BioH_2$ yields.

5.2 TYPES OF FOOD WASTE AND THEIR COMPOSITION

Food waste is generated mainly in households, in storage and handling processes in industries, and in agricultural production, with some types coming from fruits and vegetables, cereals, animals, dairy products, etc. (Mohanakrishna et al., 2023; Rajesh Banu et al., 2020). In general, food waste is mainly composed of lipids, proteins, and carbohydrates (Zeng et al., 2018). Variations in the concentration of each component may occur due to the different origins of the food residue and its content. For example, foods with vegetables and grains are normally rich in carbohydrates, while foods of animal origin are mainly made up of proteins and lipids (Paritosh et al., 2017).

Despite the particularities of each residue, in general, its composition varies between 41 and 62% of degradable carbohydrates, 13 and 30% of lipids, and 15 and 25% of proteins, in addition to a moisture content of 74–90%, a high percentage of volatile solids of around 85%, and an acidic pH of around 5.1 (Braguglia et al., 2018). Due to these characteristics, FW has an intrinsic potential to be used as substrates in microbial fermentations to generate products with high added value (Bilal & Iqbal, 2019).

Table 5.1 summarizes the composition of some FW, namely kitchen waste (Vavouraki et al., 2014), waste from restaurants (Villanueva-Galindo & Moreno-Andrade, 2021), and canteens (Alavi-Borazjani et al., 2019b), of fruits, vegetables, grains, and meat (Chu et al., 2008; Soltan et al., 2019).

5.3 FW PRETREATMENT

The production of high-value-added products using food waste as a substrate has proven to be a more ecologically viable alternative for disposing of this type of material (Rajesh Banu et al., 2020). However, its use in fermentations is limited by its composition, which can negatively affect production and degradation yields (Parthiba Karthikeyan et al., 2018). Therefore, biological processes that employ FW require a prior pretreatment step (physical, chemical, or biological), which is directly dependent on the composition of the material (Elgarahy et al., 2023).

FW has a high concentration of carbohydrates and, therefore, is an ideal substrate for biohydrogen production (Sharma et al., 2020). However, the lignocellulose fraction negatively interferes with the yields of biological processes, as components such as hemicellulose and lignin do not hydrolyze easily (Zheng et al., 2014). Knowing that biohydrogen production is directly influenced by the biomass degradation rate, greater production is achieved when the material is subjected to a prior pretreatment process (Kuang et al., 2020). This is mainly because the pretreatment step, when hydrolyzing lignocellulosic components and cellulose, contributes to greater availability of fermentable sugars and the reduction of toxic compounds (Sun & Cheng, 2002).

Table 5.2 provides examples of pretreatment using various food waste, namely fruit waste (Cieciura-Włoch et al., 2020; Mahato et al., 2020), kitchen waste (Li et al., 2014; Zhao et al., 2011),

TABLE 5.1
Physical and Chemical Characteristics of Some Food Waste Types Used for Biohydrogen Production

Food Waste	COD	pH	Total Solids	Volatile Solids	Carbohydrates	Proteins	Lipids	Reference
Kitchen food waste – 50% raw and fresh food; 15.6% cooked food; 14.5% fruit; 9.7% salad; 8% bread; and 2.2% dessert (in wet weight)	-	-	18.5±0.715 (%w/w)	94.1±0.350 (%w/w)	More than 50% of its dry weight, of which 24% is starch	16.9±0.693 (%w/w)	14.0±0.518 (%w/w) (oil and grease)	(Vavouraki et al., 2014)
Orange peel	28.38±2.21 g/g vs	3.4±0.1	10.6±0.3 % (w/w)	9.4±0.2 , % (w/w)	14.46±0,40 g/g $_{VS}$	0.7±0.04 g/g $_{VS}$	0.29±0.02 g/g $_{VS}$	(Soltan et al., 2019)
Tomato peel	13.58±1.21 g/g vs	4.4±0.32	17.6±0.3 % (w/w)	16.1±0.2 % (w/w)	3.55±0,36 g/g $_{VS}$	1.5±0.02 g/g $_{VS}$	0.12±0.01 g/g $_{VS}$	(Soltan et al., 2019)
Food waste from university canteens (vegetables and fruits, meat and fish, potatoes, bread, rice and pasta	-	4.64±0.06 (%wt)	25.32±0.63 (%wt)	23.75±0.89 (%wt)	-	-	-	(Alavi-Borazjani et al., 2019b)
Restaurant food waste (buffet)	118.6 g/kg (soluble) 316.2 g/kg (total)	4.5	0.41 g/g	0.37 g/g	155.4 g/kg (total) and 81.5 g/kg (soluble)	-	-	(Villanueva-Galindo & Moreno-Andrade, 2021)
Diluted cafeteria food waste (grains, vegetables, meat and fish)	130–155 g/L	4.79–5.08	107–128 g/L	99–117 g/L	60–72 g/L (total)	41–49 g/L (total)	-	(Chu et al., 2008)

(-) Not specified in the study.

TABLE 5.2
Examples of Pretreatments Used on Food Waste

Pretreatment	Type of Pretreatment	Food Waste	Inoculum	Conditions	Bioreactors	H2 Yield	Reference
Physical and Mechanical	Microwave (2.45 GHz for 15 min) and moist heat (121°C for 15 min)	Fruit waste	*Clostridium* BOH3	37°C, pH 6.8 and 300 rpm	250 mL bottles	2.43–2.51 mol/mol hexose)	(Mahato et al., 2020)
	Hydrothermal (200°C for 30 min)	Kitchen waste	Obtained from an anaerobic reactor of swine manure	150 rpm at 35±1°C, with pH 6	500 mL bottles	81.27 mL/g VS	(M. Li et al., 2014)
	Ultrasonication (12 min)	Cafeteria waste (rice, vegetables, fruit peels, etc.)	Obtained from a dairy industry anaerobic digester	pH 5.5 37°C and 180 rpm	125 mL serum bottles	149 mL/g VS	(Gadhe et al., 2014)
	Grinding in a crusher (obtaining particles of 3–13 mm)	Fruit and vegetable waste diluted (2:1, 3:1, 4:1)	Anaerobic sludge from municipal wastewater treatment plant	pH 3.91 ±0.02 - 4.01±0.02 under 35±1°C	Semi-continuous mode in cylindrical glass fermenters (3 dm^3)	52 cm^3/gVS	(Cieciura-Włoch et al., 2020)
Chemical	Alkaline treatment with NaOH and acid with HCl to adjust the pH	Kitchen waste (rice, meat, and vegetables)	Anaerobic digester granular sludge	35±1°C and 70 rpm	400 mL bottles	105.38 mL/g VS	(Zhao et al., 2011)
	2-Bromoethane sulfonic acid sodium salt solution (BESA, 0.2 g/L)	University canteen waste	Anaerobic activated sludge, obtained from digested sludge from a pepper treatment plant	37°C±0.5	Semi-continuous (400 mL bottles)	280.82±5.72) mL/L of H2 at pH 7	(Yasser Farouk et al., 2020)
Biological	Enzymatic (glucoamylase and protease)	Bread residue	Anaerobic sludge	35±1°C, pH >4 80-100 rpm agitation	Mixed tank reactor – MPTR (4L)	4.87 mmol hydrogen/g bread residue	(Han, Hu, et al., 2016)
	Enzymatic (glucoamylase)	Food waste from university canteen	Coming from anaerobic ETE sludge	Batch: 200 rpm and 37°C Continuous: 37°C For both, pH > 4	Batch and continuous stirred tank reactor (CSTR)	245.7 mL of hydrogen/g of glucose in the batch system and 205.8 mL of hydrogen/g glucose in the continuous system	(Han, Yan, et al., 2016)

cafeteria waste (Gadhe et al., 2014), bread residue (Han, Hu et al., 2016), and canteen waste (Yasser Farouk et al., 2020).

The use of microwave irradiation is efficient, as it acts directly on the solubilization of polymers (Chaturvedi & Verma, 2013). In a study by Mahato et al., (2020), with the application of moist heat microwave irradiation on kitchen waste (2.45 GHz for 15 min), there was an increase in the concentration of sugars (15.8–18.2 g/L) and other nutrients, which in turn contributed to the growth of *Clostridium* BOH3 cells and to the production of hydrogen (2.43–2.51 mol/mol hexose).

Another physical approach would be the use of ultrasonification, as reported in a study by Gadhe et al., (2014), which found that when applying an ultrasonication time of 12 minutes and an optimal condition of 8% total solids in cafeteria waste, a maximum yield of 149 mL/g VS was obtained. In hydrothermal pretreatment, used in the disintegration of lignocellulosic biomass, little waste is generated, and the bioavailability of nutrients is improved (Dinesh et al., 2018). Li et al., (2014) used this method at various temperatures to carry out the hydrolysis of kitchen waste. It was observed that the treatment at 200°C was the best condition to obtain $BioH_2$ (81.27 mL/g VS).

However, mechanical methods such as grinding are also effective in pretreating food waste, as reducing particle size improves absorption (Gallego-García et al., 2023). For example, Cieciura-Włoch et al., (2020), by reducing fruit and vegetable residue particles, obtained hydrogen yield of 52 cm^3/gVS. In chemical pretreatments, hydrochloric acid and sulfuric acid are often used in acid treatments (Ma et al., 2018), and sodium hydroxide and potassium hydroxide are used in alkaline treatments (Linyi et al., 2020). Although this method does not require complex installations, the use of acid and alkaline reagents is not sustainable and can result in environmental impacts (Singh et al., 2022). Sodium hydroxide and hydrochloric acid were used in the study by Zhao et al., (2011) to pretreat kitchen waste, obtaining a hydrogen yield of 105.38 mL/g VS.

Finally, another possibility of pretreating food waste biomass is through biological treatment (Bhurat et al., 2023). This method is considered a cheaper and more sustainable alternative when compared to physical and chemical methods, as it eliminates problems related to high energy consumption and costs (Yukesh Kannah et al., 2020). However, the long treatment time and the need to maintain the purity of bacterial cultures and optimized reaction conditions are some of the disadvantages of this process (Rajesh Banu et al., 2020). Han et al., (2016) used glycosamides to hydrolyze food waste, with the hydrogen yield being equal to 245.7 mL of hydrogen/g of glucose in a discontinuous system and 205.8 mL of hydrogen/g of glucose in a continuous system.

Alternatively, there are reports of combined pretreatment methods, as in the work of Vavouraki et al., (2014). In the mentioned study, acid (HCl) and alkaline (KOH) pretreatment with enzymatic hydrolysis (with α-amylase or amyloglucosidase-AMG) on kitchen waste was verified. As a result, after chemical pretreatment with acid or base followed by enzymatic treatment, the glucose concentration increased by about 300% compared to the untreated residue. In a similar study, Elbeshbishy et al., (2011) tested, in addition to individual pretreatments of food waste, the combination of ultrasonication with heat (UH), ultrasonication with acid (UA), and ultrasonication with base (UB). In the combined pretreatment tests, pretreatment with UA obtained the best result with a yield of 118 mL/g initial VS.

5.4 DARK FERMENTATION (DF) OF FOOD WASTE

Dark fermentation can be defined as the fermentative process where gases such as biohydrogen ($BioH_2$), CO_2, and other intermediate metabolites are generated, such as acetic acid, butyric acid, propionic acid, lactic acid, and some solvents such as acetone and alcohols (Martinez-Burgos et al., 2020; Rosa et al., 2020). Dark fermentation is considered one of the most promising processes to produce biological hydrogen since it is a fast process (around 24h to 72h), and the substrates used are renewable. They are generally waste (solid or liquid) that can be reused or reintegrated into the value chain (Martinez-Burgos et al., 2021b).

In dark fermentation, in addition to producing the energy vector, some pollutants are also removed. In the case of effluents, for example, phosphorus, nitrogen, and organic matter are removed, which reduces COD and BOD_5. The process can be done in batch and continuously. However, this last mode of operation is the most desirable due to the large amounts of effluents that are generated and the low price of biohydrogen. Among the substrates that can be used, the use of food waste stands out due to the large amounts that are generated in the world.

Hydrogen production via dark fermentation using food waste as a substrate takes place in several steps (Figure 5.1). In the first stage (i) non-fermentable residues such as bones and some peels are removed, then the residues are selected in such a way that there is a balance between the carbon and nitrogen content in the residues, since the C/N ratio significantly affects Bio-H_2 production (Martínez-Burgos et al., 2019). However, this parameter is influenced by the type of inoculum and even by the substrates that will be used in the production of Bio-H_2.

According to Li & Fang, (2007), the optimal C/N ratio for biohydrogen production is in the range of 130 to 3.3. Liu & Shen (2004) evaluated the C/N ratio of 130 to 3.3 in Bio-H_2 production and reported the highest YH (175 ml H_2/g hexose) at a C/N ratio of 6.7. Ueno et al., (2001) reported an optimal C/N ratio of 5 for a YH of 272 ml H_2/g hexose. Finally, Martínez-Burgos et al., (2019) reported an optimal C/N ratios of 5.0, 4.0, and 3.5 using different microbial consortia *Vac*, *Esg*, and *Lod*, respectively.

In step (ii), the residues undergo some pretreatment. Crushing and homogenization with water are common operations to obtain the desired solids size and concentration. According to Xiao et al., (2013), the solid concentration should be around 5 g/L. The pH of the medium is then adjusted; salts, bases, or acids can be used for this purpose. In general, dark fermentation takes place at neutral pH or close to this condition (Martinez-Burgos et al., 2021a), so the initial pH in fermentation is in the range of 5.5 to 8.5. The pH is one of the main variables that significantly affect the production of Bio-H_2 (Ananthi et al., 2022). In dark fermentation, the pH of the fermentation broth varies during the process, which is due to the fact that in dark fermentation some intermediate metabolites are produced, mainly organic acids, such as acetic acid and butyric acid (Equations 5.1 and 5.2, respectively) (Saady, 2013; Sydney et al., 2020). At the beginning of the dark fermentation, the pH starts to decrease, and later, when the main carbon source is depleted, the microorganisms use the organic acids as alternative carbon sources, so the pH increases (Martinez-Burgos et al., 2020).

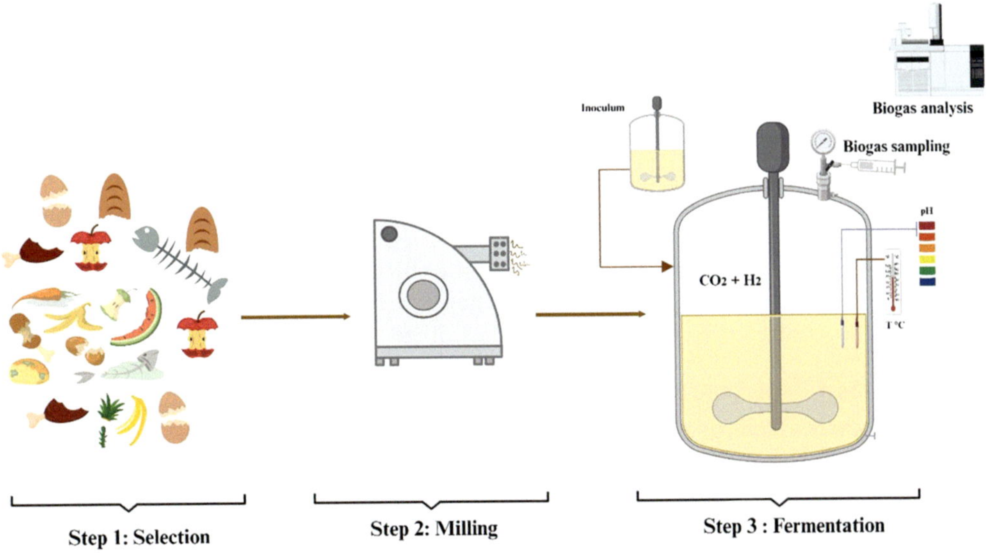

FIGURE 5.1 Steps for producing BioH_2 from food waste.

$$C_6H_{12}O_6 + 2H_2O \longrightarrow 2CH_3COOH + 4H_2 + 2CO_2 \quad (5.1)$$

$$C_6H_{12}O \longrightarrow 2CH_3CH_3CH_2COOH + 2H_2 + 2CO_2 \quad (5.2)$$

In step (iii), the medium is inoculated; the inoculum size varies between 2.0 v/v % to 10 v/v% of the medium volume. Finally, the system is incubated; the incubation temperature depends on the type of microorganisms or microbial consortia used. Temperature affects hydrogen production since it can even change the metabolic route of Bio-H_2 production. During incubation, the microorganisms grow and directly produce Bio-H_2, as this metabolite is associated with the growth of the microorganism. Thus, it has psychrophiles (20–29°C), hydrogen-producing microorganisms such as *Flavobacterium limícola*, *Pseudomonas meridiana*, and *Actimicrobium antarticum*; mesophiles (30–40°C), *Enterobacter aerogenes*, *Clostridium beijerinckii*, *Escherichia cloacae*, and *Rhodobacter sphaeroides*; thermophiles (45–65°C), *Thermotoga neapolitana*, *Clostridium thermocellum*, and *Thermoanaerobacterium thermosaccharolyticum*; and hyperthermophiles (> 65°C), *Thermotoga neapolitana* DSM4359, *Thermotoga marítima* DSM 3109, and *Pyrococcus furiosus* DSM 3638.

Okamoto et al., (2000) used seven types of solid food residues, among which are lean meat, chicken skin, rice, cabbage, carrots, egg, and fat for the production of BioH_2. The process was developed in a batch system, under mesophilic conditions (37°C) and using a microbial consortium (thermally treated sludge). The solids content in the medium was established between 2 and 3.0%, w/w. The highest hydrogen yields were obtained with cabbage, carrots, and rice, which were in the range of 6.3–61.7 mL H_2/g-VS; carrot, 44.9–70.7 mL H_2/g-VS; and 19.3–96.0 mL H_2/g-VS.

Xiao et al., (2013) produced Bio-H_2 using food waste. The residues employed were composed of 6.6% rice, 9.6% vegetables, 8.4% pork, 2.4% eggs, 1.2% corn oil, 0.2% table salt, and 66.2% water. The results showed that the highest amounts of hydrogen were obtained with *Enterobacter aerogenes* and *Escherichia cloacae* strains. The process was carried out at an initial pH of 7.0 and a temperature of 37°C. The best specific Bio-H_2 production (mL/g VS) was 28.7 and 39.6 for *E. aerogenes* and *E. cloacae*, respectively. Dong et al., (2009) obtained the maximum Bio-H_2 production yields using rice, potato, and lettuce residues, which were 134 mL H_2/g-VS, 106 mL H_2/g-VS, and 50 mL/g-VS respectively. The process was developed at 37°C, and the inoculum used was heat-treated swine manure. Other works are summarized in Table 5.3.

5.4.1 Bioreactors Used for BioH_2 Production

The production of BioH_2 and biomethane (BioCH_4) takes place in similar processes. To produce the latter, four steps are necessary (hydrolysis, acidogenesis, acetogenesis, and methanogenesis), while in dark fermentation the process ends in acetogenesis (De Carvalho et al., 2023; Karp et al., 2022), therefore, the bioreactors that have been used to produce BioCH_4 can also be used for the production of BioH_2. Among the different systems or bioreactors that have been used for the production of hydrogen are continuous stirred tank reactor (CSTR), up-flow anaerobic sludge blanket (UASB), fluidized bed reactor (FBR), anaerobic baffled reactor (ABR), and other little-used bioreactors that are derivatizations of the previous ones, such as up-flow anaerobic sludge blanket reactor (UASBR), up-flow anaerobic sludge fixed-film (UASFF), and modified anaerobic baffled bioreactor (MABB).

5.4.1.1 Up-Flow Anaerobic Sludge Blanket (UASB)

The UASB is another bioreactor widely used to treat effluents, especially those with a high organic load. The UASB can be defined as an ascending flow column, which is composed of several zones: one for digestion, one for sedimentation, and the gas-liquid-liquid phase separator device (Figure 5.2A) (Chaisri et al., 2007). Intanoo et al., 2016) produced hydrogen and methane using a two-stage UASB. The maximum hydrogen yields (HY) and methane were 39.83 L H_2/kg COD removed and 115.23 L CH_4/kg COD removed, respectively.

TABLE 5.3
Hydrogen Production from Different Food Waste

Wastes	Bioreactor Type/Inoculum	Operations Conditions	Bio-H_2 Yield	References
Animal kitchen waste, vegetable kitchen waste, cereal waste	Shake tubes/sludge	Batch (55°C, pH:6.0), purge gas: nitrogen	~28 mLH_2/gVS, ~40 mLH_2/gVS, ~150 mLH_2/gVS	(Kobayashi et al., 2012)
FW (pasta, rice, bread, fruit and vegetable husks), olive husks	Bench scale anaerobic flask/ sludge	Batch (57° C, pH: 7.0), purge gas: nitrogen	0.084 Nm^3 H_2/ $kgVS_{fed}$	(Pagliaccia et al., 2016)
Municipal food waste + Kitchen wastewater	Anaerobic baffled reactor/ heat-treated sludge	Batch (HRT: 36 h, T: 35°C, pH: 6.2–9.0), purge gas: nitrogen	164±22 mL H_2/g $COD_{removed}$	(Tawfik et al., 2015)
FW (consisting of a variety of grains, vegetables, and meats), white mud from ammonia soda process	Glass digesters bioreactor/ heat-treated sludge	Batch (T: 55°C, pH: 5.7–6.1), p urge gas: nitrogen	145.4 mL H_2/g VS	(J. Zhang & Wang, 2013)
FW, pulp and paper sludge	Glass digesters bioreactor/ heat-treated sludge	Batch (T: 37°C, pH: 5.5–8.8), p urge gas: nitrogen	64.48 mLH_2/gVS_{fed}	(Lin et al., 2013)
Synthetic food waste samples	Glass digesters/heat-treated sludge	Batch (T: 35°C, pH: 6.0), p urge gas: nitrogen	55.3 mLH_2/gVS_{fed}	(Rafieenia et al., 2017)
FW (cafeteria)	Glass digesters/heat-treated sludge	Bacth (T: 37°C, pH: 6.0), p urge gas: nitrogen	54 mL/gVS_{added}	(Jang et al., 2015)
FW	CSTR/heat-treated sludge	(T: 37°C, pH: 5.5), p urge gas: nitrogen	~40 mL/gVS_{added}	(Elbeshbishy & Nakhla, 2011)

5.4.1.2 Continuous Stirred Tank Bioreactor (CSTR)

The CSTR is one of the most studied bioreactors to produce Bio-H_2. This bioreactor operates with a constant composition flow. The CSTR is mainly composed of a tank and a mechanical stirrer, which facilitates the homogenization of the biomass in the system (Figure 5.2B), thus increasing the contact area and the production of biogas (Ohimain & Izah, 2017). Gavala et al., (2006) found that in CSTR systems, thermophilic conditions provide high rates of hydrogen production with low biomass production, reaching a maximum productivity of 104 mmol H_2/h/L in a hydraulic retention time (HRT) of 6 h.

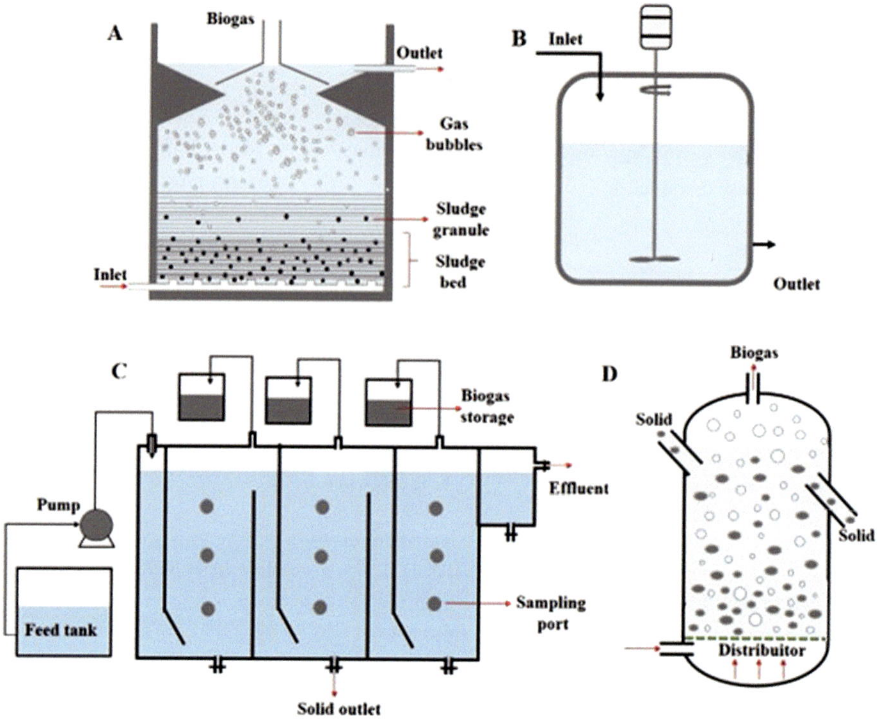

FIGURE 5.2 Bioreactors for $BioH_2$ production. A) UASB; B) CSTR; C) ABR; D) FBR.

5.4.1.3 Anaerobic Baffled Bioreactor (ABR)

The ABR is a system composed of a battery of vertical deflectors where the effluents are forced to flow over them (Figure 5.2C) (Ferraz et al., 2009). According to Ohimain & Izah (2017), these bioreactors are simple and low-cost since they do not have mechanical systems, so they are a good alternative to be used in the production of $BioH_2$. Li et al., (2007) used a 28 L three-compartment ABR for $BioH_2$ production. The maximum yield achieved from the energetic vector was 130 L H_2/Kg COD removed; the process was developed in 26 days.

5.4.1.4 Fluidized Bed Bioreactor (FBR)

The FBR is an advanced anaerobic bioreactor, which operates at a high organic loading rate (OLR). This system allows the expansion of the bed during the process (Ohimain & Izah, 2017). These bioreactors can develop several multiphase reactions and can be used mainly in high-resistance effluents (Borja & Banks, 1995). In FBR, the solid substrate is supported by a porous plate called a distributor, and the fluid passes through the voids in the solid material (Figure 5.2D). Barros et al., (2010) used a 4.2 L FBR for hydrogen production using polystyrene and expanded clay as support material. The system was operated at 30°C, pH 5.5, 2 h HRT, and the maximum HY was 1.90 mol H_2/mol glucose.

5.5 DARK FERMENTATION AND PHOTOFERMENTATION INTEGRATED TO $BIOH_2$ PRODUCTION (DF-PF)

Photofermentation is another biological route of hydrogen production, carried out by photoheterotrophic bacteria known as purple sulfur-free bacteria (PNSB), such as those of the genera *Rhodobacter* (Pandey et al., 2015), *Rhodobium* (Anam et al., 2012), *Rhodopseudomonas* (García-Sánchez et al.,

2018), and *Rhodospirillum* (Ferreira et al., 2022), that metabolize organic molecules under anaerobic conditions and the presence of light to produce H_2 and CO_2 (Niño-Navarro et al., 2020). When molecular N_2 is present in the culture medium, in addition to the anaerobic conditions and presence of light mentioned above, N_2 is reduced to NH_3 by the nitrogenase enzyme as a result of ATP expenditure. For each N_2 molecule fixed, one H_2 molecule is produced (Ghosh et al., 2017), as shown in Equation (5.3):

$$N_2 + 8H^+ + 8e- + 16\ ATP \rightarrow 2\ NH_3 + H_2 + 16\ ADP + 16\ P_i \tag{5.3}$$

In contrast, under conditions of N_2 deficiency, the enzyme nitrogenase catalyzes an alternative reaction to produce H_2 (Tiang et al., 2020), as observed in Equation (5.4):

$$8H^+ + 8e- + 16\ ATP \rightarrow 4H_2 + 16\ ADP + 16\ P_i \tag{5.4}$$

The reducing equivalents presented in Equations 5.3 and 5.4 are obtained from the oxidation of organic substrates by PNSBs, which have the capacity to produce H_2 from a variety of substrates, like simple sugars such as glucose (Tian et al., 2010) and organic acids such as pyruvate, propionate, malate, lactate, butyrate, acetate, and others (Cardeña et al., 2017). However, purple non-sulfur bacteria present these organic acids as their preferred substrate to produce H_2, and it is reported in the literature that H_2 yields are better from the use of these organic acids to the detriment of the use of pure sugars (Tao et al., 2008).

Despite the positive results extensively obtained from photofermentation and dark fermentation to produce H_2, their yields are inadequate due to the incomplete use of the energy present in the substrate (Ghimire et al., 2015). Because of this, new production approaches have been proposed with the aim of increasing H_2 yields, such as the integration of dark fermentation and photofermentation in the same bioreactor (Masihi et al., 2023) or in sequential steps (Ghimire et al., 2022).

As shown by Equation (5.5), during dark fermentation a significant fraction of the substrate is converted into volatile fatty acids such as acetic acid, butyric acid, and propionic acid, which remain in the system, causing a decrease in pH and inhibition of fermenting microorganisms, limiting the production of H_2 and maintaining COD (chemical oxygen demand) at high levels (Hitit et al., 2017; Rai & Singh, 2016; Rezaeitavabe et al., 2020). Therefore, it appears that the metabolism of microorganisms used in dark fermentation is complementary to that of purple non-sulfur bacteria (Rai & Singh, 2016), since, as mentioned previously, the volatile fatty acids produced during dark fermentation are the preferred substrates of purple non-sulfur bacteria for the production of H_2, and when integration or sequential steps of dark fermentation-photofermentation occur, these volatile fatty acids are consumed by PNSBs, resulting in the production of additional H_2 and a decrease in the COD of the culture medium (Niño-Navarro et al., 2020).

The theoretical maximum yield of H_2 obtained using dark fermentation and photofermentation is around 12 mol of H_2/mol of glucose (Equations 5.5 and 5.6), greater than the yields obtained through these two isolated processes, which are around 4 mol of hydrogen/mol of glucose obtained in dark fermentation and 8 mol of H_2 obtained in photofermentation with acetic acid as the substrate (Das & Basak, 2021).

$$C_6H_{12}O_6 + 2H_2O \rightarrow 2CH_3COOH + 4H_2 + 2CO_2 \tag{5.5}$$

$$2CH_3COOH + 4H_2O + light \rightarrow 8H_2 + 4CO_2 \tag{5.6}$$

The H_2 production processes through dark fermentation and photofermentation can be integrated into a hybrid reactor (Figure 5.3) (Masihi et al., 2023) or in sequential steps (Figure 5.4) (Ghimire et al., 2015). The production of H_2 in a hybrid reactor allows the substrate to be fermented by dark fermentation microorganisms simultaneously with the consumption of volatile fatty acids by photofermentative microorganisms, consequently controlling the acidification of the reactor and reducing fermentation time (Alian et al., 2021).

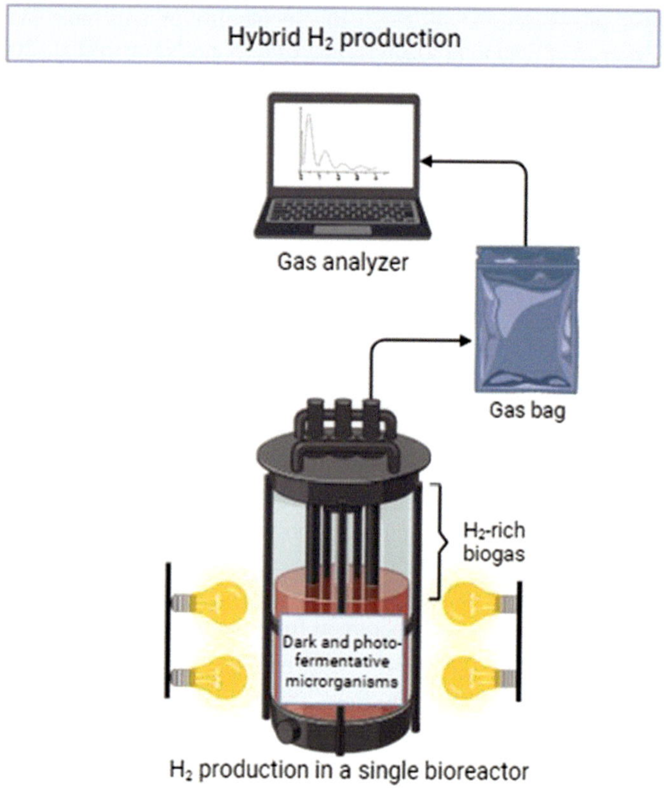

FIGURE 5.3 Hybrid hydrogen production.

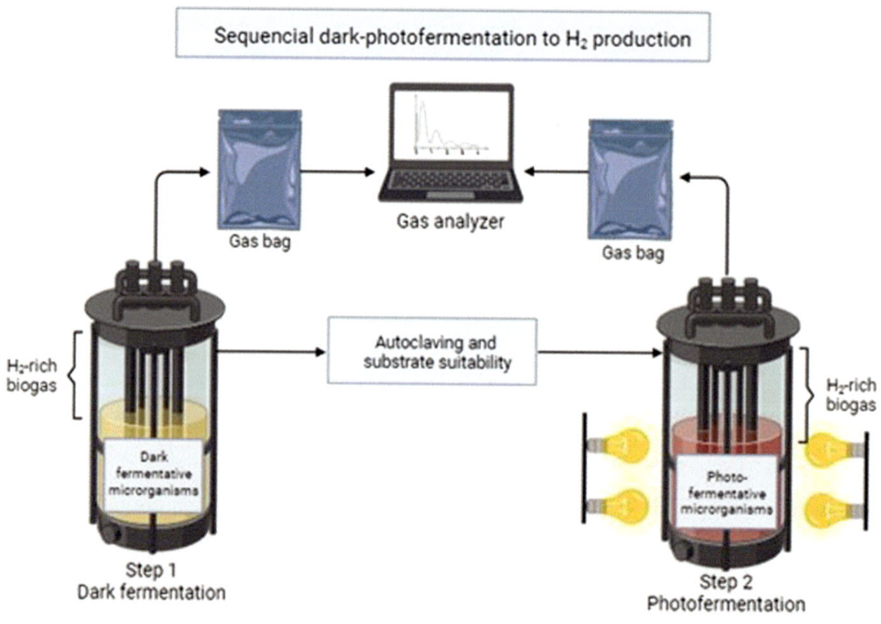

FIGURE 5.4 Sequential dark-photofermentation for hydrogen production.

In sequential fermentation, the two processes occur in separate reactors (Niño-Navarro et al., 2020). H_2 yields through these two methodologies are quite variable, and although sequential fermentation has been reported to be responsible for better yields, hybrid fermentation is simpler to operate and more economical (Ghosh et al., 2018). DF-PF integration allows the use of a wide variety of substrates during the initial dark fermentation stage, such as agro-industrial residues (Rai et al., 2014; Rodrigues et al., 2020), food processing waste (Rambabu et al., 2021b; Rao & Basak, 2022), lignocellulosic biomass (Rai et al., 2014; Zhang et al., 2020), food waste (Alian et al., 2021), and other substrates that are transformed into H_2 and volatile fatty acids that are subsequently used in the photofermentation stage.

Considering the great availability of food waste and the economic, environmental, and social problems related to food waste (Y. Zhang et al., 2023), an option for valuing this substrate is its use to produce biohydrogen through dark fermentation and photofermentation, maximizing the oxidation of the substrate and resulting in its treatment (Moreno-Andrade et al., 2023).

As seen in Table 5.4, cultivation conditions such as pH, temperature, light, and time are quite variable during the production of H_2 using food waste, both in hybrid fermentation and in two-step fermentation. In both processes, the pH usually varies in a range of 5.0–7.0, temperature around 30°C, luminosity 4000–15000 lux, and time of 24 hours to 40 days (Alian et al., 2021; Das & Basak, 2022; Ghimire et al., 2015; Masihi et al., 2023; Niño-Navarro et al., 2020; Rezaeitavabe et al., 2020). Furthermore, for anaerobiosis to be guaranteed, dark fermentation bioreactors are normally flushed with CO_2 and N_2 to ensure anaerobic conditions (Niño-Navarro et al., 2020). However, photobioreactors need to be purged with argon gas to avoid inhibition of H_2 production by the presence of N_2 (Ghimire et al., 2015; Ghosh et al., 2017).

Alian et al., (2021) used food waste to produce hydrogen in a hybrid bioreactor inoculated with the strains *Clostridium acetobutylicum* (dark fermentation) and *Rhodobacter sphaeroides* (photofermentation). The food waste, consisting of a mixture of 14.7 g of cooked rice, 2.1 g of beef, and 4.2 g of apple, was suspended in 700 mL of distilled water, which had an initial COD and pH of 25360 mg/L and 5.6, respectively. The substrate was stirred until it was homogenized, and particles with an average size of 0.5 mm were obtained. Subsequently, around 700 mL of substrate was added to the hybrid bioreactor along with 200 mL of *R. sphaeroides* and 100 mL of *C. acetobutylicum*. The hybrid bioreactor was incubated at 32°C, under agitation at 100 rpm and irradiation at 15000 lux for 24 hours. To increase biohydrogen production, iron shavings were added, which at the ideal concentration of 200 mg/L resulted in a maximum hydrogen yield of 210.2 mL/gVS and COD removal of 40.1% of the initial COD.

Das and Basak (2022) used waste potatoes from a store in India as a substrate for hybrid fermentation by *Enterobacter aerogenes* and *Rhodobacter sphaeroides*. The potatoes were dehydrated and ground into a paste, which was then dried and ground into a powder. This powder was used as a substrate to produce H_2 in a 2000 mL hybrid reactor, with a working volume of 1600 mL. The culture medium had its pH adjusted to 7.0, and 10% of the inoculum volume was added with a proportion of *Enterobacter aerogenes* to *Rhodobacter sphaeroides* of 1:7 (w/w), incubated at 30°C, 200 rpm, and light intensity of 10±2 Klux for 240 hours. The substrate conversion efficiency in this study was around 53% of the initial carbohydrate, culminating in cumulative H_2 production of 1580 mL/L of medium.

Ghimire et al., (2015) used food waste consisting of 72% fruits and vegetables, 10% cooked pasta and rice, 5% bread and bakery products, 2% dairy (cheese), 8% meat and fish, and 3% biscuits to produce H_2 in sequential stages of dark fermentation and photofermentation. The substrate was crushed in a blender and stored in the refrigerator at –20°C. Dark fermentation was carried out in a 1500 mL flask with an organic loading rate of 2.5 kg SV/m^3/d, with inoculum coming from the sludge of an anaerobic digestion reactor, at a temperature 55°C, and pH 7.0 for 96 hours. Dark fermentation resulted in an H_2 yield of 121.45±44.55 N LH_2/kg. The dark fermentation effluent was collected and centrifuged, its pH was adjusted to 6.5 and then autoclaved at 121°C for 20 min. Photofermentation was carried out by the bacteria *Rhodobacter sphaeroides* in 500 mL borosilicate glass bottles with

TABLE 5.4
Dark Fermentation and Photofermentation Integrated to H_2 Production

Composition	Dark Fermentation Microorganisms	Photofermentation Microorganisms	Dark Fermentation Operating Conditions	Photofermentation Operating Conditions	H_2 Production Dark Fermentation	H_2 Production Photofermentation	Reference
14.7 g of cooked rice, 2.1 g of beef, and 4.2 g of apple suspended in 700 mL of distilled water	*Clostridium acetobutylicum*	*Rhodobacter sphaeroides*	Hybrid reactor containing 700 mL of substrate which had an initial COD and pH of 25360 mg/L and 5.6, respectively, 200 mL of *R. sphaeroides* and 100 mL of *C. acetobutylicum* solutions, respectively. The mixture was incubated at 32°C and stirred at 100 RPM under 15000 lux irradiation for 24 h.	*	210.2 mL/g VS	*	(Alian et al., 2021)
Potato waste	*Enterobacter aerogenes*	*Rhodobacter sphaeroides*	2000 mL hybrid fermentation reactor with 1600 mL working volume, pH 7.0, 30°C, 200 rpm stirring and light intensity of 10±2 Klux for 240 h.	*	1580 mL H_2/L media	*	(S. R. Das & Basak, 2022)
Fruit and vegetables 72%, cooked pasta and rice 10%, bread and bakery 5%, dairy products (cheese) 2%, meat and fish 8%, and snacks (biscuits) 3%	-	*Rhodobacter sphaeroides*	Serum bottle with 1500 mL working volume, temperature (55±2°C), pH 7.0, 2.5 kg SV/m³/d for 96 h.	500 mL borosilicate bottle with 400 mL working volume, temperature 24±2°C, pH 6.5, luminescence 4000 lux, for 40 days.	121.45±44.55 N LH_2/kg SV	914±8 LH_2/m^3	(Ghimire, Valentino, et al., 2015)

(Continued)

TABLE 5.4 (CONTINUED)
Dark Fermentation and Photofermentation Integrated to H_2 Production

Composition	Dark Fermentation Microorganisms	Photofermentation Microorganisms	Dark Fermentation Operating Conditions	Photofermentation Operating Conditions	H_2 Production Dark Fermentation	H_2 Production Photofermentation	Reference
Grains (70%), vegetables (20%), and meats (10%)	*Clostridium acetobutulicum*	*Rhodobacter sphaeroides*	Hybrid fermentation reactor with 1000 mL working volume. The bioreactor was operated at 32°C, pH 6.85, 100 rpm, 15.000 lux lighting, for 24 h.	*	213.66 ml H_2/gVS	*	(Masihi et al., 2023)
Fruit and vegetable waste and cheese whey powder	*Klebsiella, Lactobacillus, Citrobacter, Enterobacter,* e *Clostridium*	*Rhodopseudomonas palustris*	Amber serological vials with a total volume of 122 mL were filled with 100 mL working volume, pH 5.5, incubated at 35°C, shaken at 150 rpm for 120 h.	Transparent 130 mL serological vials filled with 60 mL working volume, washed with argon, pH 6.8, incubated at 30°C with agitation at 150 rpm, and illuminated with white LED light for 240 h.	12.5 (mLH_2/gDQO)	781.1 mLH_2/gDQO)	(Niño-Navarro et al., 2020)
70% carbohydrates (rice and grains), 20% vegetables, and 10% meat	*Clostridium acetobutylicum*	*Rhodobacter sphaeroides*	Hybrid fermentation reactor, in 1500 glass flasks with 1000 mL working volume incubated at 32°C under 100 rpm shaking and 15000 lux lighting for 24 h.	*	197.15±2.9 mlg^{-1} VS	*	(Rezaeitavabe et al., 2020)

(-) Not specified in the study.
(*) Production carried out in just one bioreactor.

a working volume of 400 mL in batch, washed with argon gas to ensure anaerobic conditions, and maintained at room temperature (24±2°C). Photofermentation resulted in the production of 914±8 L/m^3 of H_2, being monitored for 40 days until it completely ceased. A decrease in COD of 60.1% and removal of volatile fatty acids of 98±1% was observed. The solid residues left after dark fermentation and photofermentation were directed to anaerobic digestion to produce biogas.

Masihi et al., (2023) carried out the hybrid production of H_2 using as substrate a mixture of grains (70%), vegetables (20%), and meat (10%). Strains of *Clostridium acetobutylicum* (dark fermentation) and *Rhodobacter sphaeroides* (photofermentation) were tested. The substrate was pretreated with acid (pH of 3.0) and heat (90°C for 20 min) to break down the polymers into simpler and more available monomers. Fermentation took place in a 1000 mL reactor, with pH 6.85, 32°C, 100 rpm, and 15000 lux lighting, for 24 hours. TiO_2 and Fe_3O_4 nanoparticles were used to increase the amount of hydrogen produced. In proportion to Fe_3O_4/TiO_2 of 100/50, cumulative hydrogen production and hydrogen yield increased from 1613 to 3392 mL. It was also observed that the addition of TiO_2 nanoparticles led to an increase in biomass concentration, indicating an improvement in bacterial growth. Furthermore, the hydrogen content and COD removal at the optimal Fe_3O_4/TiO_2 ratio increased to 71% and 40.1%, respectively.

Niño-Navarro et al., (2020) used fruit and vegetable remains and cheese whey powder to produce biohydrogen in two stages using a consortium for dark fermentation, consisting predominantly (85.4%) of *Bifidobacterium* and including *Klebsiella* (9.1%), *Lactobacillus* (0.97%), *Citrobacter* (0.21%), *Enterobacter* (0.27%), and *Clostridium* (0.18%). The bacterium *Rhodopseudomonas palustris* was used for photofermentation. Production was carried out in 122 mL amber bottles with a working volume of 100 mL. The substrate had its pH adjusted to 5.5±0.5 and was inoculated with 10 mL of the dark fermentation consortium, kept at 35°C and stirred at 150 rpm for 120 hours. In the dark fermentation stage, a production of H_2 at 12,5 mL/gDQO was obtained. Subsequently, the dark fermentation effluent was centrifuged at 13,000 rpm for 15 minutes and then sterilized in an autoclave at 120°C for 15 minutes. Photofermentation occurred in transparent 130 mL serological vials filled with 60 mL working volume, washed with argon, pH 6.8, incubated at 30°C with agitation at 150 rpm, and illuminated with white LED light for 240 hours. The sterile DF effluent was diluted 1:2, 1:5, and 1:10 with sterile distilled water. The effluent diluted at 1:10 obtained an H_2 production of 781.1 mLH_2/gCOD). Mixtures with a high concentration of organic substrates constitute substrates that are difficult to degrade, and in this case, the metabolism of microorganisms was diverted towards the production of lactate. Thus, despite the low yield in dark fermentation, around 15.2 g/L of lactate were produced and subsequently used by PNB to produce H_2 in the photofermentation stage.

Rezaeitavabe et al., (2020) used a substrate composed of 70% carbohydrates (rice and grains), 20% vegetables, and 10% meat to produce H_2 by *Clostridium acetobutylicum* and *Rhodobacter sphaeroides* in a hybrid bioreactor. The authors also investigated the effect of adding resin saturated with Ca and Mg and biochar loaded with phosphate. The residues were crushed in a blender and filtered through a 0.5 mm pore sieve, after which they were pretreated with acid and heat (90°C for 20 min and pH 3.0) to break down polymers and release monomers that are more easily usable for fermentation. Fermentation took place in 1500 mL glass bottles with a working volume of 1000 mL, inoculated with a co-culture of *R. sphaeroides* and *C. acetobutylicum* with a ratio of 2/1 (20%: 10% v/v), and subsequently incubated at 32°C under 100 rpm shaking and 15000 lux illumination for 24 hours. Maximum accumulated hydrogen production (3130±16.8 mL) and hydrogen yield (197.15±2,9 mL/ g VS) were observed when 5 g/ L of resin and 0.5 g/ L of biochar. This corresponded to a 94% increase in yield compared to the control. Thus, the application of resin saturated with Ca, Mg, and biochar loaded with phosphate in a hybrid process of dark fermentation and photofermentation increased the efficiency of converting organic waste into hydrogen, maintaining process stability.

5.6 CONCLUSIONS AND PERSPECTIVES

Biological processes are promising technologies for producing hydrogen using food waste as raw material. The amount of waste generated by the food sector and its characteristics, such as biodegradability and carbohydrate content, make these wastes particularly interesting for energy production. In addition to generating clean energy, the process reduces the polluting load of food waste that is currently disposed of in landfills.

However, many advances have been made to improve hydrogen yield, yet dark and photofermentation are still difficult to apply on a larger scale. To overcome this, studies have focused on the following aspects: selection of microbial consortia and adequate pretreatments of food wastes, reactor designs, and operating modes. It is also observed that integrated technologies, DF and PF have been attractive options. Another recent technology to obtain a higher biohydrogen yield is the use of microbial electrolysis cells.

In conclusion, DF and PF seem promising and sustainable approaches for biohydrogen production from FW, offering the potential for clean energy generation and waste treatment. Continued research and development in this field will contribute to turn hydrogen into a viable future fuel, reducing our dependence on fossil fuels and mitigating the environmental impacts of food waste.

REFERENCES

Alavi-Borazjani, S. A., Capela, I., & Tarelho, L. A. C. (2019a). Dark fermentative hydrogen production from food waste: Effect of biomass ash supplementation. *International Journal of Hydrogen Energy*, *44*(48), 26213–26225. https://doi.org/10.1016/j.ijhydene.2019.08.091

Alavi-Borazjani, S. A., Capela, I., & Tarelho, L. A. C. (2019b). Dark fermentative hydrogen production from food waste: Effect of biomass ash supplementation. *International Journal of Hydrogen Energy*, *44*(48), 26213–26225. https://doi.org/10.1016/j.ijhydene.2019.08.091

Alian, M., Saadat, S., & Rezaeitavabe, F. (2021). An investigation on the dose-dependent effect of iron shaving on bio-hydrogen production from food waste. *International Journal of Hydrogen Energy*, *46*(38), 19886–19896. https://doi.org/10.1016/j.ijhydene.2021.03.121

Anam, K., Habibi, M. S., Harwati, T. U., & Susilaningsih, D. (2012). Photofermentative hydrogen production using Rhodobium marinum from bagasse and soy sauce wastewater. *International Journal of Hydrogen Energy*, *37*(20), 15436–15442. https://doi.org/10.1016/j.ijhydene.2012.06.076

Ananthi, V., Ramesh, U., Balaji, P., & Kumar, P. (2022). A review on the impact of various factors on biohydrogen production. *International Journal of Hydrogen Energy*. https://doi.org/10.1016/j.ijhydene.2022.08.046

Barros, A. R., Cavalcante de Amorim, E. L., Reis, C. M., Shida, G. M., & Silva, E. L. (2010). Biohydrogen production in anaerobic fluidized bed reactors: Effect of support material and hydraulic retention time. *International Journal of Hydrogen Energy*, *35*(8), 3379–3388. https://doi.org/10.1016/j.ijhydene.2010.01.108

Beretta, C., & Hellweg, S. (2019). Potential environmental benefits from food waste prevention in the food service sector. *Resources, Conservation and Recycling*, *147*, 169–178. https://doi.org/10.1016/j.resconrec.2019.03.023

Bhurat, K. S., Banerjee, T., Bobde, P. V., & Sureshchandra Bhurat, S. (2023). Effect of chemical, physical, and biological pre-treatment of food wastes on bio-hydrogen production by dark anaerobic fermentation under mesophilic conditions. *Energy Sources, Part A: Recovery, Utilization and Environmental Effects*, *45*(1), 1017–1029. https://doi.org/10.1080/15567036.2023.2174615

Bilal, M., & Iqbal, H. M. N. (2019). Sustainable bioconversion of food waste into high-value products by immobilized enzymes to meet bio-economy challenges and opportunities – A review. In *Food research international* (Vol. 123, pp. 226–240). Elsevier Ltd. https://doi.org/10.1016/j.foodres.2019.04.066

Bilanovic, D., Andargatchew, A., Kroeger, T., & Shelef, G. (2009). Freshwater and marine microalgae sequestering of CO2 at different C and N concentrations - Response surface methodology analysis. *Energy Conversion and Management*, *50*(2), 262–267. https://doi.org/10.1016/j.enconman.2008.09.024

Boboescu, I. Z., Gherman, V. D., Lakatos, G., Pap, B., Bíró, T., & Maróti, G. (2016). Surpassing the current limitations of biohydrogen production systems: The case for a novel hybrid approach. *Bioresource Technology*, *204*, 192–201. https://doi.org/10.1016/J.BIORTECH.2015.12.083

Borja, R., & Banks, C. J. (1995). Response of an anaerobic fluidized bed reactor treating ice-cream wastewater to organic, hydraulic, temperature and pH shocks. *Journal of Biotechnology*, *39*, 251–259.

Braguglia, C. M., Gallipoli, A., Gianico, A., & Pagliaccia, P. (2018). Anaerobic bioconversion of food waste into energy: A critical review. In *Bioresource Technology* (Vol. 248, pp. 37–56). Elsevier Ltd. https://doi.org/10.1016/j.biortech.2017.06.145

Cardeña, R., Valdez-Vazquez, I., & Buitrón, G. (2017). Effect of volatile fatty acids mixtures on the simultaneous photofermentative production of hydrogen and polyhydroxybutyrate. *Bioprocess and Biosystems Engineering*, *40*(2), 231–239. https://doi.org/10.1007/s00449-016-1691-9

Chaisri, R., Boonsawang, P., Prasertsan, P., & Chaiprapat, S. (2007). Effect of organic loading rate on methane and volatile fatty acids productions from anaerobic treatment of palm oil mill effluent in UASB and UFAF reactors. *Journal of Science and Technology*, *29*, 311–323.

Chaturvedi, V., & Verma, P. (2013). An overview of key pretreatment processes employed for bioconversion of lignocellulosic biomass into biofuels and value added products. *3 Biotech*, *3*(5), 415–431. https://doi.org/10.1007/s13205-013-0167-8

Chu, C. F., Li, Y. Y., Xu, K. Q., Ebie, Y., Inamori, Y., & Kong, H. N. (2008). A pH- and temperature-phased two-stage process for hydrogen and methane production from food waste. *International Journal of Hydrogen Energy*, *33*(18), 4739–4746. https://doi.org/10.1016/j.ijhydene.2008.06.060

Cieciura-Włoch, W., Borowski, S., & Otlewska, A. (2020). Biohydrogen production from fruit and vegetable waste, sugar beet pulp and corn silage via dark fermentation. *Renewable Energy*, *153*, 1226–1237. https://doi.org/10.1016/j.renene.2020.02.085

Das, D., & Veziroglu, T. N. (2008). Advances in biological hydrogen production processes. *International Journal of Hydrogen Energy*, *33*(21), 6046–6057. https://doi.org/10.1016/j.ijhydene.2008.07.098

Das, S. R., & Basak, N. (2021). Molecular biohydrogen production by dark and photo fermentation from wastes containing starch: Recent advancement and future perspective. In *Bioprocess and biosystems engineering* (Vol. 44, Issue 1). Springer Science and Business Media Deutschland GmbH. https://doi.org/10.1007/s00449-020-02422-5

Das, S. R., & Basak, N. (2022). Optimization of process parameters for enhanced biohydrogen production using potato waste as substrate by combined dark and photo fermentation. *Biomass Conversion and Biorefinery*. https://doi.org/10.1007/s13399-022-02588-w

De Carvalho, J. C., De Souza Vandenberghe, L. P., Sydney, E. B., Karp, S. G., Magalhães, A. I., Martinez-Burgos, W. J., Medeiros, A. B. P., Thomaz-Soccol, V., Vieira, S., Letti, L. A. J., Rodrigues, C., Woiciechowski, A. L., & Soccol, C. R. (2023). Biomethane production from sugarcane vinasse in a circular economy: Developments and innovations. *Fermentation*, *9*(4), 1–27. https://doi.org/10.3390/fermentation9040349

Dinesh, G. K., Chauhan, R., & Chakma, S. (2018). Influence and strategies for enhanced biohydrogen production from food waste. In *Renewable and sustainable energy reviews* (Vol. 92, pp. 807–822). Elsevier Ltd. https://doi.org/10.1016/j.rser.2018.05.009

Ding, L., Cheng, J., Qiao, D., Li, H., & Zhang, Z. (2022). Continuous co-generation of biohydrogen and biomethane through two-stage anaerobic digestion of hydrothermally pretreated food waste. *Energy Conversion and Management*, *268*. https://doi.org/10.1016/j.enconman.2022.116000

Dong, L., Zhenhong, Y., Yongming, S., Xiaoying, K., & Yu, Z. (2009). Hydrogen production characteristics of the organic fraction of municipal solid wastes by anaerobic mixed culture fermentation. *Internationa Journal of Hydrogen Energy*, *34*(1), 812–820. https://doi.org/10.1016/j.ijhydene.2008.11.031

Elbeshbishy, E., Hafez, H., Dhar, B. R., & Nakhla, G. (2011). Single and combined effect of various pretreatment methods for biohydrogen production from food waste. *International Journal of Hydrogen Energy*, *36*(17), 11379–11387. https://doi.org/10.1016/j.ijhydene.2011.02.067

Elbeshbishy, E., & Nakhla, G. (2011). Comparative study of the effect of ultrasonication on the anaerobic biodegradability of food waste in single and two-stage systems. *Bioresource Technology*, *102*(11), 6449–6457. https://doi.org/10.1016/j.biortech.2011.03.082

Elgarahy, A. M., Eloffy, M. G., Alengebawy, A., El-Sherif, D. M., Gaballah, M. S., Elwakeel, K. Z., & El-Qelish, M. (2023). Sustainable management of food waste; pre-treatment strategies, techno-economic assessment, bibliometric analysis, and potential utilizations: A systematic review. In *Environmental research* (Vol. 225). Academic Press Inc. https://doi.org/10.1016/j.envres.2023.115558

Ferraz, F. M., Bruni, A. T., & Bianchi, V. L. Del. (2009). Performance of an Anaerobic Baffled Reactor (ABR) in treatment of cassava wastewater. *Brazilian Journal of Microbiology*, *40*(1), 48–53.

Ferreira, G. M. T., Moreira, F. S., Cardoso, V. L., & Batista, F. R. X. (2022). Enhancement of photo-fermentative hydrogen production with co-culture of rhodobacter capsulatus and rhodospirillum rubrum by using medium renewal strategy. *Bioenergy Research*. https://doi.org/10.1007/s12155-022-10550-x

Gadhe, A., Sonawane, S. S., & Varma, M. N. (2014). Ultrasonic pretreatment for an enhancement of biohydrogen production from complex food waste. *International Journal of Hydrogen Energy, 39*(15), 7721–7729. https://doi.org/10.1016/j.ijhydene.2014.03.105

Gallego-García, M., Moreno, A. D., Manzanares, P., Negro, M. J., & Duque, A. (2023). Recent advances on physical technologies for the pretreatment of food waste and lignocellulosic residues. In *Bioresource technology* (Vol. 369). Elsevier Ltd. https://doi.org/10.1016/j.biortech.2022.128397

García-Sánchez, R., Ramos-Ibarra, R., Guatemala-Morales, G., Arriola-Guevara, E., Toriz-González, G., & Corona-González, R. I. (2018). Photofermentation of tequila vinasses by rhodopseudomonas pseudopalustris to produce hydrogen. *International Journal of Hydrogen Energy, 43*(33), 15857–15869. https://doi.org/10.1016/j.ijhydene.2018.07.015

Gavala, H. N., Skiadas, I. V., & Ahring, B. K. (2006). Biological hydrogen production in suspended and attached growth anaerobic reactor systems. *International Journal of Hydrogen Energy, 31*(9), 1164–1175. https://doi.org/10.1016/j.ijhydene.2005.09.009

Ghimire, A., Frunzo, L., Pirozzi, F., Trably, E., Escudie, R., Lens, P. N. L., & Esposito, G. (2015). A review on dark fermentative biohydrogen production from organic biomass: Process parameters and use of by-products. In *Applied energy* (Vol. 144, pp. 73–95). Elsevier Ltd. https://doi.org/10.1016/j.apenergy.2015.01.045

Ghimire, A., Luongo, V., Frunzo, L., Lens, P. N. L., Pirozzi, F., & Esposito, G. (2022). Biohythane production from food waste in a two-stage process: Assessing the energy recovery potential. *Environmental Technology (United Kingdom), 43*(14), 2190–2196. https://doi.org/10.1080/09593330.2020.1869319

Ghimire, A., Valentino, S., Frunzo, L., Trably, E., Escudié, R., Pirozzi, F., Lens, P. N. L., & Esposito, G. (2015). Biohydrogen production from food waste by coupling semi-continuous dark-photofermentation and residue post-treatment to anaerobic digestion: A synergy for energy recovery. *International Journal of Hydrogen Energy, 40*(46), 16045–16055. https://doi.org/10.1016/j.ijhydene.2015.09.117

Ghosh, S., Chowdhury, R., & Bhattacharya, P. (2018). A review on single stage integrated dark-photo fermentative biohydrogen production: Insight into salient strategies and scopes. In *International journal of hydrogen energy* (pp. 2091–2107). Elsevier Ltd. https://doi.org/10.1016/j.ijhydene.2017.12.018

Ghosh, S., Dairkee, U. K., Chowdhury, R., & Bhattacharya, P. (2017). Hydrogen from food processing wastes via photofermentation using Purple Non-Sulfur Bacteria (PNSB) – A review. In *Energy conversion and management* (Vol. 141, pp. 299–314). Elsevier Ltd. https://doi.org/10.1016/j.enconman.2016.09.001

Han, W., Hu, Y., Li, S., Huang, J., Nie, Q., Zhao, H., & Tang, J. (2016). Simultaneous dark fermentative hydrogen and ethanol production from waste bread in a mixed packed tank reactor. *Journal of Cleaner Production, 141*, 608–611.

Han, W., Yan, Y., Shi, Y., Gu, J., Tang, J., & Zhao, H. (2016). Biohydrogen production from enzymatic hydrolysis of food waste in batch and continuous systems. *Scientific Reports, 6*. https://doi.org/10.1038/srep38395

Hawkes, F. R., Hussy, I., Kyazze, G., Dinsdale, R., & Hawkes, D. L. (2007). Continuous dark fermentative hydrogen production by mesophilic microflora: Principles and progress. *International Journal of Hydrogen Energy, 32*(2), 172–184. https://doi.org/10.1016/j.ijhydene.2006.08.014

Hitit, Z. Y., Zampol Lazaro, C., & Hallenbeck, P. C. (2017). Increased hydrogen yield and COD removal from starch/glucose based medium by sequential dark and photo-fermentation using Clostridium butyricum and rhodopseudomonas palustris. *International Journal of Hydrogen Energy, 42*(30), 18832–18843. https://doi.org/10.1016/j.ijhydene.2017.05.161

Intanoo, P., Chaimongkol, P., & Chavadej, S. (2016). Hydrogen and methane production from cassava wastewater using two-stage upflow anaerobic sludge blanket reactors (UASB) with an emphasis on maximum hydrogen production. *International Journal of Hydrogen Energy, 41*(14), 6107–6114. https://doi.org/10.1016/j.ijhydene.2015.10.125

Jafary, T., Wan Daud, W. R., Ghasemi, M., Abu Bakar, M. H., Sedighi, M., Kim, B. H., Carmona-Martínez, A. A., Jahim, J. M., & Ismail, M. (2019). Clean hydrogen production in a full biological microbial electrolysis cell. *International Journal of Hydrogen Energy, 44*(58), 30524–30531. https://doi.org/10.1016/j.ijhydene.2018.01.010

Jang, S., Kim, D. H., Yun, Y. M., Lee, M. K., Moon, C., Kang, W. S., Kwak, S. S., & Kim, M. S. (2015). Hydrogen fermentation of food waste by alkali-shock pretreatment: Microbial community analysis and limitation of continuous operation. *Bioresource Technology, 186*, 215–222. https://doi.org/10.1016/j.biortech.2015.03.031

Jia, X., Li, M., Wang, Y., Wu, Y., Zhu, L., Wang, X., & Zhao, Y. (2020). Enhancement of hydrogen production and energy recovery through electro-fermentation from the dark fermentation effluent of food waste. *Environmental Science and Ecotechnology, 1*. https://doi.org/10.1016/j.ese.2019.100006

Jung, J. H., Sim, Y. B., Ko, J., Park, S. Y., Kim, G. B., & Kim, S. H. (2022). Biohydrogen and biomethane production from food waste using a two-stage Dynamic Membrane Bioreactor (DMBR) system. *Bioresource Technology*, *352*. https://doi.org/10.1016/j.biortech.2022.127094

Karp, S. G., Burgos, W. J. M., Vandenberghe, L. P. S., Diestra, K. V., Torres, L. A. Z., Woiciechowski, A. L., Letti, L. A. J., Pereira, G. V. M., Thomaz-Soccol, V., Rodrigues, C., De Carvalho, J. C., & Soccol, C. R. (2022). Sugarcane: A promising source of green carbon in the circular bioeconomy. *Sugar Tech*, *24*(4), 1230–1245. https://doi.org/10.1007/s12355-022-01161-z

Keskin, T., & Hallenbeck, P. C. (2012). Hydrogen production from sugar industry wastes using single-stage photofermentation. *Bioresource Technology*, *112*, 131–136. https://doi.org/10.1016/j.biortech.2012.02.077

Kobayashi, T., Xu, K. Q., Li, Y. Y., & Inamori, Y. (2012). Evaluation of hydrogen and methane production from municipal solid wastes with different compositions of fat, protein, cellulosic materials and the other carbohydrates. *International Journal of Hydrogen Energy*, *37*(20), 15711–15718. https://doi.org/10.1016/j.ijhydene.2012.05.044

Kuang, Y., Zhao, J., Gao, Y., Lu, C., Luo, S., Sun, Y., & Zhang, D. (2020). Enhanced hydrogen production from food waste dark fermentation by potassium ferrate pretreatment. *Environmental Science and Pollution Research*, *27*(15), 18145–18156. https://doi.org/10.1007/s11356-020-08207-3

Li, C., & Fang, H. (2007). Fermentative hydrogen production from wastewater and solid wastes by mixed cultures. *Critical Reviews in Environmental Science and Technology*, *3389*, 1–39. https://doi.org/10.1080/10643380600729071

Li, M., Xia, T., Zhu, C., Xi, B., Jia, X., Wei, Z., & Zhu, J. (2014). Effect of short-time hydrothermal pretreatment of kitchen waste on biohydrogen production: Fluorescence spectroscopy coupled with parallel factor analysis. *Bioresource Technology*, *172*, 382–390. https://doi.org/10.1016/j.biortech.2014.09.040

Lin, Y., Wu, S., & Wang, D. (2013). Hydrogen-methane production from pulp & paper sludge and food waste by mesophilic-thermophilic anaerobic co-digestion. *International Journal of Hydrogen Energy*, *38*(35), 15055–15062. https://doi.org/10.1016/j.ijhydene.2012.01.051

Linyi, C., Yujie, Q., Buqing, C., Chenglong, W., Shaohong, Z., Renglu, C., Shaohua, Y., Lan, Y., & Zhiju, L. (2020). Enhancing degradation and biogas production during anaerobic digestion of food waste using alkali pretreatment. *Environmental Research*, *188*. https://doi.org/10.1016/j.envres.2020.109743

Liu, G., & Shen, J. (2004). Effects of culture and medium conditions on hydrogen production from starch using anaerobic bacteria. *Journal of Bioscience and Bioengineering*, *98*(4), 251–256.

Ma, C., Liu, J., Ye, M., Zou, L., Qian, G., & Li, Y. Y. (2018). Towards utmost bioenergy conversion efficiency of food waste: Pretreatment, co-digestion, and reactor type. In *Renewable and sustainable energy reviews* (Vol. 90, pp. 700–709). Elsevier Ltd. https://doi.org/10.1016/j.rser.2018.03.110

Mahato, R. K., Kumar, D., & Rajagopalan, G. (2020). Biohydrogen production from fruit waste by Clostridium strain BOH3. *Renewable Energy*, *153*, 1368–1377. https://doi.org/10.1016/j.renene.2020.02.092

Martinez-Burgos, W. J., Bittencourt, E., Rodrigues, D., Paula, D., Bianchi, A., Medeiros, P., Cesar, J., Carvalho, D., Thomaz, V., Porto, L., Vandenberghe, D. S., Lorenci, A., & Soccol, C. R. (2020). Biohydrogen production in cassava processing wastewater using microbial consortia: Process optimization and kinetic analysis of the microbial community. *Bioresource Technology*, *309*, 123331. https://doi.org/10.1016/j.biortech.2020.123331

Martinez-Burgos, W. J., Junior, J. R. do N., Medeiros, A. B. P., Herrmann, L. W., Sydney, E. B., & Soccol, C. R. (2021a). Biohydrogen production from agro-industrial wastes using clostridium beijerinckii and isolated bacteria as inoculum. *BioEnergy Research*, *15*(1), 18. https://doi.org/10.1007/s12155-021-10358-1

Martinez-Burgos, W. J., Sydney, E. B., de Paula, D. R., Medeiros, A. B. P., De Carvalho, J. C., Molina, D., & Soccol, C. R. (2021b). Hydrogen production by dark fermentation using a new low-cost culture medium composed of corn steep liquor and cassava processing water: Process optimization and scale-up. *Bioresource Technology*, *320*, 124370. https://doi.org/10.1016/j.biortech.2020.124370

Martinez-Burgos, W. J., Sydney, E. B., de Paula, D. R., Medeiros, A. B. P., de Carvalho, J. C., Soccol, V. T., de Souza Vandenberghe, L. P., Woiciechowski, A. L., & Soccol, C. R. (2020). Biohydrogen production in cassava processing wastewater using microbial consortia: Process optimization and kinetic analysis of the microbial community. *Bioresource Technology*, *309*, 123331. https://doi.org/10.1016/j.biortech.2020.123331

Martínez-Burgos, W. J., Sydney, E. B., Medeiros, A., Medeiros, P., Carvalho, J., Ricardo, C., & Soccol, C. R. (2019). The effect of hydrolysis and sterilization in biohydrogen production from cassava processing wastewater medium using anaerobic bacterial consortia. *International Journal of Hydrogren Energy*, *4*, 25551–25564. https://doi.org/10.1016/j.ijhydene.2019.08.085

Masihi, F., Rezaeitavabe, F., Karimi-Jashni, A., & Riefler, G. (2023). Optimization and enhancement of biohydrogen production in a single-stage hybrid (dark/photo) fermentation reactor using Fe3O4 and TiO2 nanoparticles. *International Journal of Hydrogen Energy*. https://doi.org/10.1016/j.ijhydene.2023.05.150

Mohanakrishna, G., Sneha, N. P., Rafi, S. M., & Sarkar, O. (2023). Dark fermentative hydrogen production: Potential of food waste as future energy needs. *Science of the Total Environment, 888*. https://doi.org/10.1016/j.scitotenv.2023.163801

Moreno-Andrade, I., Berrocal-Bravo, M. J., & Valdez-Vazquez, I. (2023). Biohydrogen production from food waste and waste activated sludge in codigestion: Influence of organic loading rate and changes in microbial community. *Journal of Chemical Technology and Biotechnology, 98*(1), 230–237. https://doi.org/10.1002/jctb.7238

Niño-Navarro, C., Chairez, I., Christen, P., Canul-Chan, M., & García-Peña, E. I. (2020). Enhanced hydrogen production by a sequential dark and photo fermentation process: Effects of initial feedstock composition, dilution and microbial population. *Renewable Energy, 147*, 924–936. https://doi.org/10.1016/j.renene.2019.09.024

Ohimain, E. I., & Izah, S. C. (2017). A review of biogas production from palm oil mill effluents using different configurations of bioreactors. *Renewable and Sustainable Energy Reviews, 70*(December 2014), 242–253. https://doi.org/10.1016/j.rser.2016.11.221

Okamoto, M., Miyahara, T., Mizuno, O., & Noike, T. (2000). Biological hydrogen potential of materials characteristic of the organic fraction of municipal solid wastes. *Water Science and Technology, 41*(3), 25–32. https://doi.org/10.2166/wst.2000.0052

Pagliaccia, P., Gallipoli, A., Gianico, A., Montecchio, D., & Braguglia, C. M. (2016). Single stage anaerobic bioconversion of food waste in mono and co-digestion with olive husks: Impact of thermal pretreatment on hydrogen and methane production. *International Journal of Hydrogen Energy, 41*(2), 905–915. https://doi.org/10.1016/j.ijhydene.2015.10.061

Pandey, A., Gupta, K., & Pandey, A. (2015). Effect of nanosized TiO2 on photofermentation by Rhodobacter sphaeroides NMBL-02. *Biomass and Bioenergy, 72*, 273–279. https://doi.org/10.1016/j.biombioe.2014.10.021

Paritosh, K., Kushwaha, S. K., Yadav, M., Pareek, N., Chawade, A., & Vivekanand, V. (2017). Food waste to energy: An overview of sustainable approaches for food waste management and nutrient recycling. In *BioMed research international* (Vol. 2017). Hindawi Limited. https://doi.org/10.1155/2017/2370927

Parthiba Karthikeyan, O., Trably, E., Mehariya, S., Bernet, N., Wong, J. W. C., & Carrere, H. (2018). Pretreatment of food waste for methane and hydrogen recovery: A review. In *Bioresource technology* (Vol. 249, pp. 1025–1039). Elsevier Ltd. https://doi.org/10.1016/j.biortech.2017.09.105

Rafieenia, R., Girotto, F., Peng, W., Cossu, R., Pivato, A., Raga, R., & Lavagnolo, M. C. (2017). Effect of aerobic pre-treatment on hydrogen and methane production in a two-stage anaerobic digestion process using food waste with different compositions. *Waste Management, 59*, 194–199. https://doi.org/10.1016/j.wasman.2016.10.028

Rai, P. K., & Singh, S. P. (2016). Integrated dark- and photo-fermentation: Recent advances and provisions for improvement. In *International journal of hydrogen energy* (Vol. 41, Issue 44, pp. 19957–19971). Elsevier Ltd. https://doi.org/10.1016/j.ijhydene.2016.08.084

Rai, P. K., Singh, S. P., Asthana, R. K., & Singh, S. (2014). Biohydrogen production from sugarcane bagasse by integrating dark- and photo-fermentation. *Bioresource Technology, 152*, 140–146. https://doi.org/10.1016/J.BIORTECH.2013.10.117

Rajesh Banu, J., Merrylin, J., Mohamed Usman, T. M., Yukesh Kannah, R., Gunasekaran, M., Kim, S. H., & Kumar, G. (2020). Impact of pretreatment on food waste for biohydrogen production: A review. *International Journal of Hydrogen Energy, 45*(36), 18211–18225. https://doi.org/10.1016/j.ijhydene.2019.09.176

Rambabu, K., Bharath, G., Banat, F., Hai, A., Show, P. L., & Phong Nguyen, T. H. (2021a). Ferric oxide/date seed activated carbon nanocomposites mediated dark fermentation of date fruit wastes for enriched biohydrogen production. *International Journal of Hydrogen Energy, 46*(31), 16631–16643. https://doi.org/10.1016/j.ijhydene.2020.06.108

Rambabu, K., Bharath, G., Thanigaivelan, A., Das, D. B., Show, P. L., & Banat, F. (2021b). Augmented biohydrogen production from rice mill wastewater through nano-metal oxides assisted dark fermentation. *Bioresource Technology, 319*. https://doi.org/10.1016/j.biortech.2020.124243

Rao, R., & Basak, N. (2022). Sequential dark-photo batch fermentation and kinetic modelling for biohydrogen production using cheese whey as a feedstock. *Applied Biochemistry and Biotechnology, 194*(9), 3930–3960. https://doi.org/10.1007/s12010-022-03958-w

Regueira-Marcos, L., García-Depraect, O., & Muñoz, R. (2023). Elucidating the role of pH and total solids content in the co-production of biohydrogen and carboxylic acids from food waste via lactate-driven dark fermentation. *Fuel*, *338*, 127238. https://doi.org/10.1016/J.FUEL.2022.127238

Rezaeitavabe, F., Saadat, S., Talebbeydokhti, N., Sartaj, M., & Tabatabaei, M. (2020). Enhancing bio-hydrogen production from food waste in single-stage hybrid dark-photo fermentation by addition of two waste materials (exhausted resin and biochar). *Biomass and Bioenergy*, *143*. https://doi.org/10.1016/j.biombioe.2020.105846

Rodrigues, C. V., Rios Alcaraz, F. A., Nespeca, M. G., Rodrigues, A. V., Motteran, F., Tallarico Adorno, M. A., Varesche, M. B. A., & Maintinguer, S. I. (2020). Biohydrogen production in an integrated biosystem using crude glycerol from waste cooking oils. *Renewable Energy*, *162*, 701–711. https://doi.org/10.1016/j.renene.2020.08.061

Rosa, D., Medeiros, A. B. P., Martinez-Burgos, W. J., do Nascimento, J. R., de Carvalho, J. C., Sydney, E. B., & Soccol, C. R. (2020). Biological hydrogen production from Palm Oil Mill Effluent (POME) by anaerobic consortia and Clostridium beijerinckii. *Journal of Biotechnology*, *323*, 17–23. https://doi.org/10.1016/j.jbiotec.2020.06.015

Saady, N. M. C. (2013). Homoacetogenesis during hydrogen production by mixed cultures dark fermentation: Unresolved challenge. *International Journal of Hydrogen Energy*, *38*(30), 13172–13191. https://doi.org/10.1016/j.ijhydene.2013.07.122

Sharma, P., Gaur, V. K., Kim, S. H., & Pandey, A. (2020). Microbial strategies for bio-transforming food waste into resources. *Bioresource Technology*, *299*. https://doi.org/10.1016/j.biortech.2019.122580

Singh, A., Singhania, R. R., Soam, S., Chen, C. W., Haldar, D., Varjani, S., Chang, J. S., Dong, C. D., & Patel, A. K. (2022). Production of bioethanol from food waste: Status and perspectives. In *Bioresource technology* (Vol. 360). Elsevier Ltd. https://doi.org/10.1016/j.biortech.2022.127651

Singh, D., Tembhare, M., Machhirake, N., & Kumar, S. (2023). Biogas generation potential of discarded food waste residue from ultra-processing activities at food manufacturing and packaging industry. *Energy*, *263*, 126138. https://doi.org/10.1016/J.ENERGY.2022.126138

Soltan, M., Elsamadony, M., Mostafa, A., Awad, H., & Tawfik, A. (2019). Nutrients balance for hydrogen potential upgrading from fruit and vegetable peels via fermentation process. *Journal of Environmental Management*, *242*, 384–393. https://doi.org/10.1016/j.jenvman.2019.04.066

Sun, Y., & Cheng, J. (2002). Hydrolysis of lignocellulosic materials for ethanol production: A review. *Bioresource Technology, 83*, 1–11. https://doi.org/10.1016/s0960-8524(01)00212-7

Sydney, E. B., Duarte, E., Martinez-Burgos, W., Carvalho, J. C., Larroche, C., & Soccol, C. R. (2020). Development of short chain fatty acid-based artificial neuron network tools applied to biohydrogen production. *Internationa Journal of Hydrogen Energy*, *45*, 4175–5181. https://doi.org/10.1016/j.ijhydene.2019.10.128

Tao, Y., He, Y., Wu, Y., Liu, F., Li, X., Zong, W., & Zhou, Z. (2008). Characteristics of a new photosynthetic bacterial strain for hydrogen production and its application in wastewater treatment. *International Journal of Hydrogen Energy*, *33*(3), 963–973. https://doi.org/10.1016/j.ijhydene.2007.11.021

Tawfik, A., El-Qelish, M., & Salem, A. (2015). Efficient anaerobic co-digestion of municipal food waste and kitchen wastewater for bio-hydrogen production. *International Journal of Green Energy*, *12*(12), 1301–1308. https://doi.org/10.1080/15435075.2014.909357

The State of Food Security and Nutrition in the World 2020. (2020). In *The State of Food Security and Nutrition in the World 2020*. FAO, IFAD, UNICEF, WFP and WHO. https://doi.org/10.4060/ca9692en

Tian, X., Liao, Q., Zhu, X., Wang, Y., Zhang, P., Li, J., & Wang, H. (2010). Characteristics of a biofilm photobioreactor as applied to photo-hydrogen production. *Bioresource Technology*, *101*(3), 977–983. https://doi.org/10.1016/j.biortech.2009.09.007

Tiang, M. F., Fitri Hanipa, M. A., Abdul, P. M., Jahim, J. M. d., Mahmod, S. S., Takriff, M. S., Lay, C. H., Reungsang, A., & Wu, S. Y. (2020). Recent advanced biotechnological strategies to enhance photofermentative biohydrogen production by purple non-sulphur bacteria: An overview. In *International Journal of Hydrogen Energy* (Vol. 45, Issue 24, pp. 13211–13230). Elsevier Ltd. https://doi.org/10.1016/j.ijhydene.2020.03.033

Ueno, Y., S, H., Ishii, M., & Igarashi, Y. (2001). Microbial community in anaerobic hydrogen-producing microflora enriched from sludge compost. *Applied Microbiology and Biotechnology*, *57*, 555–562. https://doi.org/10.1007/s002530100806

United Nations Environment Programme. (2021). Food Waste Index Report 2021. Nairobi. https://www.unep.org/resources/report/unep-food-waste-index-report-2021

Vavouraki, A. I., Volioti, V., & Kornaros, M. E. (2014). Optimization of thermo-chemical pretreatment and enzymatic hydrolysis of kitchen wastes. *Waste Management*, *34*(1), 167–173. https://doi.org/10.1016/j.wasman.2013.09.027

Villanueva-Galindo, E., & Moreno-Andrade, I. (2021). Bioaugmentation on hydrogen production from food waste. *International Journal of Hydrogen Energy*, *46*(51), 25985–25994. https://doi.org/10.1016/j.ijhydene.2020.11.092

Xiao, L., Deng, Z., Fung, K. Y., & Ng, K. M. (2013). Biohydrogen generation from anaerobic digestion of food waste. *International Journal of Hydrogen Energy*, *8*, 13907–13913. https://doi.org/10.1016/j.ijhydene.2013.08.072

Yasser Farouk, R., Li, L., Wang, Y., Li, Y., & Melak, S. (2020). Influence of pretreatment and pH on the enhancement of hydrogen and volatile fatty acids production from food waste in the semi-continuously running reactor. *International Journal of Hydrogen Energy*, *45*(6), 3729–3738. https://doi.org/10.1016/j.ijhydene.2019.07.236

Yukesh Kannah, R., Merrylin, J., Poornima Devi, T., Kavitha, S., Sivashanmugam, P., Kumar, G., & Rajesh Banu, J. (2020). Food waste valorization: Biofuels and value added product recovery. In *Bioresource technology reports* (Vol. 11). Elsevier Ltd. https://doi.org/10.1016/j.biteb.2020.100524

Zeng, Y., Xie, T., Li, P., Jian, B., Li, X., Xie, Y., & Zhang, Y. (2018). Enhanced lipid production and nutrient utilization of food waste hydrolysate by mixed culture of oleaginous yeast Rhodosporidium toruloides and oleaginous microalgae Chlorella vulgaris. *Renewable Energy*, *126*, 915–923. https://doi.org/10.1016/j.renene.2018.04.020

Zhang, J., & Wang, Q. (2013). Buffering and nutrient effects of white mud from ammonia-soda process on thermophilic hydrogen fermentation from food waste. *International Journal of Hydrogen Energy*, *38*(31), 13564–13571. https://doi.org/10.1016/j.ijhydene.2013.08.047

Zhang, T., Jiang, D., Zhang, H., Jing, Y., Tahir, N., Zhang, Y., & Zhang, Q. (2020). Comparative study on bio-hydrogen production from corn stover: Photo-fermentation, dark-fermentation and dark-photo co-fermentation. *International Journal of Hydrogen Energy*, *45*(6), 3807–3814. https://doi.org/10.1016/J.IJHYDENE.2019.04.170

Zhang, Y., van Herpen, E., Van Loo, E. J., Pandelaere, M., & Geuens, M. (2023). Save near-expired food: Does a message to avoid food waste affect food purchase and household waste prevention behaviors? *Journal of Cleaner Production*, *384*. https://doi.org/10.1016/j.jclepro.2022.135555

Zhao, M. X., Yan, Q., Ruan, W. Q., Miao, H. F., Ren, H. Y., & Xu, Y. (2011). Enhancement of substrate solubilization and hydrogen production from kitchen wastes by pH pretreatment. *Environmental Technology*, *32*(2), 119–125. https://doi.org/10.1080/09593330.2010.482596

Zheng, Y., Zhao, J., Xu, F., & Li, Y. (2014). Pretreatment of lignocellulosic biomass for enhanced biogas production. In *Progress in energy and combustion science* (Vol. 42, Issue 1, pp. 35–53). Elsevier Ltd. https://doi.org/10.1016/j.pecs.2014.01.001

6 Production of Biogas from Food Waste

Sharareh Haririchi, Mojdeh Mirshafiei, Abdullah Bilal Öztürk, Mohsen Parchami, Fatemeh Yazdian, and Mohammad J. Taherzadeh

6.1 INTRODUCTION

Food waste generation poses complex problems that require urgent attention and critical solutions. This issue arises from the unsustainable exploitation of natural resources to meet the growing global demand for food, as well as flawed consumption patterns. The necessity to mitigate the impact of environmental issues, e.g., greenhouse gas emissions (GHGs), loss of biodiversity, and ecosystem biodegradation, arising from global warming, natural resources overuse, climate change, and the depletion of fossil fuels, prompts the search for viable alternatives such as biogas (Muñiz-Márquez et al., 2023; Singh et al., 2023). Biogas offers several advantages; notably, it possesses a higher octane number and has the potential to generate significantly higher net energy compared to ethanol (Singh et al., 2023). Biogas can be produced during the anaerobic digestion of various wastes, including food waste. On the other hand, using food waste in the production of biogas is another possible solution to the problem of waste generation. This approach shows promise, as it not only reduces waste but also converts it into an energy source and contributes to the circular economy (Figure 6.1) (Muñiz-Márquez et al., 2023).

The Kyoto Protocol in 1997 and the United Nations Climate Change Conferences of the Parties (2015 and 2021) represent significant international agreements aimed at addressing climate change on a global scale. In particular, focusing on the management of food waste, around one-third of all the food produced worldwide (excluding inedible parts) is lost during its journey through the supply chain from agricultural production to the end consumer (Schirmer et al., 2023). According to the latest food waste index report in 2021, 17% of the total food produced in 2019 went to waste (Zhongming et al., 2021).

Landfilling and incineration are two traditional approaches to food waste management. But these methods have many serious drawbacks such as aquatic contamination, eutrophication, malodor, and GHG emissions that limit their widespread usage (Khanthong et al., 2023). On the other hand, the organic nature and elevated energy content of food waste make it an appropriate feedstock for biogas production; however, the bioconversion of food waste to bioenergy is costly and requires more precise research and innovative approaches to decrease costs. In general, biogas, derived from biodegradable organic materials through the process of anaerobic digestion (AD), is a fascinating and valuable product. It is mainly composed of methane and CO_2, and other gaseous compounds such as water, oxygen, nitrogen, H_2S, ammonia, siloxanes, and other hydrocarbons may be present. This gaseous fuel has multiple applications, including heating, drying, and power generation. It can replace gasoline in spark ignition engines with appropriate hardware modifications. Additionally, biogas can be utilized in dual-fuel mode to power compression ignition engines (Said et al., 2023). During AD, four essential stages, including hydrolysis, acidogenesis, acetogenesis, and methanogenesis, take place anaerobically to convert wastes into biogas (Muñiz-Márquez et al., 2023). Interestingly, different microbial communities from bacteria, archaea, and fungi to algae are identified for biogas production. In addition to biogas production from wastes, including food waste, a

DOI: 10.1201/9781032706030-6

FIGURE 6.1 A schematic overview of sustainable biogas production from food waste contributing to the circular economy.

by-product called digestate is generated that can be used for various purposes such as soil fertilizer and amendment (Kumari et al., 2021; Muñiz-Márquez et al., 2023).

Recently, progress has centered on emerging bioelectrochemical technologies like power-to-gas and microbial electrosynthesis, as well as microbial electrolysis cells. These innovations are being taken into account while designing future "cascading circular bio-systems" with the aim of generating eco-friendly advanced biofuels in a sustainable manner (Ning et al., 2021). In this chapter, we firstly discuss the properties of food waste, followed by the detailed exploration of AD traditional processes and novel advertising strategies. Additionally, we delve into the factors that impact productivity and biogas yield from food waste. Finally, recent trends and ongoing challenges are described, shedding light on the subject.

6.2 FOOD WASTES TYPES AND THEIR VALUABLE COMPOUNDS

Food waste, consisting of cooked and uncooked food, along with wasted raw materials during production, poses a significant environmental challenge across households, hotels, cafés, industries, and the entire food supply chain. Moreover, food waste includes ruined food, vegetable and fruit waste, leftover food, and corps residues from agricultural sector (López-Sandin et al., 2023). In general, food waste is mainly formed during five key stages of food supply chain, including production stage, handling and storing stage, processing and wrapping stage, distribution stage, and finally consumption stage (López-Sandin et al., 2023; Wunderlich & Martinez, 2018). In Table 6.1, some food products and their categorizations are shown (Edjabou et al., 2016).

The United Nations Food and Agriculture Organization (FAO) has highlighted the varying carbon footprint associated with different compositions of food waste (Food & Organization, 2015). Surprisingly, despite rice being the primary staple in Asian countries, there is a lack of research on the AD process of FW containing rice. Previous studies have shown that rice waste significantly

TABLE 6.1
Some Food Products and Their Categorizations

Main Source	Food and Food Products	Food Categories	Food Waste Fraction
Vegetable	Bread	Bakery	Avoidable
	Cakes		
	Vegetable pizza		
	Pizza bread		
Animal	Bread found in animal-based sandwiches		
Vegetable	Tea bags	Drinks, confectionery, and desserts	
	Coffee grounds		
	Biscuits		
	Chips		
	Beer		
	Alcohol		
	Spent tea bags		Unavoidable
	Spent coffee grounds		
	Ketchup	Condiments, sauces, herbs, and spices	Avoidable
	Peanut butter		
	Honey		
	Jam		
	Mayonnaise		
	Banana	Fresh fruit	
	Apple		
	Melon		
	Other fruits		
	Skin (e.g., pineapple)		Unavoidable
	Peals (e.g., banana)		
	Stones (e.g., avocado),		
	Fruits rinds (e.g., melon)		
	Carrots	Fresh vegetables and salads	Avoidable
	Potatoes		
	Other fresh vegetables		
	Cooked potatoes		
	Cooked yams		
	Cooked vegetable		
	Skin (e.g., potatoes, carrots, onion)		Unavoidable
	Peels (e.g., courgette, cucumber)		
	Breakfast cereal	Stable food	Avoidable
	Rice		
	Pasta		
	Flour		
	Cooked rice		
	Cooked pasta		
	Corn	Canned food	
	Bean		
	Pineapple		
	Other tinned vegetables		

(*Continued*)

TABLE 6.1 (CONTINUED)
Some Food Products and Their Categorizations

Main Source	Food and Food Products	Food Categories	Food Waste Fraction
	Brine from canned vegetables food		Unavoidable
Animal	Canned meat and fish		Avoidable
	Opened canned meat and fish food		
	Eggs	Dairy and eggs	
	Milk		
	Yoghurt		
	Cheese		
	Butter		
	Cooked or fried eggs		
	Cheese served at home		
	Cheese rinds		Unavoidable
	Eggs shells		
	Meat	Meat and fish	Avoidable
	Fish		
	Packed cold meat		
	Cut meat		
	Cooked, prepared or served at home meat, fish, etc.		
	Skin, rinds, fat, etc. of meat and fish		Unavoidable
	Fish heads		
	Shellfish shells		

affects biogas production compared to vegetable and fruit waste. Cooked rice has been found to enhance biogas production, while the presence of leafy vegetables leads to lower biogas production through AD. Moreover, due to its complex nature, AD of meat waste results in poor biogas production (Al-Wahaibi et al., 2020; Cherukuri & Parthasarathy, 2023). Food waste is primarily made up of carbohydrates (both simple and complex), proteins, lipids, and a small amount of inorganic materials. However, it's important to note that the composition of food waste can differ depending on the type and source of the food waste (López-Sandin et al., 2023).

6.3 APPROACHES FOR BIOGAS PRODUCTION FROM FOOD WASTE

Food waste management poses a significant challenge in the industrial world due to its substantial environmental impacts, including the emission of greenhouse gases, generation of leachate and odors, and the spread of pathogens, which cannot be overlooked. One of the considerable and sustainable strategies for the management of food waste is the process of AD, which is widely recognized for the production of biogas, and researchers have examined different approaches to enhance the production of methane by treating the food waste in various ways (Schirmer et al., 2023).

6.3.1 Anaerobic Mono-Digestion and Co-Digestion

During the regular AD process, there is a generation stage of volatile fatty acids (VFAs) and ammonium ions, leading to unstable conditions and a decrease in biogas production (Katuri et al., 2019). Presently, numerous studies are underway to find effective strategies for enhancing the traditional advertising process (Cerrillo et al., 2018; Malave et al., 2015). The most widely recognized and conventional technique for the AD process is anaerobic mono-digestion (AMD), wherein microorganisms break down organic compounds in the waste through biochemical reactions, resulting in the production of biogas. In anaerobic co-digestion (ACD), two or more types of substrates are simultaneously digested to address limitations associated with anaerobic mono-digestion (AMD), including instability, unbalanced reaction rates, high concentrations of inhibitors, low microbial diversity, and a weak buffering system (Karki et al., 2021; Khanthong et al., 2023; Zhang et al., 2020a). In general, ACD is a promising approach in biogas production from food wastes, as it offers various benefits such as increased gas yields, improved process stability, and enhanced waste management. Studies have shown that co-digestion can improve the performance of fermenters by providing the microorganisms with the necessary missing nutrients to increase their efficiency (Paranjpe et al., 2023). This method involves simultaneously digesting multiple organic substrates and combining food wastes with other biomass sources like agricultural residues, animal manure, or energy crops. This approach not only maximizes the use of available resources but also diversifies the nutritional composition, providing a balanced diet for the microbial community responsible for biogas production (Kasinath et al., 2021). Moreover, ACD allows efficient use of various organic materials, maximizing energy production from available sources. Food waste, often rich in organic matter, can be combined with other substrates to provide a balanced feedstock for anaerobic digestion. This diversification of inputs improves the nutritional composition and stability of the digestive system (Donoso-Bravo et al., 2021). ACD improves the overall performance and stability of the anaerobic digestion process. By combining different substrates, co-digestion overcomes the limitations associated with individual feedstocks. For instance, food wastes often contain high levels of ammonia, which can hinder microbial activity in the digestion system. However, by co-digesting food wastes with other substrates that have lower ammonia levels, the negative effects can be reduced, resulting in improved process stability and biogas production (Paranjpe et al., 2023). Furthermore, co-digestion can improve the economic feasibility of biogas production. It involves combining different organic materials, which increases the availability and reliability of the feedstock supply. This reduces the reliance on a single substrate, thereby minimizing the risk of feedstock shortages and price fluctuations. Furthermore, co-digestion allows for the diversification of revenue streams by generating income from various sources, such as waste management fees or the sale of excess digestate as a fertilizer (Tiong et al., 2021).

Due to the abundant presence of carbohydrates, lipids, and proteins and their high bioconversion rate, food waste is widely known as the most communal substrate for AD. However, the stability of the AD process declines as organic loading rates (OLRs) and carbon/nitrogen (C/N) ratios increase, along with rapid acidification and the buildup of VFAs. In order to enhance the AD of food waste, introducing an additional substrate to the digester can effectively decelerate the acidification process. One of the substrates that can be co-digested with food waste is the algal biomass, which not only decreases the acetogenesis rate but also improves system stability. However, if digested solely, algal biomass may lead to high amount of ammonia production and system failure due to its high protein and nitrogen content (Ferreira et al., 2022; Khanthong et al., 2023). For instance, in research conducted by Zhao and Ruan (2013), an effort was made to combine the ACD of algae and food waste. The authors conducted an evaluation of methane generation from algae and kitchen waste in hot conditions (55 °C). By adjusting the ACD to achieve a C/N ratio of 15, they were successful in achieving a substantial methane production of 389 mL/g-TS (total solids). Recently, Khanthong et al., (2023) found that combining algal biomass and food waste in a ratio of 8:2 produced the greatest methane yield of 334 mL CH_4/g COD_{Input}. Moreover, the metagenomic data of this study revealed a significant increase in the presence of VFA-oxidizing bacteria, as well as methylotrophic and hydrogenotrophic methanogens that designate improvement of methane production and system stability. Based on the food waste composition and accessibility, the selection of the substrate for co-digestion can be tailored to maximize biogas production. For instance, when pineapple waste is used for biogas generation, the addition of cow dung and slaughterhouse wastes can enhance the digestibility of the pineapple waste, particularly its recalcitrant lignocellulosic materials. Due to the presence of functional and degradative microorganisms in livestock wastes, these substrates are suitable for co-digestion with food waste. However, when applied solely, they are not sufficient to produce a considerable volume of biogas (Otieno et al., 2023).

6.3.2 Integrated Systems for Biogas Production

Researchers are exploring various strategies to enhance the efficiency of the AD process. These strategies include pretreating feedstocks, optimizing running conditions, and employing microaeration to effectively remove H_2S from biogas (Alagöz et al., 2015; Jenicek et al., 2008). Moreover, integrated production strategies have recently become a focal point for researchers with a keen interest in increasing the overall efficiency of AD. In addition to current approaches, the integration of other state-of-the-art technologies holds great promise for food waste management and energy production. One such technology that can be combined with the current AD process is bioelectrochemical systems (BESs). These systems such as microbial fuel cells (MFCs) and microbial electrolysis cells (MECs) have shown tremendous potential in enhancing the overall efficiency of food waste treatment and biogas production (Gao et al., 2021; Wang et al., 2022). For instance, MECs have the capability to be incorporated into the AD process in order to enhance the quality of biogas produced. Upgrading biogas is a crucial stage before injecting it into natural gas or utilizing it as vehicle fuel. In this approach, not only is CO_2 eliminated from the biogas, but it is also transformed into biomethane through the use of electrodes inserted into the AD bioreactor (Cerrillo et al., 2018; Wang et al., 2022). The significant fact is that organic-rich wastes such as food waste have the ability to generate a greater amount of biogas in microbial electrolysis cells (MECs) as opposed to diluted wastewaters. As a result, MECs are expected to be more effective for treating organic-rich waste. However, it is important to note that this approach cannot be directly compared to a firm and robust AD process. Alternatively, combining MECs with AD can optimize the retrieval of energy. For instance, an integrated MEC-AD system increased methane production by up to 59.7% compared to a standalone AD system (Katuri et al., 2019; Wang et al., 2022). Therefore, MEC has emerged as a viable and promising strategy that enhances the stability of the AD process by oxidizing high concentrations of propionate and other VFA compounds. This is achieved by facilitating mutually beneficial relationships between fermentative and exoelectrogenic microorganisms, as well as hydrogenotrophic methanogens. The successful implementation of this MEC-AD process ensures that the accumulation of VFA compounds is prevented, highlighting the important role played by the exoelectrogenic microorganisms. Additionally, this integrated process has the potential to reduce HRT (hydraulic retention time) by up to fourfold (Park et al., 2018b). By seamlessly integrating BESs, cutting-edge technology, into the AD process, we can greatly expedite the decomposition of organic matter and achieve unparalleled odor removal. In addition, BESs enables us to enhance the production of biomethane within the biogas, unlocking its potential for energy generation and various other applications (Quashie et al., 2021; Wang et al., 2022).

6.3.3 Pretreatment of Food Waste

Due to the high volume, rich organic nature, and elevated water content of food wastes compared to other types of waste, effectively handling and storing them for an extended period without microbial contamination becomes challenging. Therefore, some pretreatment methods such as drying can aid to increase food waste stability and reduce storage size (Srivastava & Malik, 2022). In addition, incorporating various pretreatment methods such as mechanical, chemical, thermal, or biological approaches can enhance the digestion rate and biogas yield derived from food waste. For example, the co-digestion of thermally pretreated yard waste and food waste increased methane production up to 431 mL/g-VS_{added} (Panigrahi et al., 2020). Implementing these pretreatments can lead to a significant reduction in volatile solids (VS) levels within the waste. Furthermore, extrusion, as a mechanical pretreatment technique, is widely employed to reduce particle size, encourage hydrolysis, enhance organic loading, and improve the stability of AD (Cherukuri & Parthasarathy, 2023). Occasionally, combining different pretreatment methods can be more beneficial than relying on a single approach. By implementing a combination of techniques, the overall effectiveness and efficiency of the pretreatment process can be improved. For example, the combination of ultrasonication and thermal pretreatment at 130 °C for 50 minutes exhibited impressive results by doubling the biogas production without causing any significant delays. Additionally, the application of both mechanical and ultrasonication pretreatment to household organic waste resulted in a noteworthy 14% increase in biodegradability and a higher methane yield compared to using only mechanical pretreatment (Menon et al., 2016).

Among various types of food waste, vegetable and fruit waste are reported to have higher levels of lignocellulosic residues, such as lignin, hemicellulose, and cellulose, which are resistant to anaerobic digestion. To enhance biogas production in anaerobic digesters, it is necessary to pretreat vegetable and fruit waste through a combination of chemical, physical, and biological methods, as recommended in technical reports (Cai et al., 2021b). By pretreating vegetable and fruit waste, it is possible to decrease the lignin content and crystallinity of cellulosic fibers, thereby increasing porosity and surface area for enzymatic attacks. The most practical approach for treating food waste is through chemical methods, involving the use of an alkaline reagent. Chemical pretreatment is commonly employed prior to the AD process due to its affordability, effectiveness, and versatility in various applications. Furthermore, the addition of an alkaline reagent such as sodium hydroxide (NaOH) not only enhances lignin removal and enzymatic hydrolysis but also effectively buffers the system during the AD process, particularly in the initial steps. However, the time of pretreatment, concentration of alkaline agent, and type of reagent should be considered carefully not to cause inhibitory effects on the methane production. Subsequently, the assessment of ideal pretreatment conditions should take into account the specific characteristics of the food waste under examination, as well as economic and sustainability factors (Schirmer et al., 2023).

6.3.4 Approaches for the Monitoring of AD of Food Waste

During the AD process, four main microbial-mediated stages (hydrolysis, acidogenesis, acetogenesis, and methanogenesis) occur that can be influenced or can even fail according to various factors, including inhibitors and unwanted operational parameters. Therefore, it is essential to monitor the system continuously through different physicochemical, biochemical and molecular methods. In Table 6.2, these monitoring methods are shown (Yaverino-Gutierrez et al., 2023).

6.4 FACTORS INFLUENCE BIOGAS PRODUCTION FROM FOOD WASTE

Biogas production from food waste is a promising and environment-friendly solution for sustainable and renewable energy generation while waste management. The process of anaerobic digestion, where microorganisms break down organic matter in the absence of oxygen, plays a crucial role in converting food waste into biogas, primarily composed of methane and carbon dioxide (Pérez-Rodríguez et al., 2019; Vijayakumar et al., 2022). However, the success of biogas production relies on various factors that can significantly influence the process. Understanding these factors is essential for optimizing biogas production processes and maximizing energy generation from food waste resources (Dos Santos et al., 2022). This section highlights some key factors including substrate composition, feedstock preparation, temperature, pH and acidity level retention time, mixing and agitation techniques, and inhibitory substances commonly found in food waste. By considering these factors in the design and operation of anaerobic digestion systems handling food wastes we can ensure efficient biogas production with higher methane content while minimizing potential issues such as inhibition or process inefficiencies (Deena et al., 2022; Yuan et al., 2022). The common influential factors in biogas production from food waste are summarized in Figure 6.2 and Table 6.3.

TABLE 6.2
Methods for the Monitoring of the AD Process

Analysis Type	Analysis	Abbreviation	Importance
Physicochemical assays	Determination of total solids	TSs analysis	Monitoring of organic materials degradation, formation of intermediates, and biomethane yield
	Determination of total suspended solids at 103–105 °C	TSSs analysis	
	Determination of fixed solids and volatile solids, ignited at 550 °C	FSs and VSs analyses	
	Determination of chemical oxygen demand	COD analysis	
	Determination of alkalinity	-	
	Determination of total volatile fatty acids	Total VFAs analysis	
	Volatile fatty acids profile and quantification	-	
	Methane measurement	-	
Enzymatic assays	Amylase, exoglucanase, endoglucanase, cellobiase, lipase, chitinase, xylanase, protease, pullulanase, benzoyl-CoA reductase, phloroglucinol reductase, resorcinol reductase, and resorcinol hydrolase	-	Monitoring of bioconversion rate and waste degradation capability
Molecular approaches	Metagenomics	-	Monitoring of microbial diversity and their dynamics at different stages
	Gene expression	*mcrA* gene analysis	Monitoring activity and functionality of methanogens
	Proteomics	-	Identifying the protein profiles, metabolic pathways, activities, and dynamics of microbial communities

6.5 FEEDSTOCK CHARACTERISTICS

The composition of the food waste substrate has a significant impact on the biogas production process. All biodegradable waste that contains carbohydrates, fats, proteins, cellulose, and hemicelluloses can be used as feedstock in bio-digesters. Different food waste materials have varying levels of organic matter, nutrient content, and biodegradability. These factors directly affect the efficiency and yield of biogas production (Ekstrand et al., 2022; Linyi et al., 2020). Previous studies conducted by researchers worldwide have found that the composition of food waste greatly influences its physicochemical characteristics. Therefore, the type and quality of food waste used as a substrate can determine the overall performance of the anaerobic digestion process (Bareha et al., 2022; Lamolinara et al., 2022).

6.5.1 Carbon/Nitrogen Ratio

Food waste is typically used as the carbon source to strengthen the cell structure of biomass. In addition to carbon, microorganisms also require nitrogen for protein synthesis. The carbon-to-nitrogen (C/N) ratio is crucial in determining the suitability of organic matter for anaerobic digestion.

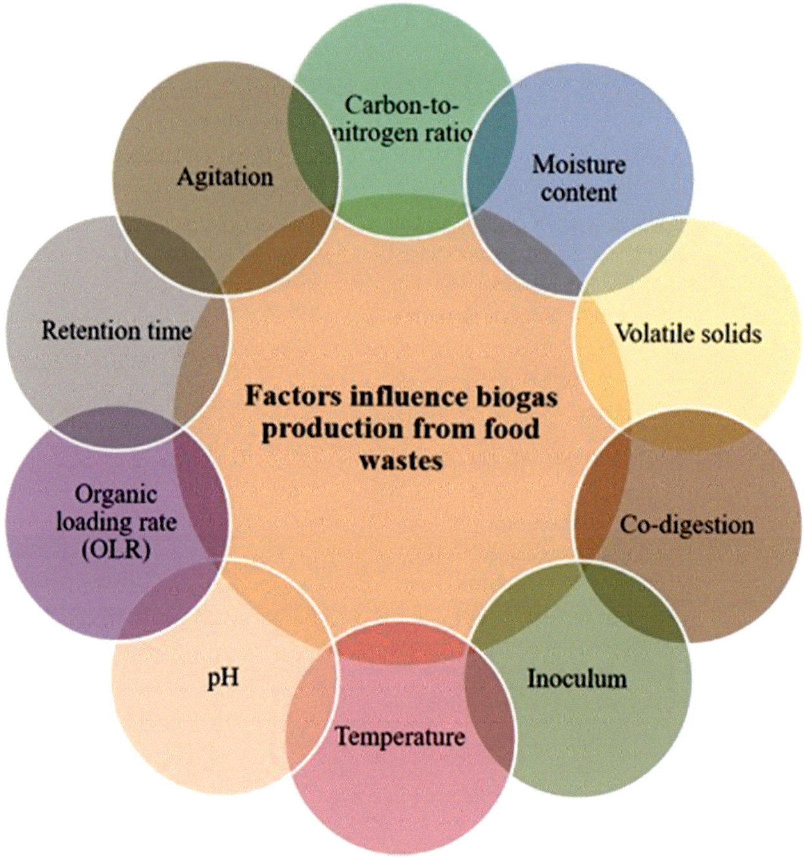

FIGURE 6.2 Schematic of common influential factors in biogas production from food waste.

The C/N ratio affects the availability and utilization of these essential nutrients, where, in turn, nitrogen limitation can lead to ammonia toxicity, inhibiting the activity of methanogens, which are responsible for methane production. A high C/N ratio indicates low nitrogen content, which is not favorable for microbial growth. Consequently, methanogens consume nitrogen for protein production, resulting in carbon wastage and ultimately leading to low biogas yield. Nitrogen limitation can lead to ammonia toxicity, inhibiting the activity of methanogens, which are responsible for methane production (Du et al., 2021; Rasool et al., 2023). On the other hand, a low C/N ratio can result in the accumulation of ammonia and nitrogen, which can inhibit the anaerobic digestion process. This limitation can cause the accumulation of volatile fatty acids and hinder the stability of the anaerobic digestion process. To optimize biogas production, it may be necessary to adjust the C/N ratio of the food waste substrate. If the C/N ratio is too high, adding a carbon-rich material, such as agricultural residues or carbonaceous waste, can help balance the ratio and provide additional carbon for the microorganisms. Conversely, if the C/N ratio is too low, adding a nitrogen-rich material, such as animal manure or nitrogen-containing organic waste, can help balance the ratio and provide additional nitrogen for microbial growth. To ensure efficient biogas production, it is recommended to maintain a balanced C/N ratio of around 20–30:1. This ratio provides a balance between carbon availability for energy generation and nitrogen supply for microbial growth. A well-balanced C/N ratio ensures that microorganisms have sufficient carbon to convert into biogas while maintaining a favorable environment for their metabolic activity. Food wastes that are rich in carbon, such as fruits and vegetables, can help maintain this ratio (Cerón-Vivas et al., 2019; da Silva et al., 2021).

TABLE 6.3
Overview of Biogas Production from Food Waste and Influential Factors

Substrate	Pretreatment	Inoculum	Influence Factors	Biogas Production	Reference
Food waste	Ionizing radiation	Granular sludge	TS, VS, pH, and soluble COD	504 mL/g-VS	(Fei et al., 2020)
	Thermal	Sewage sludge	HRT, sludge recycle ratio, water/feed ratio, sewage sludge/food waste ratio, COD, and VS	0.28 L CH4/g COD	(Cheong et al., 2022)
	-	Primary sludge under thermophilic (55°C) and mesophilic (35 °C) conditions	Food waste/primary sludge ratio, VS, OLR, and HRT	270–205 $mLCH_4$/g-VS	(Aguilar et al., 2017)
	Thermal	Pilot digester treating wastewater sludge (mesophilic)	TS, VS, time treatment, and soluble COD	68.54 ± 2,4 mg/L and 382.82 mLSTP CH_4/g-VS	(El Gnaoui et al., 2020)
	Microwave	Wastewater treatment plant sludge with A2O process	TS, VS, COD, and VFAs content	3446.3 ± 172.3 mL (35 days)	(Liu et al., 2020)
	Alkali	*Methanosaeta* and *Methanospirillum*	pH, TS, VS, DOM, SCOD, and OLR	829 mL/g-VS	(Linyi et al., 2020)
	Aerobic	*Proteobacteria* (mesophilic)	Oxygen rate, CO_2 contents, TS, VS, Van Soest fractionation, macromolecule content, VFAs, and simple sugars	500 $NLCH_4$/kg-VS	(Meng et al., 2017)
	Enzymatic	Sewage treatment plant sludge (mesophilic)	pH, TS, VS, TCOD, and SCOD	0.30, 0.28, 0.28, 0.26, 0.27 m3CH_4/kg-COD	(Zhang et al., 2020b)
	Enzymatic (Fungal mash)	Activated sludge (mesophilic)	VS, SCOD, free ammonia nitrogen (FAN) concentration, and glucose concentration	610.3–817.0 $mLCH_4$/g-VS	(Yin et al., 2016)
	Alkali	Waste-activated sludge (mesophilic)	TS, VS, and COD	271.7–339.2 $mLCH_4$/g-VS	(Naran et al., 2016)

(Continued)

TABLE 6.3 (CONTINUED)
Overview of Biogas Production from Food Waste and Influential Factors

Substrate	Pretreatment	Inoculum	Influence Factors	Biogas Production	Reference
Food waste + lipid waste	Ultrasound and microwave heating	Sludge	TS, TVS, moisture content, COD, time, and power treatment	738.63 mL/g-TVS –927.97 mL/g-TVS	(Yue et al., 2021)
Food waste + poultry manure	Autoclave, Microwave and Ultrasonication	Sludge (thermophilic)	TS, VS, FS, C/N ratio, COD, and HRT	9926 mL	(Deepanraj et al., 2017)
Catering wastes + pretreated *Parthenium* weed (as co-substrate)	Steam and microwave irradiation	*Parthenium hysterophorus*	COD, C/N ratio, solids content, pH, and mixing ratios	559 mL/L/d	(Tayyab et al., 2019)
Rice, cooked vegetables, bread, and meat + yard waste (grass, dry leaves, and wood chips)	Thermal and microwave	Waste-activated sludge	TS, VS, pH, VFA, and soluble COD	431 mL/g-VS	(Panigrahi et al., 2020)
Vegetable and fruit waste	Crushed	Fermented cow manure	TS, dissolved organic matter (DOM), ashes, Catalyst content	0.274 L/L and 0.451 L/g-DOM	(Dubrovskis & Plume, 2017)
Vegetable and fruit waste + food waste	Crushed and homogenized	Sludge	TS, VS, OLR, Vegetable and fruit waste/ food waste ratio, HRT, and pH	0.351–0.455 L/g-VS/d	(Shen et al., 2013)

By considering and managing the C/N ratio of food waste, operators of anaerobic digestion systems can optimize biogas production, improve process stability, and maximize the energy potential of the substrate (Paranjpe et al., 2023).

6.5.2 Moisture Content

The moisture content of food waste is important for the digestion process. Optimal moisture levels are necessary for efficient microbial activity and biogas production. Ideally, a moisture content of about 80–90% is desirable for maximum biogas production (Choudhary et al., 2020). Adequate moisture is necessary for the microbial community to thrive and carry out the anaerobic digestion process effectively. Microorganisms responsible for biogas production require water to dissolve and transport nutrients, enzymes, and other essential substances. Inadequate moisture levels can hinder microbial growth and metabolic activity, resulting in reduced biogas yields. In addition, moisture content plays a crucial role in facilitating the breakdown of organic matter in food waste. The presence of water helps to soften and dissolve complex compounds, making them more accessible to microbial degradation. Optimal moisture levels allow for efficient hydrolysis, the initial step in anaerobic digestion, where complex organic molecules are broken down into simpler compounds (Govarthanan et al., 2022; Paranjpe et al., 2023). However, excessive moisture levels can lead to substrate saturation, which can inhibit microbial activity and biogas production. Saturation occurs when the substrate becomes waterlogged, impeding the movement of gases and nutrients. This can create an unfavorable environment for the anaerobic bacteria involved in biogas production. Therefore, balance between providing sufficient moisture for microbial activity and avoiding substrate saturation is important (Abomohra et al., 2022; Seo et al., 2021).

It is generally accepted that anaerobic digestion processes are categorized into three types based on the total solid (TS) content in the raw materials: wet (< 10% TS), semi-dry (10% < TS < 15%), and dry (≥15% TS). Dry anaerobic digestion is a promising technology for treating and stabilizing organic wastes, such as agricultural residues and livestock waste, which have high TS contents (15%–40%). This process also allows for the recovery of energy through biogas production, primarily methane. Compared to wet anaerobic digestion, dry anaerobic digestion offers several advantages, including higher volumetric methane yield, reduced reactor volumes, lower energy requirements for heating, less digested residues, and lower moisture content in the digestate, which makes it easier to manage. Because of this, dry anaerobic digestion has gained significant attention in recent years. However, this technology still faces challenges due to its high solid content. Furthermore, dry anaerobic digestion differs significantly from wet anaerobic digestion in terms of technical operation, reactor design, and process performance. Therefore, it is essential to have a comprehensive understanding of dry anaerobic digestion and its characteristics, analyze its operational factors, and evaluate optimization methods and techniques to develop practical applications (Czekała et al., 2023; Wang et al., 2023c). The main steps and metabolic pathways in dry anaerobic digestion are illustrated in Figure 6.3.

6.5.3 Total Solids, Volatile Solids, and Their Particle Sizes

Solid content refers to the overall amount of fermentable substrate in a given volume of slurry. Higher levels of dry solids, particularly lignocellulosic content, can impact the hydrolysis process. The ideal solid content range for enhancing productivity was determined to be 7–9%. Total solids include both organic and inorganic substances. The percentage of volatile solids in the substrate is directly linked to methane production (Benti & Asfaw, 2022; Sun et al., 2022). On the other hand, the particle size affects the overall fermentation of organic matter in the digester. Smaller particles of food waste have been found to offer a larger surface area for the initial adsorption of exo-enzymes facilitating the degradation process and enhancing biogas production. Studies have demonstrated that reducing the size of the feedstock can enhance the anaerobic digestion process in two ways: (i)

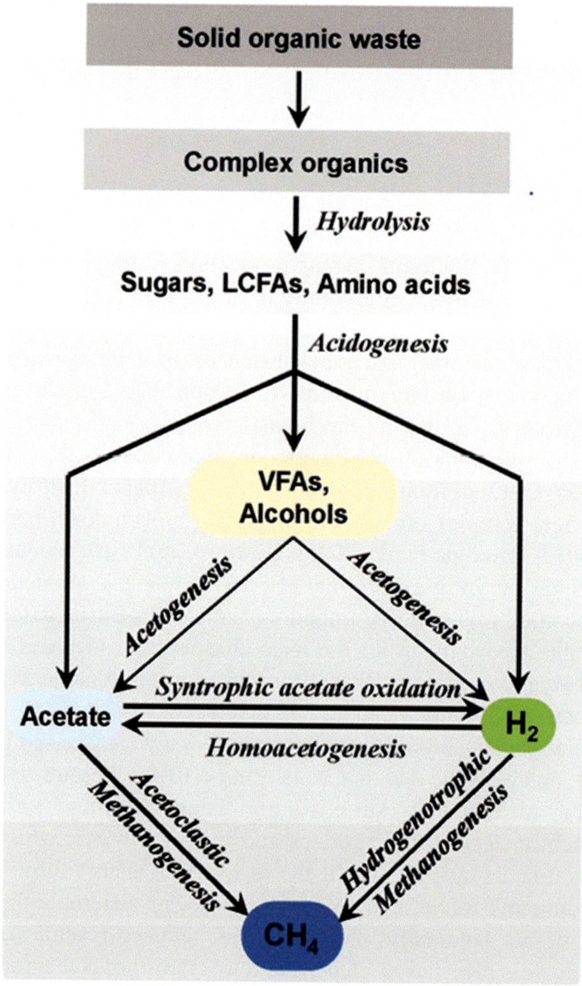

FIGURE 6.3 Primary steps and metabolic pathways involved in dry anaerobic digestion (Wang et al., 2023c).

increasing biogas production from substrates with high fiber content, and (ii) reducing the digestion time for all substrates (Mirmohamadsadeghi et al., 2019; Suhartini et al., 2022). Crushing the feedstock into smaller and uniform particle sizes is recommended before anaerobic digestion, as it can significantly reduce the volume of the digester without compromising the quantity of biogas produced. However, excessive combination of food waste can lead to the accumulation of volatile fatty acids and a decrease in methane yields (Kabaivanova et al., 2022; Shahbaz et al., 2019). This is especially harmful in solid-state anaerobic digestion processes compared to submerged processes. It has been reported that particle sizes that are too fine can cause severe foaming and process failure respectively in wet and dry digesters (Yang et al., 2021). Therefore, it is important to choose appropriate combination equipment based on the type of digester to ensure a successful anaerobic digestion process (Singh & Uppaluri, 2023).

6.5.4 Inhibitory Compounds

The presence of inhibitory compounds can hinder the efficiency and stability of biogas production. These compounds can originate from various sources, such as the composition of the food waste and the conditions of the anaerobic digestion process. Inhibitory compounds can have detrimental effects on the microbial communities that are responsible for biogas production, leading to decreased gas yields and process instability (Zhang et al., 2022b). For example, metal elements, both light (sodium, potassium, magnesium, calcium, and aluminum) and heavy (chromium, cobalt, copper, zinc, and nickel), are essential micronutrients for the survival of microorganisms. However, high concentrations of these metals can also inhibit anaerobic digestion. These compounds, unlike organic inhibitors, are not degraded and can accumulate in fermenters, reaching inhibitory levels. Heavy metals inhibit the process by disrupting the structure and function of enzymes. Nonetheless, in the anaerobic digestion of food wastes, heavy metal inhibition is usually not a concern, as their concentrations in food wastes are typically below the threshold (Demarco et al., 2019; Guo et al., 2019). In contrast, food wastes contain high concentrations of light metals, particularly Na and K ions, which could be a concern. Consequently, it is important to measure the levels of Na and K in food wastes to prevent the inhibition of anaerobic digestion (Ofili et al., 2020). On the other hand, ammonia (NH_3), carbon dioxide, long-chain fatty acids (LCFA), and H2S are also common inhibitory substances found in the substrate during the anaerobic digestion process. During protein degradation, ammonia is produced primarily in the forms of free ammonia (NH_3) and ammonium (NH_4^+). Microorganisms can consume ammonia as a macronutrient. Additionally, by neutralizing volatile fatty acids, the digester's buffering capacity can be increased, avoiding the inhibition of volatile fatty acids (Font Pomarol, 2022). However, at higher concentrations, ammonia can also inhibit the anaerobic digestion process. This inhibitory effect is caused by the free form of ammonia (NH_3), while the ammonium form is not toxic. Free ammonia can diffuse cells and disrupt the balance of H^+ and K, affecting cell function. At higher pH and temperature, as the ammonia-ammonium balance shifts toward inhibiting ammonia, the inhibitory effect of ammonia increases (Cai et al., 2021a).

Some feedstocks can also generate intermediates that have an inhibitory impact on the anaerobic digestion procedure. For instance, LCFA are found in agricultural residues, food waste, olive oil wastewater, slaughterhouse waste, and other materials, and act as inhibitor compounds during anaerobic digestion. Although lipids have a high potential for methane production, their fermentation results in the formation of LCFAs. These LCFAs can accumulate and impede the formation of methane. Food waste is typically composed of substances that are rich in lipids. When lipids are hydrolyzed, they produce LCFAs such as oleic acid, palmitoleic acid, and linoleic acid (Cai et al., 2021a; Grosser et al., 2017; Nakasaki et al., 2020). One way that LCFAs inhibit microorganisms is by adsorbing onto their cell membranes and walls, which disrupts the cell's transport system. Various solutions have been proposed to overcome the inhibition caused by LCFAs. These include adding an active inoculum, reducing the bioavailability of LCFAs through adsorption, adding co-substrates, and implementing discontinuous feeding (Xu et al., 2023).

6.5.5 Inoculum

In the anaerobic digestion process, the quality and quantity of the inoculum is a crucial factor. The quantity of inoculum specifically affects the start-up phase of anaerobic digestion. The inoculum introduces a diverse microbial community that initiates and facilitates the breakdown of organic matter into biogas. Therefore, selecting and managing an appropriate inoculum is essential for achieving efficient and stable biogas production from food waste (Luo et al., 2023). There are various sources of inoculum, but commonly used ones include digestate from existing biogas plants, animal manure, sewage sludge, or specialized microbial cultures (Koupaie et al., 2019). The choice of inoculum depends on factors such as the composition of the food waste, desired process conditions, and the availability of suitable microbial communities.

In the initial stages of anaerobic digestion, hydrolytic bacteria break down complex organic compounds into simpler molecules. Acidogenic bacteria then convert these simpler molecules into VFAs such as acetic acid, propionic acid, and butyric acid. Acetogenic bacteria convert VFAs into hydrogen, carbon dioxide, and acetate. Methanogenic archaea then use these intermediates to produce methane, which is the main component of biogas (Wainaina et al., 2019). For effective biogas production, a diverse and well-adapted microbial community in the inoculum is crucial. It ensures the proper breakdown of complex organic compounds, prevents imbalances in the process, and enhances the stability of the anaerobic digestion system (Viancelli et al., 2023). It is important to note that the inoculum itself can influence the biogas production process. The composition and activity of the introduced microbial community can affect the efficiency of degradation, gas yields, and process dynamics (Omar et al., 2018).

6.6 PROCESS PARAMETERS

In addition to feedstock characteristics, the efficiency and success of biogas production from food wastes depend on various process parameters such as temperature, pH, organic loading rate, retention time, and substrate composition. Understanding and optimizing these process parameters are essential for maximizing biogas production and ensuring the stability of the anaerobic digestion system.

6.6.1 Temperature

Temperature is a vital parameter determinant of the effectiveness of the anaerobic digestion process and influences the activity and composition of the microbial community involved in biogas production. As different microorganisms thrive at different temperature ranges during the anaerobic digestion process, the anaerobic digestion process can be carried out within three temperature ranges: (i) thermophilic (40–70 °C), (ii) mesophilic (25–40 °C), and (iii) psychrophilic (below 25 °C) (Rodríguez-Jiménez et al., 2022). Increasing the temperature can help to increase gas production but leads to lower methane content and increased CO_2 ratio, which lowers the heating value of biogas. Higher temperatures can accelerate the degradation process but also raise the possibility of destabilization or process inhibition. On the other hand, lower temperatures may prolong the digestion process and reduce gas yields (Albano et al., 2023). It is important to find the appropriate temperature range for the specific substrates and microbial community to achieve optimal biogas production (Schnürer, 2016). Accordingly, the optimal temperature for efficient and continuous biogas production was reported to be between 32 °C and 35 °C (Duong & Lim, 2023).

6.6.2 pH

The pH is a critical process parameter in anaerobic digestion and biogas production from food wastes. Maintaining an appropriate pH range ensures the optimal growth and activity of the microbial community responsible for anaerobic digestion and biogas production (Li et al., 2022a). The pH and temperature are interdependent, and the optimal pH range for enhancing biogas yield is typically between 6.5 and 7.2 (Bhajani & Pal, 2022). Within this pH range, the microbial community, including methanogenic bacteria, can thrive and efficiently convert organic matter into biogas (Li et al., 2022a). If the pH deviates from this optimal range, it can negatively impact the anaerobic digestion process. Acidic conditions that occur due to the accumulation of VFAs can hinder the growth and activity of methanogenic bacteria, resulting in reduced biogas production. The accumulation of VFAs, which are intermediate products of the digestion process, can occur due to high organic loading rates, short retention times, or an imbalance in the microbial community. High levels of VFAs can lower the pH and create an unfavorable environment for methanogens (Wainaina et al., 2019). On the other hand, the presence of alkalinity can promote the growth of specific microbial groups, such as acetogenic bacteria. These bacteria compete with methanogens for the available organic matter. Consequently, this competition can decrease the efficiency of methane production and lead to lower yields of biogas. Therefore, to ensure the optimal pH range in biogas production from food wastes, it is crucial to monitor and control the pH levels in the digestion system, which can achieve through regularly measuring the pH and making adjustments using buffering agents or alkalinity supplements. Furthermore, managing the organic loading rate, retention time, and substrate composition properly can help prevent excessive acidification or alkalization (Jaman et al., 2023; Yuan et al., 2019). As demonstrated in Figure 6.4, the productivity of biogas is influenced by changes in pH with respect to the retention time.

6.6.3 Organic Loading Rate (OLR)

OLR refers to the amount of organic matter that is introduced into the anaerobic digestion system within a specific volume and time. OLR plays a significant role in determining the efficiency, stability, and biogas production potential of the anaerobic digestion process. The organic matter in food waste serves as the substrate for the microbial community responsible for biogas production (Yang et al., 2023). The OLR directly affects the availability and concentration of organic matter within the system, which in turn influences the yield of biogas. Therefore, gas production is mainly influenced by the OLR. So, it is important to find the optimal OLR that allows for the efficient utilization of the substrate without overwhelming the microorganisms (Borowski & Cieciura-Włoch, 2021). A high OLR can cause imbalances in the process and hinder biogas production. When the organic matter exceeds the capacity of the microbial community to degrade it, several issues can arise. Excessive acidification may occur due to the accumulation of VFAs. These VFAs can prevent the growth and activity of methanogenic bacteria, resulting in reduced biogas yields. High OLRs can also lead to the accumulation of inhibitory compounds, such as ammonia, which further hinder microbial activity and biogas production (Lianhua et al., 2020). On the other hand, a low OLR may not fully utilize the available substrate, limiting the potential for biogas production. In such cases, the anaerobic digestion system may not reach its maximum capacity in converting organic matter into biogas. So, it is important to strike a balance between maximizing biogas production and avoiding process imbalances (Duan et al., 2019; Fuentes et al., 2021). Determining the optimal OLR for biogas production from food wastes requires considering various factors, such as substrate characteristics, microbial community capacity, and anaerobic digestion system design and size. Regular monitoring and adjustment of the OLR are essential for maintaining the stability and efficiency of the anaerobic digestion process (Mills et al., 2023).

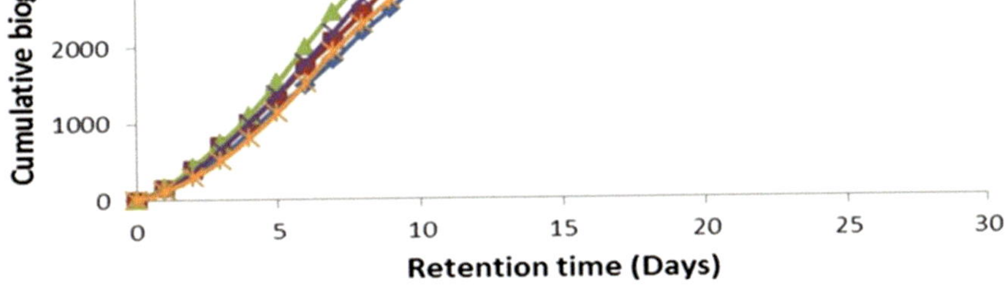

FIGURE 6.4 Effect of pH on biogas production (Bhajani & Pal, 2022).

6.6.4 Retention Time

Microorganisms require an adequate retention time to convert organic materials into biogas. Retention time refers to the duration that the substrate remains in the anaerobic digestion system, and it affects the extent of organic matter degradation and biogas production. The duration of retention time is crucial for efficient biogas production. Increasing the retention time leads to a higher reduction of volatile solids, a larger digester volume requirement, and better adaptation to pH variations and toxic compounds (Mirmohamadsadeghi et al., 2019). Longer retention times allow for more complete degradation of complex organic compounds present in food waste, resulting in higher gas yields. The microbial community responsible for anaerobic digestion requires sufficient time to break down organic matter into simpler compounds such as VFAs and convert them into biogas, primarily methane, and carbon dioxide. On the other hand, shorter retention times can reduce the required volume of the digester and therefore lower the investment cost while still producing biogas of the same quality and quantity. Therefore, striking a balance between retention time and gas production is essential for optimizing the anaerobic digestion process. By gradually adjusting the retention time and monitoring the rates and quality of biogas production, the optimal duration can be determined to maximize gas production while maintaining process stability (Novakovic et al., 2021). It is important to note that retention time can also vary depending on the composition and biodegradability of the food waste. Some food waste types may require longer retention times to ensure complete degradation, while others with higher biodegradability may achieve satisfactory results with shorter retention times (Pau et al., 2023).

6.6.5 Agitation

Agitation is essential for efficient digestion, as it helps to ensure a consistent mixture of the substrate in the anaerobic digestion system. It also aids the homogeneous distribution of nutrients and microorganisms, facilitating better interaction between the organic matter and the microbial community. This interaction promotes the degradation of complex organic compounds into simpler ones, ultimately enhancing the production of biogas (Harirchi et al., 2022). Agitation prevents the formation of stagnant zones or dead zones in the digestive system. Stagnant zones can lead to localized accumulation of organic matter, inhibitory compounds, or gas bubbles, which can hinder microbial activity and reduce biogas production. By ensuring proper mixing and circulation of the substrate, agitation helps to minimize these issues and maintain a favorable environment for the microbial community (Singh et al., 2021). In addition, agitation prevents foaming, disturbs temperature gradient, and eliminates floating and sinking layers which can hinder the exchange of gases between the substrate and the microbial community (Mirmohamadsadeghi et al., 2019). The intensity and frequency of agitation depend on various factors such as the characteristics of the substrate, the design of the digestion system, and the specific requirements of the microbial community. It is important to strike a balance between providing sufficient mixing to promote digestion and avoiding excessive agitation that may lead to the loss of volatile gases or excessive energy consumption (Harirchi et al., 2022). Various agitation methods can be used in biogas production from food waste including mechanical stirring, circulation pumps, or gas recirculation systems. The choice of agitation method depends on the design, scale, and specific requirements of the process (Singh et al., 2019). Agitation is essential for efficient digestion, as it helps to ensure a consistent mixture of the substrate in the anaerobic digestion system. It also aids the homogeneous distribution of nutrients and microorganisms, facilitating better interaction between the organic matter and the microbial community. This interaction promotes the degradation of complex organic compounds into simpler ones, ultimately enhancing the production of biogas (Harirchi et al., 2022). Agitation prevents the formation of stagnant zones or dead zones in the digestive system. Stagnant zones can lead to localized accumulation of organic matter, inhibitory compounds, or gas bubbles, which can hinder microbial activity and reduce biogas production. By ensuring proper mixing and circulation of the substrate, agitation helps to minimize these issues and maintain a favorable environment for the microbial community (Singh et al., 2021). In addition, agitation prevents foaming, disturbs temperature gradient, and eliminates floating and sinking layers which can hinder the exchange of gases between the substrate and the microbial community (Mirmohamadsadeghi et al., 2019). The intensity and frequency of agitation depend on various factors such as the characteristics of the substrate, the design of the digestion system, and the specific requirements of the microbial community. It is important to strike a balance between providing sufficient mixing to promote digestion and avoiding excessive agitation that may lead to the loss of volatile gases or excessive energy consumption (Harirchi et al., 2022). Various agitation methods can be used in biogas production from food waste including mechanical stirring, circulation pumps, or gas recirculation systems. The choice of agitation method depends on the design, scale, and specific requirements of the process (Singh et al., 2019).

6.7 TRENDS IN BIOGAS PRODUCTION FROM FOOD WASTE

Food waste has significant socioeconomic and environmental costs, and the recovery and valorization of these resources could have a positive impact on the environment and society. Food waste not only affects the economy negatively but also contributes to the emission of greenhouse gases, particularly methane. To reduce production costs and promote environmental sustainability, it is crucial to improve the efficiency of the food chain and minimize food loss and waste (Palaniveloo et al., 2020; Vigneshwar et al., 2022). Given the rising expenses related to energy supply and waste disposal, as well as growing public concerns about environmental quality, converting food waste into energy is becoming an attractive and environmentally beneficial solution. Utilizing food waste for bioenergy offers numerous benefits, including mitigating climate change, promoting the circular economy, and fostering sustainable development (Lin et al., 2013). Food waste can be transformed into various types of biofuels, such as biomethane through anaerobic digestion, biohydrogen through dark fermentation, biohythane (a combination of biohydrogen and biomethane gas) through two-stage anaerobic fermentation, and biodiesel through transesterification (Dahiya et al., 2015; Kannah et al., 2020; Kiran et al., 2015; Mahyari et al., 2021).

The process of converting food waste into biogas through anaerobic digestion is a promising and sustainable solution for waste management and renewable energy production. The process of converting food waste into biogas involves several stages that work together to transform organic waste into a valuable renewable energy source. Understanding these stages is essential for stakeholders involved in waste management, renewable energy, and sustainability, as they provide insights into the intricacies of the conversion process and its potential for a more circular and resource-efficient future (Mahyari et al., 2021).

The first stage is the collection and sorting of food waste, where it is separated from other waste streams to ensure purity. The next stage is pre-treatment, which may include shredding or grinding the waste to increase its surface area and facilitate microbial digestion. In some cases, pre-treatment of food waste is necessary to enhance the efficiency of biogas production. Pre-treatment can also involve adjusting the pH or moisture content of the waste to optimize the anaerobic digestion process (Le Pera et al., 2023). The main stage is anaerobic digestion, where the pre-treated food waste is placed in an oxygen-free environment. Microorganisms break down the organic matter in the waste, resulting in the production of biogas. This biogas is mainly composed of methane and carbon dioxide, along with smaller amounts of other gases. Following the digestion process, the remaining material, digestate, can be used as a nutrient-rich fertilizer. The biogas produced can be directly used for heating or generating electricity. It can also be upgraded to biomethane and injected into the natural gas grid or used as fuel for transportation. Understanding these stages is essential for effective waste management and the utilization of renewable energy from food waste (Govarthanan et al., 2022).

As shown in Figure 6.5, the metabolic reactions in anaerobic digestion involve four main stages: hydrolysis, acidogenesis, acetogenesis, and methanogenesis (Wang et al., 2023a).

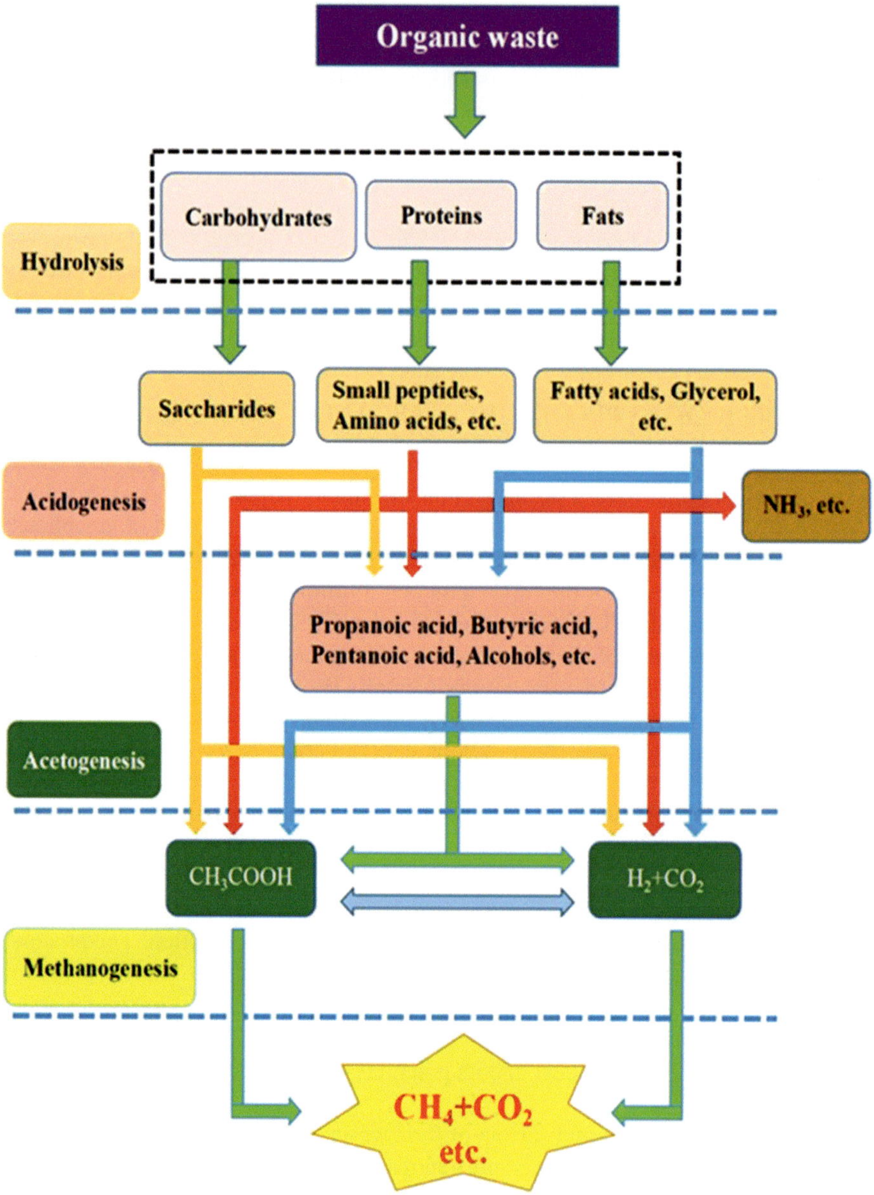

FIGURE 6.5 The anaerobic digestion process of organic waste (Xiong et al., 2023).

The overall reaction of the bioconversion of food waste to biogas is shown in Equation (6.1):

$$C_cH_hO_oN_nS_s + wH_2O \longrightarrow mCH_4 + nNH_3 + sH2S + (c\text{-}m)\ CO_2 \tag{6.1}$$

Where, m=1/8(4c+h-2o-3n-2s) and w=1/4(4c-h-2o+3n+3s).

The biodegradable portion of food waste mainly consists of carbohydrates ($C_6H_{12}O_6$), proteins ($C_{13}H_{25}O_7N_3S$), and lipids ($C_{12}H_{24}O_6$) (Deublein & Steinhauser, 2011). The organic content includes various particles as well as water-insoluble polymers, so microorganisms cannot directly access these polymers. Therefore, hydrolyzing bacteria expel exo-enzymes like amylase, cellulase, xylanase, lipase, and protease to initiate hydrolysis. In the hydrolysis stage, hydrolytic enzymes are adsorbed onto the surface of the substrate and gradually convert the complex organic compounds found in food waste such as carbohydrates, proteins, and lipids into monomers and oligomers (such as glucose, fatty acids, glycerol, and amino acids). Through this process, insoluble organic matter is converted into substances that are easily consumed by microorganisms (Mirmohamadsadeghi et al., 2019). The hydrolysis stage is typically the rate-limiting stage in the production of biogas from waste streams that contain highly resistant materials, such as lignocellulosic materials (Mutungwazi et al., 2023). Certain bacteria, including *Clostridium*, *Micrococci*, *Bacteroides*, *Butyrivibrio*, *Fusobacterium*, and *Selenomonas*, participate in this rate-limiting stage (Agrawal et al., 2023).

Following the conversion of carbohydrates, lipids, and proteins into sugars, long-chain fatty acids, and amino acids, respectively, in the next step, i.e., acidogenesis, the products of hydrolysis are further broken down by acidogenic bacteria. These bacteria convert the soluble compounds into VFAs, alcohols, and other organic acids (García-Depraect et al., 2022). Apart from acetic acid, propionic acid, and ethanol, other acids are produced by *Actinomyces*, *Peptostreptococcus anaerobius*, *Clostridium*, and *Lactobacillus*, respectively (Kalaiselvan et al., 2022). This stage produces intermediate products that will be further metabolized in the subsequent stages. With the support of a proton-reducing agent, long VFAs and alcohols are oxidized to produce acetic acid and H_2 during the acetogenesis stage. Acetogenic bacteria convert the intermediate products of acidogenesis, such as VFAs, into acetic acid, hydrogen, and carbon dioxide. These compounds serve as substrates for the final step in methane production, known as methanation. Acetogenesis is an important step in the production of methane, as these compounds serve as substrates for the final step in methane production, namely methanation (Detman et al., 2021). The final stage of anaerobic digestion is methanogenesis, where methanogenic archaea convert the acetic acid, hydrogen, and carbon dioxide produced in the previous stages into methane (CH_4) and carbon dioxide (CO_2). During this stage, two types of methanogens are produced: hydrogenotrophic and acidogenic. This is due to the reduction of CO_2 to H_2 as well as the scrubbing of the sliced acetic acid formed in the acetogenesis stage. Methanogenesis is the key step that produces biogas, which primarily consists of methane and carbon dioxide, along with trace amounts of other gases. This stage is the rate-limiting stage in the anaerobic digestion of readily degradable feedstocks with low buffering capacity (Khan et al., 2022).

Anaerobic digestion is a complex biochemical process that involves a diverse consortium of bacterial species. The variety of microbial consortia are responsible for each step of the process and maintain synergistic relationships with the microbial consortiums involved in the other steps to break down organic material into biogas through hydrolysis and fermentation. Although most of these bacterial species are strict anaerobes, such as *Bifidobacteria*, *Bacteroides*, and *Clostridia*, facultative anaerobes such as *Enterobacteriaceae* and *Streptococci* could also be present. *Acetobacterium woodii* and *Clostridium aceticum* are acetogenic bacteria that produce hydrogen. Methane is produced by methanogenic bacteria from acetate, hydrogen, and carbon dioxide. Only a few species (*Methanosarcina barkeri*, *Methanococcus mazei*, and *Methanotrix soehngenii*) of methanogenic bacteria can break down acetate into CH_4 and CO_2, though all methanogenic bacteria are capable of producing methane from CO_2 and hydrogen (Xiong et al., 2023). The process can be achieved in two stages because all the microbes in the consortium are closely linked to each other. Methanogenic bacteria have the slowest growth rates and are the most sensitive to environmental changes of the four microbial groups. As a result, the rate-limiting step in the entire anaerobic digestion process is methanogenesis. However, hydrolysis might be a limiting step in the degradation of lignocellulosic feedstock because of its recalcitrant structure. Therefore, a pretreatment step of lignocellulosic materials is required before anaerobic digestion to achieve a more efficient cellulose and hemicellulose microbial breakdown, which ultimately leads to an increased biogas yield (Olatunji et al., 2021).

The production of biogas from food waste is gaining popularity worldwide due to its many advantages. A major trend is the increasing use of biogas production as a sustainable solution for waste management. Numerous countries and regions are realizing the potential of converting food waste into biogas, not only to decrease greenhouse gas emissions but also to tackle the problems of waste disposal and limited landfill space (Chidi et al., 2022). Policy measures are crucial in promoting the growth of the biogas sector. Governments are acknowledging the potential advantages of producing biogas from food waste, such as reducing greenhouse gas emissions and decreasing dependence on fossil fuels, which align with their sustainability objectives (Obaideen et al., 2022). Supportive policies provide incentives for investing in infrastructure development for digesters of different sizes, whether it is small-scale community-based facilities or larger industrial installations. These policies encourage both public-private partnerships and individual initiatives that aim to create circular economies within local communities (Zebra et al., 2021).

Moreover, businesses are becoming more aware of the importance of corporate social responsibility and the economic benefits of resource optimization. The idea of "waste-to-wealth" is particularly appealing to companies looking for innovative ways to cut waste disposal costs and generate extra income by producing renewable energy (Xu et al., 2019). Biotechnology industries also play an essential role in developing specialized microbial cultures that can effectively degrade different types of feedstock. These advancements guarantee the production of high-quality output gases that can be directly used or further processed into compressed natural gas (CNG) or biomethane and can then be injected into existing gas grids (Schnürer, 2016). Furthermore, the integration of high-rate anaerobic co-digesters, which facilitate faster decomposition of organic materials, has significantly improved the biogas production process. Co-digestion, the practice of combining food wastes with other organic materials, has also proven to be beneficial in enhancing biogas yield, diversifying feedstock availability and process stability (Lisowyj & Wright, 2020). In addition, pre-treatment techniques such as thermal hydrolysis and mechanical disruption have been implemented to break down complex organic compounds, resulting in increased biogas production. These include the development of high-rate anaerobic digesters, co-digestion with other organic materials, and the use of pre-treatment techniques to enhance the biodegradability of food waste (Zhang et al., 2022b).

It is essential to prioritize research and development to enhance the efficiency and scalability of biogas production systems. Collaboration among academia, industry stakeholders, and policymakers will promote knowledge sharing and integration across sectors, such as agriculture and hospitality. This collaboration will create valuable partnerships that utilize wasted resources effectively. Therefore, the growing trend of biogas production from food waste is revolutionizing waste management and offering sustainable solutions for environmental issues. Technological advancements are enhancing efficiency and productivity within anaerobic digestion processes while policies incentivize investment at various scales.

6.7 CHALLENGES AND OPPORTUNITIES

It is known that food wastes can possess crucial components for producing biogas, and the higher production amount could be based on the optimum digestion parameters. Considering the higher titer production goals, biogas production from food waste can offer several benefits. However, similar to other bioprocesses, these benefits may come with varied and unique challenges. Herein, the prospects and challenges associated with biogas production from food waste are discussed comprehensively. They can be categorized in various aspects, but in this study, the key opportunities and challenges are described as listed in Table 6.4.

Considering the GHG-emitting power itself, the utilization of use of food waste for the production of value-added products and biofuels is one of the promising approaches in terms of sustainability and environmental friendliness (Flanagan & Priyadarshini, 2021). However, the chemical composition of this feedstock can differ enormously based on geopolitical location, cultural differences, and agriculture and diet habits of locals, and consequently, the feedstock variation of food wastes can be accepted as one of the critical challenges (de Abreu et al., 2023). Apart from location dependence, food waste can be from a wide variety of sources, including households, restaurants, hotels, university cafeterias, and mostly from food processing industries (Chen et al., 2017; Yang et al., 2019). This

TABLE 6.4
Opportunities and Challenges of Food Waste Utilization on Biogas Production

Challenges	Opportunities
Feedstock variation and contaminants	Renewable energy production
Acidification and microorganism inhibition	Waste management and utilization
Hydraulic retention time (HRT)	Digestate as a biological fertilizer
Process stability	Circular economy and GHG emission reduction
Microbial community management	Lower-cost potential through dry digestion
Concerns on hygiene and odour	Process integration for bioproducts

source diversity can result in composition variations in the beneficial part of the food wastes such as carbohydrates, fats, proteins, and other organic compounds, and finally, can imbalance the microbial populations responsible for anaerobic digestion (Braguglia et al., 2018). Based on the composition of food wastes, the amount of contaminants is another vital parameter of biogas production. The contaminants, including heavy metals, microplastics, pesticides, and other compounds, can accumulate in the digester over time, and the inhibition potential of the biogas-producing microorganisms can be elevated. Also, it can negatively affect the quality of the biogas produced and pose a risk to the environment if not handled properly (Kuo & Dow, 2017; Porterfield et al., 2023).

6.7.1 Acidification and Microorganism Inhibition

Acidification and microorganism inhibition are another critical challenge that can arise during the biogas production process from food waste (Alavi-Borazjani et al., 2020). This negative issue can impact the digestion process stability and efficiency, which results in lower biogas titer. More specifically, acidification refers to a decline in the medium pH during the process through the higher production of VFAs by methanogenic microorganisms. Typically, VFAs are intermediate products of anaerobic digestion and must be utilized for methane conversion. If this accumulation exceeds to toxic levels due to inhibition or compromise of microorganisms, a longer retention time, lower biogas production, and imbalanced microbial community can be observed (Li et al., 2010; Li & Jin, 2015; Sheng et al., 2023).

6.7.2 Hydraulic Retention Time (HRT)

HRT is a critical parameter for efficient biogas production from food waste as a feedstock. Since it refers to the amount of time that substances remain in the anaerobic digester, it is crucial to establish optimum process parameters to minimize the HRT. Higher HRT is one of the economic burdens for industrial-scale biogas production due to higher operational expenses, higher equipment volumes, and lower annual plant capacity (Atelge et al., 2020; Buitrón et al., 2014). Therefore, it's critical to consider the feedstock composition, establish optimal operating conditions, and monitor digester composition simultaneously to adjust the HRT as required. This can necessitate pretreatment techniques in combination to remove inhibitors, co-digestion with other organic materials to set attainable composition, and maintaining a balance between lower HRT and higher digestion by a stable operation (Kumar et al., 2016; Thi et al., 2016).

6.7.3 Process Stability

Process stability is a decisive parameter for an industrial-scale biotechnology plant. To ensure consistent microbial activity and optimal conditions throughout the operation, several parameters are associated with maintaining process stability (Nemestóthy et al., 2018). Foaming, agitation, pH and temperature fluctuations, concentration variations in the digester, and inhibitory components' existence in the medium can be promoters of the unstable microbial community performance on biogas production from food waste (Park et al., 2018a; Rajesh Banu et al., 2020; Yukesh Kannah et al., 2020; Zhen et al., 2016).

6.7.4 Microbial Community Management

The microbial community that is evaluated in food waste digestion is considered a mixture of populations responsible for biogas production. Therefore, it is vital to maintain higher activity in these cultures by managing detrimental process factors, particularly food waste composition (Choudhary et al., 2023; Goswami et al., 2016). Food waste composition is a location-dependent factor, and the nutrient availability may deviate significantly, which can result in competition of strains for limited beneficial resources. In other words, specific types of nutrients can require suitable microorganism strains in the digester inoculum. Therefore, the characterization of food waste and inoculum, applying proper pretreatment methods, and adjusting optimum process parameters are critical to the management of the microbial community (Li et al., 2022b; Zhang et al., 2022a).

6.7.5 Hygiene and Odor

Most of the food wastes contain relatively higher moisture, which elevates the spoilage tendency of wastes to higher levels. For that reason, hygiene and odor concerns of food waste management and utilization of food wastes in biogas production are significant (Dhungana et al., 2022). Food waste can emit disgusting and unpleasant odors as it is processed through anaerobic digestion by releasing VOCs, and the existence of these compounds can result in disruptive effects on the nearby area and residents. Moreover, these emissions can lead to negative perceptions by locals, and complaints may arise against the biogas plants. In order to overcome these challenges, proper pretreatment methods, proper filtration processes, and ensuring regulatory compliance are critical steps (Bai et al., 2022; Wang et al., 2023b).

6.7.6 Opportunities

6.7.6.1 Renewable Energy Production

Biogas is considered one of the most promising renewable energy alternatives, mainly consisting of methane and carbon dioxide, and food waste valorization on biogas production offers several opportunities. Biogas can be used in electricity production and transportation fuel by methane upgrade, and this can facilitate reliance on conventional fossil fuel utilization (Iqbal et al., 2014). Overall, biogas production from food waste offers a multifaceted approach to addressing and contributing to the transition to more sustainable and renewable energy in a globalized future (Mahmudul et al., 2022; Negri et al., 2020).

6.7.6.2 Waste Management and Utilization

Food waste management and utilization through biogas production is gaining attention, particularly due to the most efficient waste-handling route to contribute to environmental sustainability and resource recovery (Ferdeș et al., 2022). Methane emission during the decomposition of organic wastes at landfills and incineration facilities can be eliminated by biogas production facilities and the utilization of methane. Ultimately, biogas production can encourage waste reduction efforts and help stakeholders, including municipalities, businesses, and consumers, to meet waste reduction goals more efficiently (Ajay et al., 2021; Eriksson et al., 2016; Woon & Lo, 2016).

6.7.6.3 Digestate as Biological Fertilizer

Digestate is the nutrient-rich end product of the anaerobic digestion process in biogas production from food waste. It is mainly composed of unconverted organic matter by biogas-producing microorganisms (Stoknes et al., 2016). It provides several application areas and opportunities as a biofertilizer into biogas, along with the nutrients and minerals present in the original feedstock. Digestate offers several valuable opportunities as a source of nitrogen (N), phosphorus (P), and potassium (K), as well as micronutrients. Thus, digestate is a promising candidate for replacing synthetic fertilizers to contribute to the carbon balance, water retention, and nutrient-holding capacity of soil (Lee et al., 2020; Lu & Xu, 2021; Song et al., 2021). However, one of the most important benefits of using digestate as a fertilizer is promoting sustainable agriculture by reducing reliance on non-renewable resources and closing the circular economy cycle by returning nutrients and organic matter to the soil (Xiao et al., 2022).

6.7.6.4 Circular Economy and GHG Emission Reduction

Using biogas instead of conventional fossil fuels is crucial for reducing greenhouse gas emissions. Biofuels can complete the carbon cycle within a significantly shorter time compared to non-renewable sources (Devi et al., 2023; Yong et al., 2021). Therefore, utilizing food waste in biogas production offers an environmentally friendly and economically viable solution for waste management and gas emission prevention. Overall, circular economy approaches and solutions often result in a range of positive externalities beyond GHG emission reduction, including cleaner air, better water quality, and enhanced biodiversity (Bowman et al., 2022; Kapoor et al., 2020).

6.7.6.5 Lower-Cost Potential through Dry Digestion

One of the promising approaches to constructing an economically feasible industrial-scale biogas production plant is the use of dry digestion. Dry digestion is a type of anaerobic digestion process that operates with relatively higher solid content (more than 15% w/w) in the feedstock compared to traditional wet digestion (Jiang et al., 2018; Matheri et al., 2018; Xiao et al., 2022). This approach presents a number of significant opportunities for lower potential operational and capital costs in biogas production from food waste due to lower volumes and water requirements. These advantages are reduced water usage and handling expenditures, lower energy consumption, lower need for dilution, fewer equipment and infrastructural requirements, lower transportation expenses, more straightforward feedstock handling, shorter retention times, enhanced biogas composition, flexibility in facility site, cost-effective biogas upgrading, and applicability for certain feedstocks (Maragkaki et al., 2018; Muhammad Nasir et al., 2012; Seo et al., 2021).

6.7.6.6 Process Integration for Bioproducts

Herein, process integration for bioproducts refers to the concept of maximizing the value that can be derived from food waste by incorporating additional processes beyond biogas production. Several value-added products, including bio-based chemicals, biofuels, bio-plastics, and enzymes can be produced as further revenue streams (Battista et al., 2020). Also, it is possible to evaluate the waste stream in this type of integration. Ultimately, delivering multiple products from a single feedstock can improve the economic viability of the process and facilitate the efficient usage of resources by minimizing waste (Atabani et al., 2022).

6.8 CONCLUSIONS AND PERSPECTIVES

Today, as fossil energy becomes increasingly scarce, it is imperative to develop renewable energy sources in order to address the energy dilemma. Furthermore, the rapid pace of industrialization and urbanization exacerbates the problem of food waste and the generation of other types of waste. The recovery of these resources not only serves as an alternative energy source but also presents an effective solution for waste management, especially food waste. The AD process stands out as a viable method for generating clean energy in the form of biogas. By employing AD, valuable resources in food waste can be recovered as high-value added materials such as biogas, fertilizer, and bioactive compounds. However, traditional AD suffers from drawbacks like low gas production rates and lengthy digestion times, which limit its widespread adoption. Consequently, investigating the highly efficient and cost-effective technologies for AD has emerged as an exciting avenue for research.

REFERENCES

Abomohra, A., Faisal, S., Ebaid, R., Huang, J., Wang, Q., Elsayed, M. 2022. Recent advances in anaerobic digestion of lipid-rich waste: Challenges and potential of seaweeds to mitigate the inhibitory effect. *Chemical Engineering Journal*, **449**, 137829.

Agrawal, A., Chaudhari, P.K., Ghosh, P. 2023. Anaerobic digestion of fruit and vegetable waste: A critical review of associated challenges. *Environmental Science and Pollution Research*, **30**(10), 24987–25012.

Aguilar, M.C., Wang, Y.D., Roskilly, T., Pathare, P.B., Lamidi, R.O. 2017. Biogas from anaerobic co-digestion of food waste and primary sludge for cogeneration of power and heat. *Energy Procedia*, **142**, 70–76.

Ajay, C.M., Mohan, S., Dinesha, P. 2021. Decentralized energy from portable biogas digesters using domestic kitchen waste: a review. *Waste Management*, **125**, 10–26.

Alagöz, B.A., Yenigün, O., Erdinçler, A. 2015. Enhancement of anaerobic digestion efficiency of wastewater sludge and olive waste: Synergistic effect of co-digestion and ultrasonic/microwave sludge pre-treatment. *Waste Management*, **46**, 182–188.

Alavi-Borazjani, S.A., Capela, I., Tarelho, L.A.C. 2020. Over-acidification control strategies for enhanced biogas production from anaerobic digestion: A review. *Biomass and Bioenergy*, **143**, 105833.

Albano, M., Madeira, L.M., Miguel, C.V. 2023. Use of Pd-Ag membrane reactors for low-temperature dry reforming of biogas—A simulation study. *Membranes*, **13**(7), 630.

Al-Wahaibi, A., Osman, A., Al-Muhtaseb, A., Alqaisi, O., Baawain, M., Fawzy, S., Rooney, D. 2020. Techno-economic evaluation of biogas production from food waste via anaerobic digestion. *Scientific Reports*, **10**, 15719.

Atabani, A.E., Tyagi, V.K., Fongaro, G., Treichel, H., Pugazhendhi, A., Hoang, A.T. 2022. Integrated biorefineries, circular bio-economy, and valorization of organic waste streams with respect to bio-products. *Biomass Conversion and Biorefinery*, **12**(3), 565–565.

Atelge, M.R., Krisa, D., Kumar, G., Eskicioglu, C., Nguyen, D.D., Chang, S.W., Atabani, A.E., Al-Muhtaseb, A.H., Unalan, S. 2020. Biogas production from organic waste: Recent progress and perspectives. *Waste and Biomass Valorization*, **11**(3), 1019–1040.

Bai, D., Jain, V., Tripathi, M., Ali, S.A., Shabbir, M.S., Mohamed, M.A.A., Ramos-Meza, C.S. 2022. Performance of biogas plant analysis and policy implications: evidence from the commercial sources. *Energy Policy*, **169**, 113173.

Bareha, Y., Faucher, J.-P., Michel, M., Houdon, M., Vaneeckhaute, C. 2022. Evaluating the impact of substrate addition for anaerobic co-digestion on biogas production and digestate quality: The case of deinking sludge. *Journal of Environmental Management*, **319**, 115657.

Battista, F., Frison, N., Pavan, P., Cavinato, C., Gottardo, M., Fatone, F., Eusebi, A.L., Majone, M., Zeppilli, M., Valentino, F., Fino, D., Tommasi, T., Bolzonella, D. 2020. Food wastes and sewage sludge as feedstock for an urban biorefinery producing biofuels and added-value bioproducts. *Journal of Chemical Technology & Biotechnology*, **95**(2), 328–338.

Benti, N.E., Asfaw, A.A. 2022. Challenges and solutions in biogas technology adoption in ethiopia: A mini review. *Ethiopian Journal of Science and Sustainable Development*, **9**(2), 78–95.

Bhajani, S.S., Pal, S.L. 2022. Review: Factors affecting biogas production. *International Journal for Research in Applied Science and Engineering Technology*, **10**(11), 78–88.

Borowski, S., Cieciura-Włoch, W. 2021. Enzymatic pretreatment of byproducts from soapstock splitting and glycerol processing for improvement of biogas production. *Molecules*, **26**(22), 6782.

Bowman, G., Ayed, L., Burg, V. 2022. Material and energy flows of industrial biogas plants in Switzerland in the context of the circular economy. *Bioresource Technology Reports*, **20**, 101273.

Braguglia, C.M., Gallipoli, A., Gianico, A., Pagliaccia, P. 2018. Anaerobic bioconversion of food waste into energy: A critical review. *Bioresource Technology*, **248**, 37–56.

Buitrón, G., Kumar, G., Martinez-Arce, A., Moreno, G. 2014. Hydrogen and methane production via a two-stage processes (H2-SBR + CH4-UASB) using tequila vinasses. *International Journal of Hydrogen Energy*, **39**(33), 19249–19255.

Cai, Y., Gallegos, D., Zheng, Z., Stinner, W., Wang, X., Pröter, J., Schäfer, F. 2021a. Exploring the combined effect of total ammonia nitrogen, pH and temperature on anaerobic digestion of chicken manure using response surface methodology and two kinetic models. *Bioresource Technology*, **337**, 125328.

Cai, Y., Zheng, Z., Schaefer, F., Stinner, W., Yuan, X., Wang, H., Cui, Z., Wang, X. 2021b. A review about pretreatment of lignocellulosic biomass in anaerobic digestion: Achievement and challenge in Germany and China. *Journal of Cleaner Production*, **299**, 126885.

Cerón-Vivas, A., Cáceres, K.T., Rincón, A., Cajigas, Á. 2019. Influence of pH and the C/N ratio on the biogas production of wastewater. *Revista Facultad de Ingeniería Universidad de Antioquia,* **92**, 70–79.

Cerrillo, M., Viñas, M., Bonmatí, A. 2018. Anaerobic digestion and electromethanogenic microbial electrolysis cell integrated system: Increased stability and recovery of ammonia and methane. *Renewable Energy*, **120**, 178–189.

Chen, H., Jiang, W., Yang, Y., Yang, Y., Man, X. 2017. State of the art on food waste research: A bibliometrics study from 1997 to 2014. *Journal of Cleaner Production*, **140**, 840–846.

Cheong, W.L., Chan, Y.J., Tiong, T.J., Chong, W.C., Kiatkittipong, W., Kiatkittipong, K., Mohamad, M., Daud, H., Suryawan, I.W.K., Sari, M.M. 2022. Anaerobic co-digestion of food waste with sewage sludge: Simulation and optimization for maximum biogas production. *Water*, **14**(7), 1075.

Cherukuri, N.R., Parthasarathy, P. 2023. Impact of sequential hybrid pretreatment in anaerobic digestion of food waste and garden waste co-digestion on waste characteristics and biogas production. *Journal of Material Cycles and Waste Management*, **25**, 1–14.

Chidi, B.S., Okudoh, V.I., Hutchinson, U.F., Ngongang, M.M., Maphanga, T., Madonsela, B.S., Shale, K., Lim, J.W., Ntwampe, S.K.O. 2022. A perspective on emerging inter-disciplinary solutions for the sustainable management of food waste. *Applied Sciences*, **12**(22), 11399.

Choudhary, A., Kumar, A., Kumar, S. 2020. Techno-economic analysis, kinetics, global warming potential comparison and optimization of a pilot-scale unheated semi-continuous anaerobic reactor in a hilly area: For north Indian hilly states. *Renewable Energy*, **155**, 1181–1190.

Choudhary, A., Kumar, A., Kumar, S., Govil, T., Sani, R.K. 2023. Enhanced biogas production from pine litter codigestion with food waste, microbial community, kinetics, and technoeconomic feasibility. *Journal of Environmental Engineering*, **149**(1), 04022089.

Czekała, W., Nowak, M., Bojarski, W. 2023. Characteristics of substrates used for biogas production in terms of water content. *Fermentation*, **9**(5), 449.

Dahiya, S., Sarkar, O., Swamy, Y., Mohan, S.V. 2015. Acidogenic fermentation of food waste for volatile fatty acid production with co-generation of biohydrogen. *Bioresource Technology*, **182**, 103–113.

da Silva, G.H., Barros, N.O., Santana, L.A.R., Carneiro J.d.C., Otenio, M.H. 2021. Shifts of acidogenic bacterial group and biogas production by adding two industrial residues in anaerobic co-digestion with cattle manure. *Journal of Environmental Science and Health, Part A*, **56**(14), 1503–1511.

de Abreu, Í.B.S., de Sousa, M.H., da Silva, A.P., de Araújo Padilha, C.E., Sales, A.T., da Silva, A.S.A., Dutra, E.D., Menezes, R.S.C. 2023. Global variability of food waste chemical composition and its consequences on the production of biofuels and chemical compounds. *Journal of Material Cycles and Waste Management*, **25**(3), 1309–1324.

Deena, S.R., Vickram, A., Manikandan, S., Subbaiya, R., Karmegam, N., Ravindran, B., Chang, S.W., Awasthi, M.K. 2022. Enhanced biogas production from food waste and activated sludge using advanced techniques–a review. *Bioresource Technology*, **355**, 127234.

Deepanraj, B., Sivasubramanian, V., Jayaraj, S. 2017. Effect of substrate pretreatment on biogas production through anaerobic digestion of food waste. *International Journal of Hydrogen Energy*, **42**(42), 26522–26528.

Demarco, C.F., Afonso, T.F., Pieniz, S., Quadro, M.S., Camargo, F.A.d.O., Andreazza, R. 2019. Phytoremediation of heavy metals and nutrients by the Sagittaria montevidensis into an anthropogenic contaminated site at Southern of Brazil. *International Journal of Phytoremediation*, **21**(11), 1145–1152.

Detman, A., Bucha, M., Treu, L., Chojnacka, A., Pleśniak, Ł., Salamon, A., Łupikasza, E., Gromadka, R., Gawor, J., Gromadka, A. 2021. Evaluation of acidogenesis products' effect on biogas production performed with metagenomics and isotopic approaches. *Biotechnology for Biofuels*, **14**(1), 125.

Deublein, D., Steinhauser, A. 2011. *Biogas from waste and renewable resources: An introduction*. Mörlenbach, Germany: John Wiley & Sons.

Devi, M.K., Manikandan, S., Kumar, P.S., Yaashikaa, P.R., Oviyapriya, M., Rangasamy, G. 2023. A comprehensive review on current trends and development of biomethane production from food waste: Circular economy and techno economic analysis. *Fuel*, **351**, 128963.

Dhungana, B., Lohani, S.P., Marsolek, M. 2022. Anaerobic co-digestion of food waste with livestock manure at ambient temperature: A biogas based circular economy and sustainable development goals. *Sustainability*, **14**(6), 3307.

Donoso-Bravo, A., Toledo-Alarcón, J., Ortega, V., Barría, V., Lesty, Y., Fontana, J., Bossche, H.V. 2021. New findings on the anaerobic co-digestion of thermally pretreated sludge and food waste: Laboratory and pilot-scale studies. *Water Science and Technology*, **84**(9), 2530–2540.

Dos Santos, L.A., Silva, T.H.L., de Melo Oliveira, C.R., Jucá, J.F.T., Santos, A.F.d.M.S. 2022. Silage as a pretreatment of orange bagasse waste to increase the potential for methane generation. *Science of The Total Environment*, **823**, 153613.

Du, N., Li, M., Zhang, Q., Ulsido, M.D., Xu, R., Huang, W. 2021. Study on the biogas potential of anaerobic digestion of coffee husks wastes in ethiopia. *Waste Management & Research*, **39**(2), 291–301.

Duan, N., Zhang, D., Lin, C., Zhang, Y., Zhao, L., Liu, H., Liu, Z. 2019. Effect of organic loading rate on anaerobic digestion of pig manure: Methane production, mass flow, reactor scale and heating scenarios. *Journal of Environmental Management*, **231**, 646–652.

Dubrovskis, V., Plume, I. 2017. Biogas from wastes of pumpkin, marrow and apple. *Agronomy Research*, **15**(1), 69–78.

Duong, C.M., Lim, T.-T. 2023. Use of regression models for development of a simple and effective biogas decision-support tool. *Scientific Reports*, **13**(1), 4933.

Edjabou, M.E., Petersen, C., Scheutz, C., Astrup, T.F. 2016. Food waste from Danish households: Generation and composition. *Waste Management*, **52**, 256–268.

Ekstrand, E.-M., Björn, A., Karlsson, A., Schnürer, A., Kanders, L., Yekta, S.S., Karlsson, M., Moestedt, J. 2022. Identifying targets for increased biogas production through chemical and organic matter characterization of digestate from full-scale biogas plants: What remains and why? *Biotechnology for Biofuels and Bioproducts*, **15**(1), 1–22.

El Gnaoui, Y., Karouach, F., Bakraoui, M., Barz, M., El Bari, H. 2020. Mesophilic anaerobic digestion of food waste: Effect of thermal pretreatment on improvement of anaerobic digestion process. *Energy Reports*, **6**, 417–422.

Eriksson, O., Bisaillon, M., Haraldsson, M., Sundberg, J. 2016. Enhancement of biogas production from food waste and sewage sludge – Environmental and economic life cycle performance. *Journal of Environmental Management*, **175**, 33–39.

Fei, X., Chen, T., Jia, W., Shan, Q., Hei, D., Ling, Y., Feng, J., Feng, H. 2020. Enhancement effect of ionizing radiation pretreatment on biogas production from anaerobic fermentation of food waste. *Radiation Physics and Chemistry*, **168**, 108534.

Ferdeș, M., Zăbavă, B.Ș., Paraschiv, G., Ionescu, M., Dincă, M.N., Moiceanu, G. 2022. Food waste management for biogas production in the context of sustainable development. *Energies*, **15**(17), 6268.

Ferreira, L.O., Astals, S., Passos, F. 2022. Anaerobic co-digestion of food waste and microalgae in an integrated treatment plant. *Journal of Chemical Technology & Biotechnology*, **97**(6), 1545–1554.

Flanagan, A., Priyadarshini, A. 2021. A study of consumer behaviour towards food-waste in Ireland: Attitudes, quantities and global warming potentials. *Journal of Environmental Management*, **284**, 112046.

Font Pomarol, J. 2022. Assessment and design of a vermifilter for the post-treatment of digestate from low-tech digesters implemented in rural areas of Colombia, Universitat Politècnica de Catalunya. http://dx.doi.org/10.2139/ssrn.4446980

Food, Organization, A. 2015. *Food wastage footprint & climate change*. Italy: FAO Rome.

Fuentes, K.L., Torres–Lozada, P., Chaparro, T.R. 2021. Beverage wastewater treatment by anaerobic digestion in two-stages for organic matter removal and energy production. *Biomass and Bioenergy*, **154**, 106260.

Gao, T., Zhang, H., Xu, X., Teng, J. 2021. Integrating microbial electrolysis cell based on electrochemical carbon dioxide reduction into anaerobic osmosis membrane reactor for biogas upgrading. *Water Research*, **190**, 116679.

García-Depraect, O., Martínez-Mendoza, L.J., Diaz, I., Muñoz, R. 2022. Two-stage anaerobic digestion of food waste: Enhanced bioenergy production rate by steering lactate-type fermentation during hydrolysis-acidogenesis. *Bioresource Technology*, **358**, 127358.

Goswami, R., Chattopadhyay, P., Shome, A., Banerjee, S.N., Chakraborty, A.K., Mathew, A.K., Chaudhury, S. 2016. An overview of physico-chemical mechanisms of biogas production by microbial communities: A step towards sustainable waste management. *3 Biotech*, **6**(1), 72.

Govarthanan, M., Manikandan, S., Subbaiya, R., Krishnan, R.Y., Srinivasan, S., Karmegam, N., Kim, W. 2022. Emerging trends and nanotechnology advances for sustainable biogas production from lignocellulosic waste biomass: A critical review. *Fuel*, **312**, 122928.

Grosser, A., Neczaj, E., Singh, B., Almås, Å., Brattebø, H., Kacprzak, M. 2017. Anaerobic digestion of sewage sludge with grease trap sludge and municipal solid waste as co-substrates. *Environmental research*, **155**, 249–260.

Guo, Q., Majeed, S., Xu, R., Zhang, K., Kakade, A., Khan, A., Hafeez, F.Y., Mao, C., Liu, P., Li, X. 2019. Heavy metals interact with the microbial community and affect biogas production in anaerobic digestion: A review. *Journal of environmental management*, **240**, 266–272.

Harirchi, S., Wainaina, S., Sar, T., Nojoumi, S.A., Parchami, M., Parchami, M., Varjani, S., Khanal, S.K., Wong, J., Awasthi, M.K. 2022. Microbiological insights into anaerobic digestion for biogas, hydrogen or volatile fatty acids (VFAs): A review. *Bioengineered*, **13**(3), 6521–6557.

Iqbal, S.A., Rahaman, S., Rahman, M., Yousuf, A. 2014. Anaerobic Digestion of Kitchen Waste to Produce Biogas. *Procedia Engineering*, **90**, 657–662.

Jaman, K., Idrus, S., Wahab, A.M.A., Harun, R., Daud, N.N.N., Ahsan, A., Shams, S., Uddin, M.A. 2023. Influence of molasses residue on treatment of cow manure in an anaerobic filter with perforated weed membrane and a conventional reactor: Variations of organic loading and a machine learning application. *Membranes*, **13**(2), 159.

Jenicek, P., Keclik, F., Maca, J., Bindzar, J. 2008. Use of microaerobic conditions for the improvement of anaerobic digestion of solid wastes. *Water Science and Technology*, **58**(7), 1491–1496.

Jiang, Y., Dennehy, C., Lawlor, P.G., Hu, Z., McCabe, M., Cormican, P., Zhan, X., Gardiner, G.E. 2018. Inhibition of volatile fatty acids on methane production kinetics during dry co-digestion of food waste and pig manure. *Waste Management*, **79**, 302–311.

Kabaivanova, L., Petrova, P., Hubenov, V., Simeonov, I. 2022. Biogas production potential of thermophilic anaerobic biodegradation of organic waste by a microbial consortium identified with metagenomics. *Life*, **12**(5), 702.

Kalaiselvan, N., Glivin, G., Bakthavatsalam, A., Mariappan, V., Premalatha, M., Raveendran, P.S., Jayaraj, S., Sekhar, S.J. 2022. A waste to energy technology for Enrichment of biomethane generation: A review on operating parameters, types of biodigesters, solar assisted heating systems, socio economic benefits and challenges. *Chemosphere*, **293**, 133486.

Kannah, R.Y., Merrylin, J., Devi, T.P., Kavitha, S., Sivashanmugam, P., Kumar, G., Banu, J.R. 2020. Food waste valorization: Biofuels and value added product recovery. *Bioresource Technology Reports*, **11**, 100524.

Kapoor, R., Ghosh, P., Kumar, M., Sengupta, S., Gupta, A., Kumar, S.S., Vijay, V., Kumar, V., Kumar Vijay, V., Pant, D. 2020. Valorization of agricultural waste for biogas based circular economy in India: A research outlook. *Bioresource Technology*, **304**, 123036.

Karki, R., Chuenchart, W., Surendra, K., Shrestha, S., Raskin, L., Sung, S., Hashimoto, A., Khanal, S.K. 2021. Anaerobic co-digestion: Current status and perspectives. *Bioresource Technology*, **330**, 125001.

Kasinath, A., Fudala-Ksiazek, S., Szopinska, M., Bylinski, H., Artichowicz, W., Remiszewska-Skwarek, A., Luczkiewicz, A. 2021. Biomass in biogas production: Pretreatment and codigestion. *Renewable and Sustainable Energy Reviews*, **150**, 111509.

Katuri, K.P., Ali, M., Saikaly, P.E. 2019. The role of microbial electrolysis cell in urban wastewater treatment: integration options, challenges, and prospects. *Current Opinion in Biotechnology*, **57**, 101–110.

Khan, A., Akbar, S., Okonkwo, V., Smith, C., Khan, S., Shah, A.A., Adnan, F., Ijaz, U.Z., Ahmed, S., Badshah, M. 2022. Enrichment of the hydrogenotrophic methanogens for, in-situ biogas up-gradation by recirculation of gases and supply of hydrogen in methanogenic reactor. *Bioresource Technology*, **345**, 126219.

Khanthong, K., Kadam, R., Kim, T., Park, J. 2023. Synergetic effects of anaerobic co-digestion of food waste and algae on biogas production. *Bioresource Technology*, **382**, 129208.

Kiran, E.U., Trzcinski, A.P., Liu, Y. 2015. Enhancing the hydrolysis and methane production potential of mixed food waste by an effective enzymatic pretreatment. *Bioresource Technology*, **183**, 47–52.

Koupaie, E.H., Azizi, A., Lakeh, A.B., Hafez, H., Elbeshbishy, E. 2019. A comprehensive dataset on anaerobic digestion of cattle manure, source separated organics, and municipal sludge using different inoculum sources. *Data in brief*, **24**, 103913.

Kumar, G., Sivagurunathan, P., Park, J.-H., Kim, S.-H. 2016. Anaerobic digestion of food waste to methane at various organic loading rates (OLRs) and hydraulic retention times (HRTs): Thermophilic vs. mesophilic regimes. *Environmental Engineering Research*, **21**(1), 69–73.

Kumari, P., Varma, A.K., Shankar, R., Thakur, L.S., Mondal, P. 2021. Phycoremediation of wastewater by Chlorella pyrenoidosa and utilization of its biomass for biogas production. *Journal of Environmental Chemical Engineering*, **9**(1), 104974.

Kuo, J., Dow, J. 2017. Biogas production from anaerobic digestion of food waste and relevant air quality implications. *Journal of the Air & Waste Management Association*, **67**(9), 1000–1011.

Lamolinara, B., Pérez-Martínez, A., Guardado-Yordi, E., Fiallos, C.G., Diéguez-Santana, K., Ruiz-Mercado, G.J. 2022. Anaerobic digestate management, environmental impacts, and techno-economic challenges. *Waste Management*, **140**, 14–30.

Lee, M.E., Steiman, M.W., St. Angelo, S.K. 2020. Biogas digestate as a renewable fertilizer: Effects of digestate application on crop growth and nutrient composition. *Renewable Agriculture and Food Systems*, **36**(2), 173–181.

Le Pera, A., Sellaro, M., Sicilia, F., Ciccoli, R., Sceberras, B., Freda, C., Fanelli, E., Cornacchia, G. 2023. Environmental and economic impacts of improper materials in the recycling of separated collected food waste through anaerobic digestion and composting. *Science of The Total Environment*, **880**, 163240.

Li, R., Chen, S., Li, X. 2010. Biogas production from anaerobic co-digestion of food waste with dairy manure in a two-phase digestion system. *Applied Biochemistry and Biotechnology*, **160**(2), 643–654.

Li, Y., Chen, Z., Peng, Y., Huang, W., Liu, J., Mironov, V., Zhang, S. 2022a. Deeper insights into the effects of substrate to inoculum ratio selection on the relationship of kinetic parameters, microbial communities, and key metabolic pathways during the anaerobic digestion of food waste. *Water Research*, **217**, 118440.

Li, Y., Jin, Y. 2015. Effects of thermal pretreatment on acidification phase during two-phase batch anaerobic digestion of kitchen waste. *Renewable Energy*, **77**, 550–557.

Li, Y., Jing, Z., Pan, J., Luo, G., Feng, L., Jiang, H., Zhou, H., Xu, Q., Lu, Y., Liu, H. 2022b. Multi-omics joint analysis of the effect of temperature on microbial communities, metabolism, and genetics in full-scale biogas reactors with food waste. *Renewable and Sustainable Energy Reviews*, **160**, 112261.

Lianhua, L., Shuibin, H., Yongming, S., Xihui, K., Junfeng, J., Zhenhong, Y., Dingfa, L. 2020. Anaerobic co-digestion of Pennisetum hybrid and pig manure: A comparative study of performance and microbial community at different mixture ratio and organic loading rate. *Chemosphere*, **247**, 125871.

Lin, C.S.K., Pfaltzgraff, L.A., Herrero-Davila, L., Mubofu, E.B., Abderrahim, S., Clark, J.H., Koutinas, A.A., Kopsahelis, N., Stamatelatou, K., Dickson, F. 2013. Food waste as a valuable resource for the production of chemicals, materials and fuels. Current situation and global perspective. *Energy & Environmental Science*, **6**(2), 426–464.

Linyi, C., Yujie, Q., Buqing, C., Chenglong, W., Shaohong, Z., Renglu, C., Shaohua, Y., Lan, Y., Zhiju, L. 2020. Enhancing degradation and biogas production during anaerobic digestion of food waste using alkali pretreatment. *Environmental Research*, **188**, 109743.

Lisowyj, M., Wright, M.M. 2020. A review of biogas and an assessment of its economic impact and future role as a renewable energy source. *Reviews in Chemical Engineering*, **36**(3), 401–421.

Liu, J., Zhao, M., Lv, C., Yue, P. 2020. The effect of microwave pretreatment on anaerobic co-digestion of sludge and food waste: Performance, kinetics and energy recovery. *Environmental Research*, **189**, 109856.

López-Sandin, I., Rodríguez-Jasso, R.M., Ruiz, H.A. 2023. Energy integration of the hydrothermal pretreatment of food waste in terms of a sustainable biorefinery. In: Aguilar Gonzalez, C.N., Gómez-García, R., Kuddus, M. (Eds.). Food Waste Conversion. Methods and Protocols in Food Science. Humana, New York. https://doi.org/10.1007/978-1-0716-3303-8_9.

Lu, J., Xu, S. 2021. Post-treatment of food waste digestate towards land application: A review. *Journal of Cleaner Production*, **303**, 127033.

Luo, C., He, T., Wang, G., Tian, M., Dai, L., Pu, T., Tian, G. 2023. Up-flow anaerobic sludge blanket treatment of swine wastewater: Effect of heterologous and homologous inocula on anaerobic digestion performance and the microbial community. *Bioresource Technology*, **386**, 129463.

Mahmudul, H.M., Rasul, M.G., Akbar, D., Narayanan, R., Mofijur, M. 2022. Food waste as a source of sustainable energy: Technical, economical, environmental and regulatory feasibility analysis. *Renewable and Sustainable Energy Reviews*, **166**, 112577.

Mahyari, Z.F., Khorasanizadeh, Z., Khanali, M., Mahyari, K.F. 2021. Biodiesel production from slaughter wastes of broiler chicken: A potential survey in Iran. *SN Applied Sciences*, **3**(1), 57.

Malave, A.C.L., Bernardi, M., Fino, D., Ruggeri, B. 2015. Multistep anaerobic digestion (MAD) as a tool to increase energy production via H2+ CH4. *International Journal of Hydrogen Energy*, **40**(15), 5050–5061.

Maragkaki, A.E., Vasileiadis, I., Fountoulakis, M., Kyriakou, A., Lasaridi, K., Manios, T. 2018. Improving biogas production from anaerobic co-digestion of sewage sludge with a thermal dried mixture of food waste, cheese whey and olive mill wastewater. *Waste Management*, **71**, 644–651.

Matheri, A.N., Sethunya, V.L., Belaid, M., Muzenda, E. 2018. Analysis of the biogas productivity from dry anaerobic digestion of organic fraction of municipal solid waste. *Renewable and Sustainable Energy Reviews*, **81**, 2328–2334.

Meng, Y., Luan, F., Yuan, H., Chen, X., Li, X. 2017. Enhancing anaerobic digestion performance of crude lipid in food waste by enzymatic pretreatment. *Bioresource Technology*, **224**, 48–55.

Menon, A., Ren, F., Wang, J.-Y., Giannis, A. 2016. Effect of pretreatment techniques on food waste solubilization and biogas production during thermophilic batch anaerobic digestion. *Journal of Material Cycles and Waste Management*, **18**, 222–230.

Mills, S., Nguyen, T.P.Y., Ijaz, U.Z., Lens, P.N. 2023. Process stability in expanded granular sludge bed bioreactors enhances resistance to organic load shocks. *Journal of Environmental Management*, **342**, 118271.

Mirmohamadsadeghi, S., Karimi, K., Tabatabaei, M., Aghbashlo, M. 2019. Biogas production from food wastes: A review on recent developments and future perspectives. *Bioresource Technology Reports*, **7**, 100202.

Muhammad Nasir, I., Mohd Ghazi, T.I., Omar, R. 2012. Production of biogas from solid organic wastes through anaerobic digestion: A review. *Applied Microbiology and Biotechnology*, **95**(2), 321–329.

Muñiz-Márquez, D.B., Cano-Gómez, C.I., Wong-Paz, J.E., Balderas-Hernández, V.E., Veana, F. 2023. Microbial biogas production: Challenges and opportunities. In S.P. Singh and S.K. Upadhyay (Eds.). *Microbial Bioreactors for Industrial Molecules*, Chandigarh India. https://doi.org/10.1002/9781119874096.ch16.

Mutungwazi, A., Awosusi, A., Matambo, T.S. 2023. Comparative functional microbiome profiling of various animal manures during their anaerobic digestion in biogas production processes. *Biomass and Bioenergy*, **170**, 106728.

Nakasaki, K., Nguyen, K.K., Ballesteros Jr, F.C., Maekawa, T., Koyama, M. 2020. Characterizing the microbial community involved in anaerobic digestion of lipid-rich wastewater to produce methane gas. *Anaerobe*, **61**, 102082.

Naran, E., Toor, U.A., Kim, D.-J. 2016. Effect of pretreatment and anaerobic co-digestion of food waste and waste activated sludge on stabilization and methane production. *International Biodeterioration & Biodegradation*, **113**, 17–21.

Negri, C., Ricci, M., Zilio, M., D'Imporzano, G., Qiao, W., Dong, R., Adani, F. 2020. Anaerobic digestion of food waste for bio-energy production in China and Southeast Asia: A review. *Renewable and Sustainable Energy Reviews*, **133**, 110138.

Nemestóthy, N., Bakonyi, P., Szentgyörgyi, E., Kumar, G., Nguyen, D.D., Chang, S.W., Kim, S.-H., Bélafi-Bakó, K. 2018. Evaluation of a membrane permeation system for biogas upgrading using model and real gaseous mixtures: The effect of operating conditions on separation behaviour, methane recovery and process stability. *Journal of Cleaner Production*, **185**, 44–51.

Ning, X., Lin, R., O'Shea, R., Wall, D., Deng, C., Wu, B., Murphy, J.D. 2021. Emerging bioelectrochemical technologies for biogas production and upgrading in cascading circular bioenergy systems. *Iscience*, **24**(9). https://doi.org/10.1016/j.isci.2021.102998

Novakovic, J., Kontogianni, N., Barampouti, E.M., Mai, S., Moustakas, K., Malamis, D., Loizidou, M. 2021. Towards upscaling the valorization of wheat straw residues: Alkaline pretreatment using sodium hydroxide, enzymatic hydrolysis and biogas production. *Environmental Science and Pollution Research*, **28**, 24486–24498.

Obaideen, K., Abdelkareem, M.A., Wilberforce, T., Elsaid, K., Sayed, E.T., Maghrabie, H.M., Olabi, A. 2022. Biogas role in achievement of the sustainable development goals: Evaluation, challenges, and guidelines. *Journal of the Taiwan Institute of Chemical Engineers*, **131**, 104207.

Ofili, I., Al Ali, O., Emeka, E., Ugwoke, E., Eze, E. 2020. Biogas production from anaerobic digestion of food waste generated in Enugu Metropolis. *Pacific Journal of Science and Technology*, **21**(1), 30–37.

Olatunji, K.O., Ahmed, N.A., Ogunkunle, O. 2021. Optimization of biogas yield from lignocellulosic materials with different pretreatment methods: A review. *Biotechnology for Biofuels*, **14**(1), 1–34.

Omar, B., Abou-Shanab, R., El-Gammal, M., Fotidis, I.A., Kougias, P.G., Zhang, Y., Angelidaki, I. 2018. Simultaneous biogas upgrading and biochemicals production using anaerobic bacterial mixed cultures. *Water Research*, **142**, 86–95.

Otieno, E.O., Kiplimo, R., Mutwiwa, U. 2023. Optimization of anaerobic digestion parameters for biogas production from pineapple wastes co-digested with livestock wastes. *Heliyon*, **9**(3). https://doi.org/10.1016/j.heliyon.2023.e14041

Palaniveloo, K., Amran, M.A., Norhashim, N.A., Mohamad-Fauzi, N., Peng-Hui, F., Hui-Wen, L., Kai-Lin, Y., Jiale, L., Chian-Yee, M.G., Jing-Yi, L. 2020. Food waste composting and microbial community structure profiling. *Processes*, **8**(6), 723.

Panigrahi, S., Sharma, H.B., Dubey, B.K. 2020. Anaerobic co-digestion of food waste with pretreated yard waste: A comparative study of methane production, kinetic modeling and energy balance. *Journal of Cleaner Production*, **243**, 118480.

Paranjpe, A., Saxena, S., Jain, P. 2023. A review on performance improvement of anaerobic digestion using co-digestion of food waste and sewage sludge. *Journal of Environmental Management*, **338**, 117733.

Park, J.-H., Kumar, G., Yun, Y.-M., Kwon, J.-C., Kim, S.-H. 2018a. Effect of feeding mode and dilution on the performance and microbial community population in anaerobic digestion of food waste. *Bioresource Technology*, **248**, 134–140.

Park, J.-H, Lee, B., Tian, D., Jun, H. 2018b. Bioelectrochemical enhancement of methane production from highly concentrated food waste in a combined anaerobic digester and microbial electrolysis cell. *Bioresource Technology*, **247**, 226–233.

Pau, S., Tan, L.C., Arriaga, S., Lens, P.N. 2023. Lactic acid fermentation of food waste in a semicontinuous SBR system: Influence of the influent composition and hydraulic retention time: Environmental technology. *Environmental Technology* (Just-accepted), 1–21. https://doi.org/10.1080/09593330.2023.2202824

Pérez-Rodríguez, M., Cano, A., Durán, U., Barrios, J. 2019. Solubilization of organic matter by electrochemical treatment of sludge: Influence of operating conditions. *Journal of Environmental Management*, **236**, 317–322.

Porterfield, K.K., Scarborough, M.J., Roy, E.D. 2023. Organics recycling tradeoffs: Biogas potential and microplastic content of mechanically depackaged food waste. *ACS Sustainable Chemistry & Engineering*, **11**(28), 10422–10429.

Quashie, F.K., Fang, A., Wei, L., Kabutey, F.T., Xing, D. 2021. Prediction of biogas production from food waste in a continuous stirred microbial electrolysis cell (CSMEC) with backpropagation artificial neural network. *Biomass Conversion and Biorefinery*, 1–12. https://doi.org/10.1007/s13399-020-01179-x

Rajesh Banu, J., Merrylin, J., Mohamed Usman, T.M., Yukesh Kannah, R., Gunasekaran, M., Kim, S.-H., Kumar, G. 2020. Impact of pretreatment on food waste for biohydrogen production: A review. *International Journal of Hydrogen Energy*, **45**(36), 18211–18225.

Rasool, K., Hussain, S., Shahzad, A., Miran, W., Mahmoud, K.A., Ali, N., Almomani, F. 2023. Comprehensive insights into sustainable conversion of agricultural and food waste into microbial protein for animal feed production. *Reviews in Environmental Science and Bio/Technology*, 1–36. https://doi.org/10.1007/s11157-023-09651-6

Rodríguez-Jiménez, L.M., Pérez-Vidal, A., Torres-Lozada, P. 2022. Research trends and strategies for the improvement of anaerobic digestion of food waste in psychrophilic temperatures conditions. *Heliyon*. https://doi.org/10.1016/j.heliyon.2022.e11174

Said, Z., Sharma, P., Bora, B.J., Bui, T.A.E., Nguyen, D.T., Dinh, X.T., Nguyen, X.P. 2023. Modeling-optimization of performance and emission characteristics of dual-fuel engine powered with pilot diesel and agricultural-food waste-derived biogas. *International Journal of Hydrogen Energy*, **48**(18), 6761–6777.

Schirmer, W.N., dos Santos, L.A., Martins, K.G., Gueri, M.V.D., Jucá, J.F.T. 2023. The effect of alkaline pretreatment on the anaerobic digestion of fruit and vegetable wastes from a central food distribution market. *Journal of Material Cycles and Waste Management*, 1–13. https://doi.org/10.1007/s10163-023-01722-8

Schnürer, A. 2016. Biogas production: Microbiology and technology. In *Anaerobes in biotechnology* (pp. 195–234). Berlin: Springer. https://doi.org/10.1007/10_2016_5

Seo, K.W., Seo, J., Kim, K., Lim, S.J., Chung, J. 2021. Prediction of biogas production rate from dry anaerobic digestion of food waste: Process-based approach vs. recurrent neural network black-box model. *Bioresource Technology*, **341**, 125829.

Shahbaz, M., Ammar, M., Zou, D., Korai, R.M., Li, X. 2019. An insight into the anaerobic co-digestion of municipal solid waste and food waste: Influence of co-substrate mixture ratio and substrate to inoculum ratio on biogas production. *Applied Biochemistry and Biotechnology*, **187**, 1356–1370.

Shen, F., Yuan, H., Pang, Y., Chen, S., Zhu, B., Zou, D., Liu, Y., Ma, J., Yu, L., Li, X. 2013. Performances of anaerobic co-digestion of fruit & vegetable waste (FVW) and food waste (FW): Single-phase vs. two-phase. *Bioresource Technology*, **144**, 80–85.

Sheng, Q., Lu, Y., Yuan, S., Li, X., Dai, X., Guo, Y., Dong, B. 2023. Effect of nitrite on hydrolysis-acidification, biogas production and microbial community in semi-continuous two-phase anaerobic digestion of sewage sludge. *Journal of Environmental Sciences*, **126**, 434–444.

Singh, B., Kovács, K.L., Bagi, Z., Nyári, J., Szepesi, G.L., Petrik, M., Siménfalvi, Z., Szamosi, Z. 2021. Enhancing efficiency of anaerobic digestion by optimization of mixing regimes using helical ribbon impeller. *Fermentation*, **7**(4), 251.

Singh, B., Szamosi, Z., Siménfalvi, Z. 2019. State of the art on mixing in an anaerobic digester: A review. *Renewable Energy*, **141**, 922–936.

Singh, H., Padhi, T., Kashyap, A., Taneja, S. 2023. Recent advances in biogas production using various bio-waste's and its potential application: An overview. *Materials Today: Proceedings*. http://dx.doi.org/10.1016/j.matpr.2023.04.484

Singh, T., Uppaluri, R.V. 2023. Optimizing biogas production: A novel hybrid approach using anaerobic digestion calculator and machine learning techniques on Indian biogas plant. *Clean Technologies and Environmental Policy*, 1–25. https://doi.org/10.1007/s10098-023-02584-2

Song, S., Lim, J.W., Lee, J.T.E., Cheong, J.C., Hoy, S.H., Hu, Q., Tan, J.K.N., Chiam, Z., Arora, S., Lum, T.Q.H., Lim, E.Y., Wang, C.-H., Tan, H.T.W., Tong, Y.W. 2021. Food-waste anaerobic digestate as a fertilizer: The agronomic properties of untreated digestate and biochar-filtered digestate residue. *Waste Management*, **136**, 143–152.

Srivastava, N., Malik, M.A. 2022. *Food waste to green fuel: Trend & development*. Singapore: Springer.

Stoknes, K., Scholwin, F., Krzesiński, W., Wojciechowska, E., Jasińska, A. 2016. Efficiency of a novel "Food to waste to food" system including anaerobic digestion of food waste and cultivation of vegetables on digestate in a bubble-insulated greenhouse. *Waste Management*, **56**, 466–476.

Suhartini, S., Rohma, N.A., Elviliana, Santoso, I., Paul, R., Listiningrum, P., Melville, L. 2022. Food waste to bioenergy: Current status and role in future circular economies in Indonesia. *Energy, Ecology and Environment*, **7**(4), 297–339.

Sun, C., Meng, X., Sun, F., Zhang, J., Tu, M., Chang, J.-S., Reungsang, A., Xia, A., Ragauskas, A.J. 2022. Advances and perspectives on mass transfer and enzymatic hydrolysis in the enzyme-mediated lignocellulosic biorefinery: A review. *Biotechnology Advances*, 108059. https://doi.org/10.1016/j.biotechadv.2022.108059

Tayyab, A., Ahmad, Z., Mahmood, T., Khalid, A., Qadeer, S., Mahmood, S., Andleeb, S., Anjum, M. 2019. Anaerobic co-digestion of catering food waste utilizing Parthenium hysterophorus as co-substrate for biogas production. *Biomass and Bioenergy*, **124**, 74–82.

Thi, N.B.D., Lin, C.-Y., Kumar, G. 2016. Waste-to-wealth for valorization of food waste to hydrogen and methane towards creating a sustainable ideal source of bioenergy. *Journal of Cleaner Production*, **122**, 29–41.

Tiong, J.S.M., Chan, Y.J., Lim, J.W., Mohamad, M., Ho, C.-D., Rahmah, A.U., Kiatkittipong, W., Sriseubsai, W., Kumakiri, I. 2021. Simulation and optimization of anaerobic co-digestion of food waste with palm oil mill effluent for biogas production. *Sustainability*, **13**(24), 13665.

Viancelli, A., Schneider, T.M., Demczuk, T., Delmoral, A.P., Petry, B., Collato, M.M., Michelon, W. 2023. Unlocking the value of biomass: Exploring microbial strategies for biogas and volatile fatty acids generation. *Bioresource Technology Reports*, 101552. https://doi.org/10.1016/j.biteb.2023.101552

Vigneshwar, S.S., Swetha, A., Gopinath, K.P., Goutham, R., Pal, R., Arun, J., SundarRajan, P., Bhatnagar, A., Chi, N.T.L., Pugazhendhi, A. 2022. Bioprocessing of biowaste derived from food supply chain side-streams for extraction of value added bioproducts through biorefinery approach. *Food and Chemical Toxicology*, **165**, 113184.

Vijayakumar, P., Ayyadurai, S., Arunachalam, K.D., Mishra, G., Chen, W.-H., Juan, J.C., Naqvi, S.R. 2022. Current technologies of biochemical conversion of food waste into biogas production: A review. *Fuel*, **323**, 124321.

Wainaina, S., Lukitawesa, Kumar Awasthi, M., Taherzadeh, M.J. 2019. Bioengineering of anaerobic digestion for volatile fatty acids, hydrogen or methane production: A critical review. *Bioengineered*, **10**(1), 437–458.

Wang, L., Li, Y., Yi, X., Yang, F., Wang, D., Han, H. 2023a. Dissimilatory manganese reduction facilitates synergistic cooperation of hydrolysis, acidogenesis, acetogenesis and methanogenesis via promoting microbial interaction during anaerobic digestion of waste activated sludge. *Environmental Research*, **218**, 114992.

Wang, W., Lee, D.-J., Lei, Z. 2022. Integrating anaerobic digestion with microbial electrolysis cell for performance enhancement: A review. *Bioresource Technology*, **344**, 126321.

Wang, Y., Fang, J., Lü, F., Zhang, H., He, P. 2023b. Food waste anaerobic digestion plants: Underestimated air pollutants and control strategy. *Science of The Total Environment*, **903**, 166143.

Wang, Z., Hu, Y., Wang, S., Wu, G., Zhan, X. 2023c. A critical review on dry anaerobic digestion of organic waste: Characteristics, operational conditions, and improvement strategies. *Renewable and Sustainable Energy Reviews*, **176**, 113208.

Woon, K.S., Lo, I.M.C. 2016. A proposed framework of food waste collection and recycling for renewable biogas fuel production in Hong Kong. *Waste Management*, **47**, 3–10.

Wunderlich, S.M., Martinez, N.M. 2018. Conserving natural resources through food loss reduction: Production and consumption stages of the food supply chain. *International Soil and Water Conservation Research*, **6**(4), 331–339.

Xiao, H., Zhang, D., Tang, Z., Li, K., Guo, H., Niu, X., Yi, L. 2022. Comparative environmental and economic life cycle assessment of dry and wet anaerobic digestion for treating food waste and biogas digestate. *Journal of Cleaner Production*, **338**, 130674.

Xiong, X., Zhang, W., Ha, X., Li, N., Chen, S., Xing, H., Yang, J. 2023. The preparation processes and influencing factors of biofuel production from kitchen waste. *Fermentation*, **9**(3), 247.

Xu, C., Nasrollahzadeh, M., Selva, M., Issaabadi, Z., Luque, R. 2019. Waste-to-wealth: Biowaste valorization into valuable bio (nano) materials. *Chemical Society Reviews*, **48**(18), 4791–4822.

Xu, W., He, X., Wang, C., Zhao, Z. 2023. Effect of granular activated carbon adsorption and size of microbial aggregates in inoculum on stimulating direct interspecies electron transfer during anaerobic digestion of fat, oil, and grease. *Bioresource Technology*, **368**, 128289.

Yang, P., Peng, Y., Tan, H., Liu, H., Wu, D., Wang, X., Li, L., Peng, X. 2021. Foaming mechanisms and control strategies during the anaerobic digestion of organic waste: A critical review. *Science of the Total Environment*, **779**, 146531.

Yang, S., Luo, F., Yan, J., Zhang, T., Xian, Z., Huang, W., Zhang, H., Cao, Y., Huang, L. 2023. Biogas production of food waste with in-situ sulfide control under high organic loading in two-stage anaerobic digestion process: Strategy and response of microbial community. *Bioresource Technology*, **373**, 128712.

Yang, Y., Bao, W., Xie, G.H. 2019. Estimate of restaurant food waste and its biogas production potential in China. *Journal of Cleaner Production*, **211**, 309–320.

Yaverino-Gutierrez, M.A., Flores-Iga, J.G., Velez-Mercado, M.I., Sosa-Herrera, A., Esparza-Perusquia, M.d.l.M., Luévanos Escareño, M.P., Hernández Almanza, A.Y., Terán, F.H., Hernández-Beltrán, J.U., Balagurusamy, N. 2023. Monitoring methods for anaerobic digestion of food waste: Physicochemical and molecular analysis. In Cristobal N. Aguilar Gonzalez, Ricardo Gómez-García, & Mohammed Kuddus (Eds.), *Food waste conversion*. New York,: Springer, pp. 187–227.

Yin, Y., Liu, Y.-J., Meng, S.-J., Kiran, E.U., Liu, Y. 2016. Enzymatic pretreatment of activated sludge, food waste and their mixture for enhanced bioenergy recovery and waste volume reduction via anaerobic digestion. *Applied Energy*, **179**, 1131–1137.

Yong, Z.J., Bashir, M.J.K., Hassan, M.S. 2021. Biogas and biofertilizer production from organic fraction municipal solid waste for sustainable circular economy and environmental protection in Malaysia. *Science of The Total Environment*, **776**, 145961.

Yuan, T., Zhang, Z., Lei, Z., Shimizu, K., Lee, D.-J. 2022. A review on biogas upgrading in anaerobic digestion systems treating organic solids and wastewaters via biogas recirculation. *Bioresource Technology*, **344**, 126412.

Yuan, Y., Hu, X., Chen, H., Zhou, Y., Zhou, Y., Wang, D. 2019. Advances in enhanced volatile fatty acid production from anaerobic fermentation of waste activated sludge. *Science of the Total Environment*, **694**, 133741.

Yue, L., Cheng, J., Tang, S., An, X., Hua, J., Dong, H., Zhou, J. 2021. Ultrasound and microwave pretreatments promote methane production potential and energy conversion during anaerobic digestion of lipid and food wastes. *Energy*, **228**, 120525.

Yukesh Kannah, R., Merrylin, J., Poornima Devi, T., Kavitha, S., Sivashanmugam, P., Kumar, G., Rajesh Banu, J. 2020. Food waste valorization: Biofuels and value added product recovery. *Bioresource Technology Reports*, **11**, 100524.

Zebra, E.I.C., van der Windt, H.J., Nhumaio, G., Faaij, A.P. 2021. A review of hybrid renewable energy systems in mini-grids for off-grid electrification in developing countries. *Renewable and Sustainable Energy Reviews*, **144**, 111036.

Zhang, C., Sun, Y., Cao, T., Wang, W., Huo, S., Liu, Z.-H. 2022a. Influence of organic load on biogas production and response of microbial community in anaerobic digestion of food waste. *International Journal of Hydrogen Energy*, **47**(77), 32849–32860.

Zhang, L., Li, F., Kuroki, A., Loh, K.-C., Wang, C.-H., Dai, Y., Tong, Y.W. 2020a. Methane yield enhancement of mesophilic and thermophilic anaerobic co-digestion of algal biomass and food waste using algal biochar: Semi-continuous operation and microbial community analysis. *Bioresource Technology*, **302**, 122892.

Zhang, S., Xiao, M., Liang, C., Chui, C., Wang, N., Shi, J., Liu, L. 2022b. Multivariate insights into enhanced biogas production in thermophilic dry anaerobic co-digestion of food waste with kitchen waste or garden waste: Process properties, microbial communities and metagenomic analyses. *Bioresource Technology*, **361**, 127684.

Zhang, S., Zou, L., Wan, Y., Ye, M., Ye, J., Li, Y.-Y., Liu, J. 2020b. Using an expended granular sludge bed reactor for advanced anaerobic digestion of food waste pretreated with enzyme: The feasibility and its performance. *Bioresource Technology*, **311**, 123504.

Zhao, M.-X., Ruan, W.-Q. 2013. Biogas performance from co-digestion of Taihu algae and kitchen wastes. *Energy Conversion and Management*, **75**, 21–24.

Zhen, G., Lu, X., Kobayashi, T., Kumar, G., Xu, K. 2016. Anaerobic co-digestion on improving methane production from mixed microalgae (Scenedesmus sp., Chlorella sp.) and food waste: Kinetic modeling and synergistic impact evaluation. *Chemical Engineering Journal*, **299**, 332–341.

Zhongming, Z., Linong, L., Xiaona, Y., Wangqiang, Z., Wei, L. 2021. *UNEP food waste index report 2021*. Nairobi, Kenya: UNEP.

7 Production of Biopolymers from Food Waste

Ariane Fátima Murawski de Mello, Luciana Porto de Souza Vandenberghe, Clara Matte Borges Machado, Priscilla Zwiercheczewski de Oliveira, Thamarys Scapini, Rafael Novaes Barros, Adenise Lorenci Woiciechowski, and Carlos Ricardo Soccol

7.1 INTRODUCTION

Food is an essential commodity for society's development, being generated in large quantities to reach demands. However, humanity still faces some challenges regarding food distribution and waste. According to the Food Agriculture Organization (FAO), approximately 1.3 billion tons of food are wasted each year. Generally, these by-products are discarded in landfills or incinerated, which can cause a severe environmental burden (Ranganathan et al., 2020). In addition, this indiscriminate disposal fits in a linear economy model, in which natural resources are taken from the environment, transformed by physicochemical and/or biological methods, products are used, and by-products are discarded with little regard to the processes' impact.

However, this mode of economy has led to the present harsh environmental scenario: the greenhouse effect with rising temperatures and heat waves, plastic waste found in remote places, microplastics pollution reaching water and food chains, and countless other effects (Sariatli 2017). Therefore, humanity is now facing an economic transition to a circular bioeconomy, in which biomass in all its different forms is applied to the production of bulk chemicals and other important molecules, especially through bioprocesses, with the main objective of generating a no-waste cycle (Carus and Dammer 2018). Thus, applying food waste for biopolymer extraction and/or production, and bringing more value to biomass that is considered a residue and would be inappropriately discarded, directly aligns with the precepts of circular bioeconomy.

Biopolymers are also aligned with the circular economy. Being biodegradable and bio-based, they can serve as potential substitutes for conventional plastics without generating the environmental burden of plastic waste. Different biopolymers present unique characteristics that make them suitable for specific applications. For example, starch blends can serve as packaging materials (Agarwal et al., 2021), while polyhydroxyalkanoates (PHAs) can be applied in drug delivery systems in the pharmaceutical area (Prakash et al., 2022). Thus, biopolymers are versatile molecules with a constantly growing market. It is estimated that the biopolymers' market size in 2018 was 12 billion dollars, with a growth rate of 19% (CAGR) in the period of 2019–2025 (Baranwal et al., 2022). In addition, research on both the production and application of biopolymers has been rising over the past few years as the search for alternatives to conventional plastics becomes urgent.

One of the main research topics regarding biopolymers is the substrate used for their manufacturing. The source of these molecules should be cheap and easy to obtain. In addition, there has been a trend in the development of second and third-generation biomolecules, including bioplastics, using waste materials as a resource for high-added value biomolecules (de Mello et al., 2023). This

DOI: 10.1201/9781032706030-7

reiterates the possibility of using food waste for the obtainment of several biopolymers. In this case, the biopolymers can be either directly extracted from biomass or produced by microorganisms (Figure 7.1). This chapter will discuss the different food waste categories, how this substrate can be treated either to extract biopolymers or to make it susceptible to microbial action, the different biopolymers that can be obtained, and the future prospects regarding the theme.

7.2 FOOD WASTE CATEGORIES

Food waste can be labeled based on technical, environmental, or social aspects. Edible food waste is the discarded food that is still safe to eat, often due to factors like overproduction, expiration dates, or aesthetic imperfections. Inedible food waste includes parts of food items that are not meant for consumption, such as peels, bones, and shells. Household food waste encompasses the food discarded by individuals and families, often due to spoilage or excess. Commercial food waste is generated by businesses, including restaurants, supermarkets, and catering services. Agricultural food waste concerns the food lost along the production process, in harvesting and post-harvest handling due to factors like pests, weather, or transportation issues. Industrial food waste is generated during food processing and manufacturing (Corrado et al., 2019). The main fundamental factors for food waste classification are spoilage – food goes bad due to improper storage, handling, or temperature control; excess production – overproduction of food items that exceed market demand; aesthetic standards – food discarded due to their disconformity with visual appearance standards set by consumers or retailers; expiration dates – food discarded considering their ideal period for consumption; and portion sizes – food waste resulting from oversized portions at restaurants or when cooking at home (Garcia-Garcia et al., 2017).

7.2.1 Dairy

Globally in the dairy industry, is estimated that about 4–11 million tons of effluents are generated per year (Ahmad et al., 2019). Whey is the liquid by-product of cheese and yogurt production. It consists primarily of water, proteins (63–70%), with α-lactalbumin and β-lactoglobulin being the predominant fractions, and lactose (4–5% w/w), while the lipid content (approximately 4%) is primarily in the form of milk fat globule membranes. Minerals like calcium, potassium, and phosphorus are present in whey (0.6–0.8%), although in relatively low percentages, as well as vitamins such as riboflavin and B12 (Sebastián-Nicolás et al., 2020; Tsermoula et al., 2021). Its rich lactose and protein content makes it an attractive resource for biopolymer production, where proteins can be isolated and used as building blocks for the development of whey protein-based biopolymers. These biopolymers are characterized by their biodegradability and biocompatibility, making them suitable for applications in food packaging, implants, and pharmaceuticals (Sebastián-Nicolás et al., 2020). The global production of whey is substantial, estimated at over 180 million tons annually (Sebastián-Nicolás et al., 2020). This volume varies based on regional dairy processing practices and the types of dairy products manufactured.

Casein is the primary protein in milk, corresponding to approximately 80% of the total protein content in dairy products (Carter et al., 2021). It can be isolated from dairy effluents and used as a key component in the formulation of casein-based biopolymers. These biopolymers possess excellent film-forming properties, mechanical strength, and water resistance, making them suitable for applications in food packaging and coatings (Ahmad et al., 2019).

Lactose, a disaccharide composed of glucose and galactose, is the primary carbohydrate in milk and its by-products (46%–52%), including whey and milk permeate. Lactose can be enzymatically hydrolyzed into its monomers, which can subsequently undergo microbial fermentation to produce lactic acid. In turn, lactic acid is a versatile precursor for the production of polylactic acid (PLA) (Sebastián-Nicolás et al., 2020). Milk permeate is generated during ultrafiltration or reverse osmosis processes, where water is removed from milk aimed at concentrating dairy components. It is primarily composed of water (90–95%) and contains lactose (4–6%), minerals (0.2–0.4%), and minor proteins (0.1–0.3%). Similar to lactose, milk permeate can serve as a substrate for microbial fermentation, leading to the production of biopolymers (Carter et al., 2021; Sebastián-Nicolás et al., 2020).

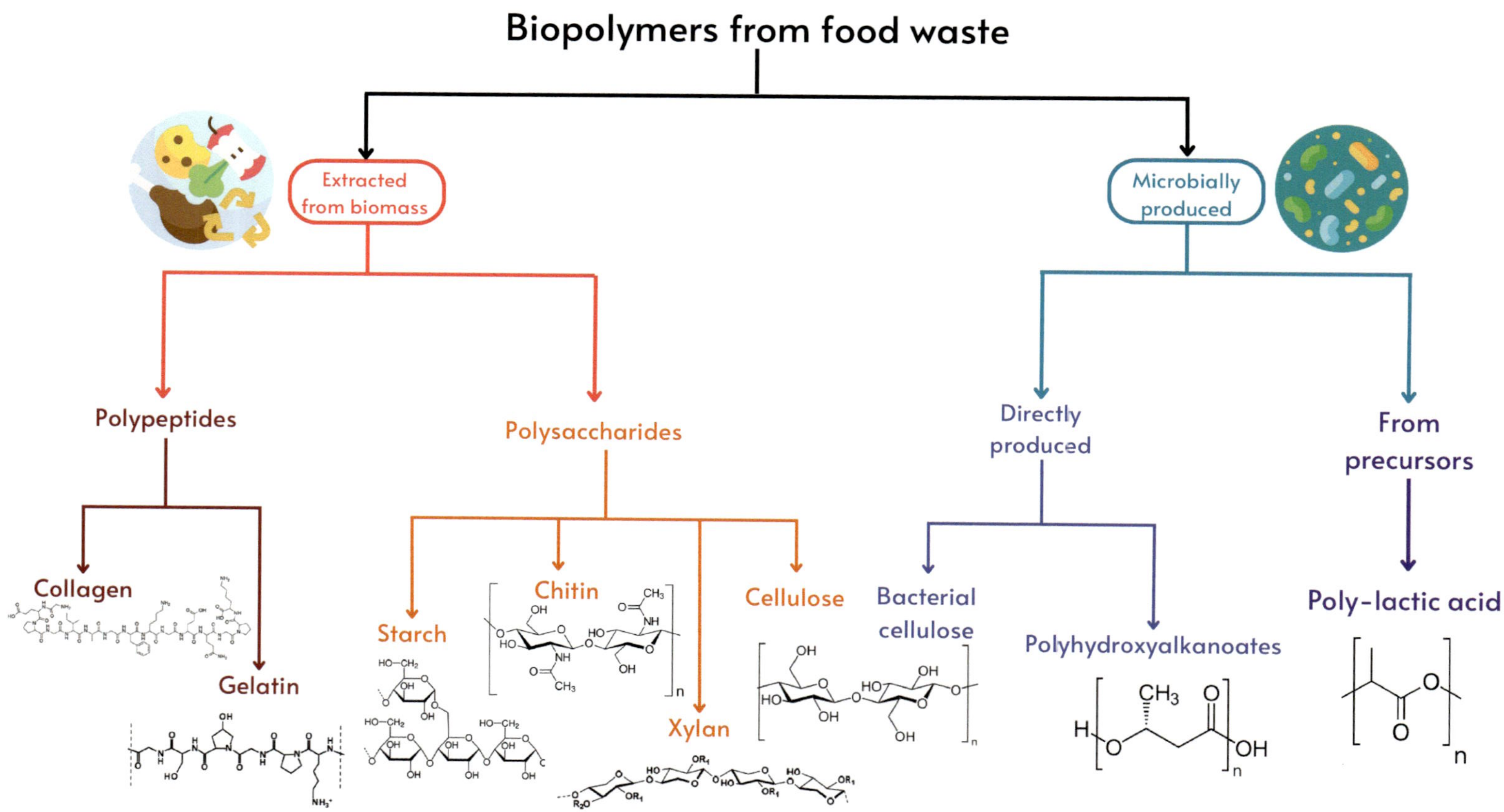

FIGURE 7.1 Different routes for biopolymer extraction. Adapted from (Banu J and Sharmila V 2023).

Dairy wastewater may contain water in the majority (95–99%), organic matter (1–5%) comprising organic constituents, including residual milk components, and inorganic components (0.2–0.5%) comprising minerals and trace elements. Other components can also be found in dairy effluents, such as residual enzymes generated during cheese production (rennet or microbial enzymes) used in the curdling and fermentation processes, and microbial biomass resulting from fermentation processes (yogurt or probiotic production) can include live bacterial cultures and their metabolic by-products (Ahmad et al., 2019; Adesra et al., 2021; Sar et al., 2022).

7.2.2 Oil

Annually, the oil industry produces approximately 350.9 million tons of effluents (Sharma et al., 2020). Wastewater from oil refineries is often characterized by its complex composition, containing a diverse range of organic compounds, including sugars, organic acids, and glycerol (Atasoy et al., 2018). The composition may vary depending on factors such as the crude oil source, refining processes, and product specifications. Nevertheless, the presence of organic compounds makes this effluent a promising feedstock for biopolymer production. Sugars present in refinery wastewater, originating from the degradation of carbohydrates in crude oil, can be metabolized by microorganisms to synthesize biopolymers, particularly polyhydroxyalkanoates (PHAs) (Aremu et al., 2021).

Crude glycerol is produced during transesterification, the process of converting vegetable oils or animal fats into biodiesel. Glycerol is a key component of these oils and fats, and during transesterification, it is separated from the fatty acid esters that form biodiesel. The resulting crude glycerol contains impurities and excess methanol, making it unsuitable for most applications (Chatzifragkou and Papanikolaou 2012). However, with appropriate purification and conversion processes, glycerol can be used as a carbon source for the synthesis of various biopolymers, including PHA and polyglycerols (Moita et al., 2014). The quantities of crude glycerol generated are substantial and primarily linked to biodiesel production volumes; the global total amount produced is estimated at 15 million tons annually (Wang et al., 2019a).

Crude palm oil mill effluent (POME) from the extraction and processing of palm oil is characterized by its high organic content, consisting of free fatty acids (40–60% w/w) and organic matter (10–20% w/w) mainly in the form of carbohydrates, proteins, and lipids. Total solids, including suspended solids and dissolved solids, comprise around 2–5% (w/w) of POME, as well as nitrogen (1–3% w/w), phosphorus (0.5–2% w/w), and inorganic salts (2–5% w/w) (Binma-ae, Prasertsan, and Choorit 2020). POME has garnered attention as a potential feedstock for biopolymer production due to its high organic load composition, and efforts have been made to treat and recover valuable components from this residue. Its carbohydrate content, such as sugars and starches, can be subjected to microbial fermentation to produce biopolymers, including PHAs (Hassan et al., 2013). The quantities of POME generated are substantial, with large palm oil mills located in Malaysia which produced 17.5 million tons in 2009 (Ahmad et al., 2023).

Waste cooking oil is a ubiquitous by-product generated by households, restaurants, and food processing facilities. This effluent consists of triglycerides, monoglycerides, diglycerides, and variable quantities of free fatty acids (11%–56% w/w), which may be contaminated with food particles and impurities (Awogbemi et al., 2019). The lipid-rich nature of waste cooking oil makes it a valuable feedstock for biopolymer production (Mandic et al., 2019). Lipids, in the form of triglycerides, can be converted into biodegradable polymers through transesterification and polymerization. The quantities generated are significant, especially in urban areas and regions with high food consumption; the annual production is estimated at 350.9 million tons (Sharma et al., 2020).

7.2.3 Fruits and Vegetables

Juice production generates significant quantities of waste, particularly in the form of pulp and peel residues. The amount of juice processing waste is estimated to be up to 30% or more of the raw fruit/

vegetable weight that may become waste during juice extraction. The composition of this waste stream is characterized by a high content (approximately 80%) of carbohydrates and total solids (Khalil et al., 2022). The peels and pomace resulting from peeling and pulping operations constitute a substantial portion of the solid waste. Their composition is rich in cellulose (10–35%), hemicellulose (5–20%), a mix of various sugar monomers (8–50%) such as glucose, xylose, and mannose, and pectin (Patidar et al., 2018). Around 15 to 25% of fruits and vegetables are wasted at the end of seasonal production (Edwiges et al., 2018). Considering this availability and the abundant sources of carbohydrates, including cellulose, hemicellulose, and pectin, it is suitable for biopolymer production through microbial fermentation to produce biopolymers such as PHA and PLA (Ranganathan et al., 2020). Pectin is another valuable component that can be used for the production of biodegradable films and coatings, making it a promising candidate for biopolymer applications (Sharmila et al., 2020). Citrus fruit processing, in particular, generates distinctive waste streams, such as citrus peel waste, which is rich in pectin (10–20%), making it suitable for the production of pectin-based bioplastics. The quantity of citrus processing waste is substantial, as citrus fruits are processed in large quantities worldwide (Zema et al., 2018).

Certain fruit and vegetable seeds, such as those from avocados, mangos, and olives, are rich in oils (Patidar et al., 2018; Sharma et al., 2020). These seeds contain valuable lipids that can be extracted and used as feedstocks for biopolymer production, particularly in the synthesis of polyurethane biopolymers. Their lipid composition can include oleic acid (C18:1) ranging from 40 to 70%, linoleic acid (C18:2) between 5 and 30%, palmitic acid (C16:0) between 7 and 20%, stearic acid (C18:0) between 0.5 and 5%, and other fatty acids in variable quantities (Ranganathan et al., 2020). Starch, in turn, is prevalent in bananas and potatoes (Sharma et al., 2020). It can be extracted from residues and used as a raw material for the production of starch-based biodegradable plastics (Sharmila et al., 2020). Starch is a carbohydrate composed of amylose and amylopectin, where amylose is a linear polysaccharide of glucose units linked by α-1,4-glycosidic bonds, while amylopectin is a branched polysaccharide with additional α-1,6-glycosidic linkages. Starch residues can be processed to extract amylose or amylopectin, both of which can be used in the production of biodegradable films and coatings (Sharmila et al., 2020). The quantity of starch residues generated is substantial, driven by the large-scale processing of starchy vegetables and fruits. For example, potato processing plants generate significant volumes of starch-rich wastewater and residues, often exceeding 30% of the original potato mass (Sharmila et al., 2020; Ranganathan et al., 2020).

7.2.4 Grains

Wheat milling involves the separation of wheat grains into their constituent parts, including the endosperm, bran, and germ. The bran, a fibrous outer layer, is typically rich in cellulose (20–25%) and hemicellulose (approximately 20%), both of which are potential feedstocks for biopolymer production (Sharmila et al., 2020; Ranganathan et al., 2020). Cellulose can be converted into cellulose acetate, a biodegradable plastic, while hemicellulose can be used in the synthesis of xanthan gum, a biopolymer with various industrial applications (Sharmila et al., 2020; Sharma et al., 2020). Rice milling generates rice bran as a by-product, which is also a source of potential feedstock for biopolymers. Rice bran typically contains lipids (15–20%), cellulose (20–25%), and a significant portion of hemicellulose (Sharmila et al., 2020; Ranganathan et al., 2020). The cellulose can be hydrolyzed and processed into cellulose-based biopolymers (e.g., cellulose acetate), while hemicellulose can be converted into xylan-based biopolymers (Sharmila et al., 2020).

Corn milling produces corn stover, consisting of cellulose and hemicellulose, collectively comprising around 70% of the dry weight. Lignin content is approximately 20 to 25%. Cellulose and hemicellulose can serve as feedstocks for various biopolymers, while lignin can potentially be used for lignin-based biopolymers (Sharmila et al., 2020; Ranganathan et al., 2020). Brewery spent grains generated during brewing are primarily composed of carbohydrates (approximately 75% of the dry weight), including unfermented sugars, cellulose, and hemicellulose, as well as residual proteins. These carbohydrates can be used as a source for microbial fermentation to produce PHA (Sharmila et al., 2020; Ranganathan et al., 2020).

Ethanol distillation processes generate distillery waste that may contain residual starches and sugars from the grains used in the fermentation process, as well as small quantities of lipids and proteins (Ranganathan et al., 2020). These carbohydrates can potentially be utilized as feedstocks for PHA production, particularly through microbial fermentation (Yadav et al., 2020). Grains can be grown for oil extraction, such as soybeans, canola, or sunflower seeds. Residues from oil extraction are known as oilseed meal and typically contain around 30 to 40% protein, depending on the grain type (Sharmila et al., 2020). Cellulose content is also around 40% (w/w) and is a current potential source for biopolymer production. Enzymatic or chemical processes can be employed to convert cellulose into fermentable sugars, which can then be used for biopolymer synthesis. Hemicellulose, another carbohydrate polymer, constitutes approximately 20% (w/w) of the effluents (Sharmila et al., 2020; Ranganathan et al., 2020; Sharma et al., 2020).

7.2.5 Meat Derived and Seafood

Blood is a substantial by-product of the meat industry and is often considered waste material. Its composition comprises water (75–85%), proteins (12–18%) of which hemoglobin and serum albumin are the significant protein components, lipids in form of triglycerides (6–8%), and mineral and organic compounds (1–2%) (Ranganathan et al., 2020; Sharma et al., 2020). In large meat processing facilities, thousands of liters of blood can be produced daily. Blood-based biopolymers have gained attention due to their potential for biodegradability and biocompatibility. Hemoglobin-derived biopolymers exhibit properties that make them suitable for applications in drug delivery and tissue engineering (Ranganathan et al., 2020).

Bones are another significant waste product; they primarily consist of hydroxyapatite (65–70%), a mineral rich in calcium and phosphorus. Hydroxyapatite can be used as a filler in biopolymer formulations, enhancing their mechanical properties and biocompatibility (Sharmila et al., 2020). This makes it a valuable resource for the development of biocompatible and biodegradable medical implants and scaffolds. Organic matter, including collagen and other proteins (20–25%), and water (5–10%) are also constituents of bone composition (Sharmila et al., 2020). The skin and hides of animals, such as cows and pigs, are typically rich in collagen, a fibrous protein. Collagen-based biopolymers have gained prominence in various applications, including tissue engineering, wound dressings, and cosmetics. Collagen corresponds to 25–35%, water, and organic compounds such as lipids and amino acids (30–40%) (Ranganathan et al., 2020). Collagen extracted from skin and hides can be used to produce biopolymers with excellent biocompatibility. These biopolymers have found applications in the medical and cosmetic industries, contributing to the development of sustainable materials (Sharmila et al., 2020).

Animal fat and grease are common by-products of meat processing, and they can be converted into bio-based feedstocks for PHA through microbial fermentation. Triglycerides (90–95%) are composed of fatty acids bound to glycerol molecules; water (2–5%) and other compounds (<1%) such as cholesterol and phospholipids are found in smaller quantities. Fish oil is a lipid-rich effluent generated especially in the processing of fatty fish species like salmon, mackerel, and sardines. The major composition is lipids (90–95%), primarily consisting of triglycerides containing long-chain polyunsaturated fatty acids (PUFAs), where omega-3 fatty acids, including eicosapentaenoic acid (EPA) and docosahexaenoic acid (DHA), are found (García-Moreno et al., 2018). Fish oil can serve as a valuable feedstock for the production of biodegradable plastics, particularly for PHAs (Ranganathan et al., 2020).

Feathers and hair are often considered waste products in the meat industry. However, they contain keratin (90–95%), a protein that can be used for biopolymer production. Keratin-based biopolymers have applications in cosmetics, agriculture, and environmental remediation (Ranganathan et al., 2020). Shellfish by-products, such as shrimp and crab shells, are abundant in the seafood processing industry. These shells are primarily composed of chitin (15–40%), a natural biopolymer, that can be converted into chitosan, a biodegradable and versatile biopolymer (Sharmila et al., 2020). Chitosan's properties make it suitable for a wide range of applications, including food packaging, wound healing, drug delivery, and water treatment. Proteins (20–40% w/w), minerals (20–50% w/w), and water (10–20%) are also found (Ranganathan et al., 2020). Processing shrimp and crab involves the removal of shells, and these shells can account for up to 40% of the total weight of the seafood (Sharma et al., 2020).

While not directly generated as effluents in the seafood industry, algae and seaweed are closely associated with marine environments. Their structure presents proteins (18–46% w/w) and lipids (12–48%) as well as their very rich composition in polysaccharides (18–46% w/w) (Devadas et al., 2021) – including alginate (mannuronic and guluronic acid units), agar (agarose and agaropectin), and carrageenan (linear sulfated polysaccharides) (Dang et al., 2022). They have diverse applications: in foods they are used as thickening agents, stabilizers, and gelling agents. In pharmaceuticals and cosmetics, they serve as encapsulation materials and emulsifiers. Thus, biodegradable films and coatings produced from these polysaccharides can replace synthetic materials (Dang et al., 2022).

7.3 FOOD WASTE TREATMENTS FOR BIOPOLYMERS' EXTRACTION AND PRODUCTION

The challenge associated with utilizing food waste is linked to the complexity of the matrix, which requires a stage where desired compounds can be extracted for conversion into biopolymers (such as polysaccharides and polypeptides) or the pretreatment of the waste to enhance the susceptibility of the macro-components to microbial fermentation. Effectively utilizing pretreatment and extraction unit operations is essential for achieving sustainable biopolymer production using low-cost raw materials.

7.3.1 Extraction Methods

An alternative for the biotechnological exploitation of food waste involves employing extraction methods to isolate biopolymers, which can be derived from carbohydrates, proteins, and lipids. These methods may encompass chemical, thermal, or biological approaches. Carbohydrate-based biopolymers extracted from food waste include starch, cellulose, dextran, chitosan, and chitin. Due to their abundant source of raw material, a range of bioproducts has been explored utilizing these compounds (Gautam et al., 2022). For instance, cellulose is a biopolymer that can be extracted from food waste rich in fruit and vegetable peels and seeds, and agro-industrial waste such as grains and straw (Sharmila et al., 2020; Chandrasekar et al., 2023). On the other hand, chitin, another biopolymer, is abundantly present in seafood debris, and food waste containing shrimp shells, fish, lobster, and krill can be an interesting source of chitin (Azelee et al., 2023). When considering protein-based biopolymers, the focus should be on waste streams rich in animal, milk, and plant residues, as these contain significant content of collagen, keratin, and gelatin (Sharmila et al., 2020; Gautam et al., 2022). Regarding lipid extraction, food waste has substantial concentrations of these compounds, especially from domestic and urban waste, and can be considered a renewable alternative for obtaining biomolecules (Gautam et al., 2022).

Methods for extracting biopolymers from food waste typically involve chemical technologies (Table 7.1) since it requires an affinity to remove non-target structures and isolate the biomolecule. For instance, when extracting chitosan or collagen, it is necessary to remove proteins, acetyl groups, and pigments, which can require a sequence of chemical processes (Muley et al., 2018; Azelee et al., 2023; Bhuimbar et al., 2019). Muley and collaborators (2018) employed an alkaline process with NaOH to remove proteins, followed by acid treatment using HCl for demineralization, and subsequent removal of pigments using acetone. Similarly, in the extraction of collagen from fish skin, a sequence of chemical processes was utilized, including an alkaline solution (NaOH) to remove non-collagenous proteins, deproteinized with butyl alcohol, and treatment with lactic acid (Bhuimbar et al., 2019).

Chemical processes have also been explored for lignocellulosic-based food waste, which is rich in residues from straws, pods, food grains, among other sources, to isolate polysaccharides, mainly cellulose and hemicellulose. Due to the recalcitrance of lignocellulosic biomass, high-pressure and high-temperature processes can be combined with chemical methods to enhance extraction yields. Moreover, the combination of temperature and chemicals can improve the extraction of xylan, an industrial biopolymer (Manzanares et al., 2020). Reaction systems such as ultrasound and microwaves have shown promise in improving extraction yields, as demonstrated by Thi Thuy Van et al., (2022) in a microwave-assisted reaction with NaOH-H_2O_2, achieving cellulose extraction yields of over 75% even in short reaction times (10 min). Ultrasound and microwave-based systems have also been evaluated for different biopolymer extraction processes, including pectin extraction from mango peels and potato waste (Thakare et al., 2023; Arrutia et al., 2020), and polysaccharides extraction from kinnow peels and mung bean peels (Kaur et al., 2023; Zhang et al., 2023).

TABLE 7.1
Recent Studies on Processes for Extracting Biopolymers from Food Waste

Waste Food Source	Biopolymer Extracted	Sequences Extraction Conditions	Reference
Banana peels	Starch and cellulose	Chemical treatment: solvent extraction with ethanol and hexane at room temperature for 16 h (removal extractives); ethanol 70% for 5 min; NaOH 0.1 M for 5 min (cellulose); supernatant was decanted, filtrated, and washed with NaOH 0.1 M and water (starch)	(Chandrasekar et al., 2023)
Banana stem	Cellulose	Microwave and NaOH-H_2O_2 treatment: 2% NaOH for 20 min of microwave radiation time following by addition of 15% H_2O_2	(Thi Thuy Van et al., 2022)
Cassava peel and bagasse	Starch	Physical treatment: sodium acetate buffer at 120°C for 10 min; sonication for 20 min; centrifugated and freeze dried	(Thuppahige et al., 2023)
Chicken feathers	Keratin	Chemical treatment: 1 M NaOH for 24 h following by regular stirring at 50°C for 5h; neutralization with HCl 2N	(A. Ahmad et al., 2022)
Prawn shells	Chitosan	Alkali-acid treatment: 5% NaOH at 90°C for 24 h (deproteinate); 4% HCl and 36 h (demineralize); acetone treatment (decolorize); 50% NaOH treatment (deacetylation)	(Muley et al., 2018)
Shrimp shells	Chitin and chitosan	Biological treatment: successive culture: 1- *Lactobacillus delbrueckii* + glucose (50 $g.L^{-1}$) for 24 h, and 2- a pulse with inoculum of *Bifidobacterium lactis* +glucose (35 $g.L^{-1}$) - 40°C, 200 rpm for 72 h (deproteinate and demineralize); crude deacetylases cocktail (*Aspergillus niger*)– 40°C, 200 rpm for 24 h (deacetylation)	(Sixto-Berrocal et al., 2023)
Skin of Medusa fish	Collagen type I	Chemical treatment (the process was conducted at cold): 0.1 M NaOH for 36 h (removal non-collagenous proteins); 10%butyl alcohol for 48 h (deproteinate); 0.5 M lactic acid for 72 h (precipitation)	(Bhuimbar, Bhagwat, and Dandge 2019)

Biopolymer extraction processes can also be conducted through biological means, employing enzyme cocktails or microorganisms. For instance, a sequential co-cultivation strategy involving *Lactobacillus delbrueckii* and *Bifidobacterium lactis* for demineralization and deproteinization, followed by deacetylation using an enzymatic cocktail of *Aspergillus niger*, achieved a degree of deacetylation of over 70% (Sixto-Berrocal et al., 2023). Biological processes offer advantages such as reducing the use of chemicals, being environmentally friendly, and allowing for easy product recovery without complex unit operations requiring high temperatures and pressure. In addition to conventional technologies, there is an emerging interest in isolating compounds from food waste that can be converted into biopolymers through fermentation.

Food waste was subjected to anaerobic digestion followed by a lactic acid extraction step from a polymeric resin (BA765). Integrating anaerobic digestion and lactic acid recovery processes with the aim of energy and biopolymer conversion showed significance improvement in the overall process efficiency. However, the main challenge of this bioprocess is associated with impurities present in the fermentation broth (Bühlmann et al., 2022).

7.3.2 Pretreatments for Further Microbial Production

The heterogeneity of food waste can be a significant challenge in developing an efficient process for converting them into biopolymers through microbial processes. This is why different pretreatment techniques are being explored to fractionate the macro-structures of these raw materials. In contrast to extraction processes, pretreatment methods have the goal of enhancing the susceptibility of these structures to microbial action. This is achieved by increasing surface area, improving access to specific structures (e.g., polysaccharides), and increasing conversion efficiency (Treichel et al., 2020).

Pretreatment processes are commonly categorized into physical, chemical, and biological. These include mechanical, acidic, alkaline, enzymatic, and microbial methods, and can be applied individually or in combination with one another. Conversion processes are considered successful when there is partial or total release of the monomers present in the waste, increasing the accessibility of proteins, lipids, and carbohydrates for subsequent transformation into biopolymers (Tsang et al., 2019). Recent studies on the pretreatment of FW for biopolymer conversion are presented in Table 7.2.

Mechanical milling processes are often employed in the conversion of food waste due to the impact of particle size on the efficiency of subsequent processes. Grinding can increase the surface area and promote greater uniformity in the raw material without altering the macro-structural composition of the substrate (Kwan et al., 2016; Ji et al., 2023). It is worth considering separation techniques for non-biodegradable waste before grinding to mitigate overloading and the potential failure of pretreatment and bioconversion reactors.

Ultrasonic pretreatment techniques have also found application in the conversion of food waste into PHA (Dan et al., 2023). This technology, already widely used in various biotechnological applications, operates by subjecting the biomass structure to the effect of ultrasound waves through the cavitation effect, which results from the propagation of high-frequency acoustic waves in the liquid in continuous cycles (Treichel et al., 2020). Recently, a combination of ultrasonic pretreatment with alkaline was experimented on food waste before undergoing anaerobic digestion followed by PHA production. The results demonstrate that the combined system could significantly enhance the soluble organic matter content in the raw material (Dan et al., 2023).

The chemical process approach mainly involves the use of acid and alkaline reagents. Chemical pretreatment technology alters the characteristics of the waste by breaking down the macro-components of organic matter, such as cellulose, starch, and hemicellulose, which is typically measured by the release of free sugars like fructose and glucose (Treichel et al., 2020; Soni et al., 2023). Despite the advantages of rapid and efficient macromolecule reaction and economic viability, chemical processes require pH adjustment for the bioconversion process (Soni et al., 2023; Fu et al., 2023). In addition, it is important that the process conditions are suitably adjusted to the biomass matrix, as harsh

TABLE 7.2
Recent Studies on the Pretreatment of Food Waste for Fermentation Processes to Obtain Biopolymers

Waste Food Source	Pretreatment		Biopolymer Production			Reference
	Type	Conditions	Microorganism	Reactor Type and Volume	Biopolymer	
Banana peel	Acid	5% (w.v^{-1}) of sulfuric acid (H_2SO_4) at 121°C for 15 min	*Limnothrix planktonica*	Shake flask (50 mL)	PHB[a]	Mishra and Panda (2023)
Banana peel	Alkali	1% (w.v^{-1}) of sodium hydroxide (NaOH) at 121°C for 15 min	*Limnothrix planktonica*	Shake flask (50 mL)	PHB	Mishra and Panda (2023)
Banana peel and mustard cake	Acid	Loading ratio of banana peel 10% (w.v^{-1}) and mustard cake 1% (w.v^{-1}); 5% (w.v^{-1}) of H_2SO_4 at 121°C	*Bacillus badiuus*	Shake flasks (250 mL)	PHB	(Soni et al., 2023)
Canteen	Enzymatic hydrolysis	Enzymes cocktail: cellulases, alpha-amylases, proteases, glucan 1,4-alpha-glucosidase. Temperature at 55°C until the concentration of glucose ceased to rise	*Halomonas bluephagenesis* WZY278	Bioreactor (7 L)	PHB	(Ji et al., 2023)
Canteen	Ultrasonic waves	Ultrasonic waves for 10 min at pH 9.0 and 10 g.L^{-1} Fe_3O_4 followed by acidification (VFA production) by anaerobic digestion	*Rhodopseudomonas palustris*	Serum bottle (150 mL)	PHA[b]	(Dan et al., 2023)
Cafeteria	Enzymatic hydrolysis*	Enzymes cocktail (1% w.v^{-1}): glucoamylase, cellulase, proteases, lipases. Loading ratio of 30% (w.v^{-1}) 450 rpm, 50°C, 6 hours	*Haloferax mediterranei*	Bench reactor (1.5 L)	PHA	(Priya et al., 2022)

* Before the pretreatment described above, the waste was milled to homogenize it
[a] Polyhydroxybutyrate

conditions can generate compounds inhibiting microbiological activity, reducing biopolymer yield. Although chemical pretreatments are widely used industrially, mainly for the bio-utilization of lignocellulosic biomass, the high organic matter content in food waste has aroused greater interest in bio-based pretreatment methods. These technologies are more frequently found in recent literature for conversion into biopolymers, since biological processes involve the addition of microorganisms and enzymes or the stimulation of the indigenous microbiome present in the raw material, thereby reducing the risk of generating microbiological inhibitors (Yousuf et al., 2018; Kwan et al., 2016).

Biological processes for converting food waste into biopolymers also deal with the challenge of material heterogeneity of the waste material. This characteristic is reflected in the necessity of using enzymes cocktails, which may include cellulases, amylases, hemicellulases, lipases, and proteases (Table 7.2), among other enzymes capable of hydrolyzing the structural macro-components of waste into monomeric structures that can be further converted into biopolymers by specialized microorganisms (Yousuf et al., 2018; Priya et al., 2022). The use of enzymes for waste pretreatment remains a topic of debate among researchers, as they can be expensive to apply, often accounting for more than 50% of the total cost of the process. Nevertheless, exploring operational strategies to enable the utilization of biological technologies in food waste pretreatment can be economically advantageous and can show results in reducing the amount of waste stream generated after the production of the molecule of interest or biopolymer (Peinemann et al., 2019).

One strategy that has been explored is the combination of two biological steps, usually starting with anaerobic digestion followed by the conversion of the resulting products into biopolymers. These alternative approaches offer several advantages, as they do not require the addition of chemicals and the isolation of an individual compound for biopolymer conversion. Moreover, the heterogeneity of food waste requires robust processes to make the technology economically viable. One possibility is to employ anaerobic digestion, interrupting it at the acidogenesis stage, where the digestate contains a high concentration of volatile fatty acids (VFAs). These VFAs can then be transformed into biopolymers. In this pretreatment process, the microbiological diversity of anaerobic digestion breaks down the structures of food waste during the hydrolysis stage, converting them into VFAs in the acetogenesis and acidogenesis stages. Subsequently, these compounds are inoculated with microorganisms capable of accumulating biopolymers (Khatami et al., 2022; Gong et al., 2021; Dan et al., 2023). Gong et al., (2021) studied the combination of anaerobic digestion steps and PHA production using food waste. In this scenario, the stream generated after anaerobic digestion (rich in VFAs) was used as feedstock for PHA accumulation, resulting in more than 50% of dry cell weight.

7.4 BIOPOLYMERS DERIVED FROM FOOD WASTE

7.4.1 Polysaccharides

Polysaccharides are macromolecules composed of monosaccharides, linked together by glycosidic bonds, and they can be formed by long, linear, or branched chains. They are a type of complex carbohydrate and perform several important functions in living organisms (Saleh et al., 2021). Some polysaccharides, such as starch found in plants, are used to store energy. Others, such as cellulose in plants and chitin in arthropods and fungi, have a structural function, as they compose the walls of cells, and provide support and rigidity to organisms (Kongala and Kondreddy 2023). In addition to their importance to living organisms, polysaccharides have a variety of uses and applications in industry due to their unique properties. Industrial production of polysaccharides can face several challenges depending on the type of polysaccharide and the specific application. The availability and quality of raw materials used in the production of polysaccharides is a challenge. The use of food waste for the production of polysaccharides is a sustainable and interesting approach that contributes to the reduction of food waste and the valorization of by-products from the food industry (Berenguer et al., 2023). It is important to ensure that these wastes are properly prepared and processed to make them safe and suitable for application. Continuous research and development of new technologies to optimize production processes are still necessary. Table 7.3 presents some of the advances published in recent years in food waste reuse processes for the production of polysaccharides-based biopolymers.

7.4.1.1 Cellulose

Cellulose $(C_6H_{10}O_5)_n$ is a fundamental structural component in plants and is one of the most abundant substances in nature. It is composed of glucose monomers that form long chains, linked together through glycosidic β(1-4) bonds. It is known for its high strength and rigidity, which makes it an essential component of plant cell walls (Amorim et al., 2020). In addition to its structural role in plants, cellulose is also an important raw material for industry. It is used in the manufacture of paper, cardboard, fabrics, and a variety of derivative products. Cellulose is mainly obtained from plant sources, such as trees and cultivated plants, and goes through extraction and purification processes before being transformed into final products (Seddiqi et al., 2021).

Several residues contain significant amounts of cellulose and can be used as raw material for extraction, such as cereal straw, rice husks, and sugar cane bagasse (Deshwal et al., 2021). The cellulose found in these wastes can be extracted, purified, and used in a variety of industrial applications. The extraction process includes steps of collecting and preparing the food waste, and enzymatic or chemical hydrolyzing to separate the cellulose from other components, such as lignin and hemicellulose (Melikoğlu et al., 2019). Cellulose obtained from food waste can be used in various industrial applications. Depending on the characteristics of the cellulose, it can be applied in the manufacture of paper, cardboard, textiles, chemicals, food, pharmaceuticals, and biodegradable materials.

7.4.1.2 Xylan

Xylan is part of the hemicellulose group and is mainly composed of xylose units, a simple sugar. Its structure may vary depending on the plant source and extraction conditions. For example, xylan can be found in linear forms with xylose residues linked by β(1-4) bonds or with branches in its backbone, containing glucuronic acid (GlcA), (methyl) glucuronic acid ([Me]GlcA), and/or arabinose residues (de Vries et al., 2021). It is found in various plant materials, being especially abundant in plant cell walls, where it plays an important role in the structure of plant cells, providing rigidity and strength. It is often used in industry as a raw material for the production of xylose, which can be used in the manufacture of food products, chemical products, and bioplastics, among others. Furthermore, xylan also has prebiotic properties and can be beneficial for intestinal health when consumed as part of the diet (Bouiche et al., 2023). Xylan-rich food waste often includes plant parts such as corn husks and cobs, and fruit peels such as apples, pineapple, and bananas. These residues can be generated during the processing of these fruits and subsequently used in the production of xylan in a process that involves stages of extraction, purification, and isolation of xylan.

7.4.1.3 Starch

Starch $(C_6H_{10}O_5)_n$ is one of the main sources of energy for many organisms, including humans. It is composed of long chains of glucose linked by glycosidic bonds. It is found mainly in plants, in storage organs such as roots (potatoes) and seeds (wheat grains), and it is an energy reserve for the plant. There are two main polymeric forms of starch: amylose and amylopectin. Amylose is mainly composed of linear glucose chains formed by α-1,4 linkages, while amylopectin is highly branched, with α (1-4) and α (1-6) glycosidic linkages (Martins et al., 2022). These differences in structure affect how starch is digested and how it behaves in culinary and industrial applications. Starch is extracted mainly from corn, cassava, and potatoes, and is widely used in the food industry as a thickener, stabilizer, and gelling agent, as well as having various industrial applications in chemicals, paper, and textiles. Various residues can be used for the production of starch: residues from processing starchy foods, such as residues from cassava factories (Tuppahige et al., 2023) or corn, or even waste from fruit processing, such as avocado pits (Martins et al., 2022).

TABLE 7.3
Production of Different Polysaccharides from Food Waste

Food Waste	Biopolymers	Pre-Treatment/Productions Method	Yield/Purity	References
Tomato pomace (*Solanum lycopersicum*)	Cellulosic nanocrystals	Pre-treatment with alkaline solution and bleached. Acid hydrolysis with H_2SO_4	Yield 19.8%	(Moradi and Fathi 2023)
Walnut shell (*Juglans Regia L.*)	Nanocellulose	Pre-treatment with alkaline solution and bleached. Acid hydrolysis with H_2SO_4	Cellulose content 89%	(Zheng et al., [s.d.])
Rice bran (*Oryza sativa*)	Xylan	Extraction of lipids with petroleum ether. Extraction with water 25°C	5.8 ± 0.03 g $100g^{-1}$	(Palaniappan et al., 2017)
Corn seed coat (*Zea mays*)	Xylan	Extraction of lipids with petroleum ether. Extraction with water 25°C	4.8 ± 0.07 g $100g^{-1}$	(Palaniappan et al., 2017)
Pea pod (*Pisum sativum*)	Xylan xylooligosaccharides	Delignification with $NaClO_2 + CH_3COOH$. Alkaline extraction with KOH	Xylan conversion 35.8%	(Cebin et al., 2021)
Avocado seeds (*Persea americana*)	Starch	Infusion in water with $Na_2S_2O_5$. Crushed and sieved through mesh	Yield 19.54%	(Martins et al., 2022)
Cassava bagasse and peel (*Manihot esculenta*)	Starch	Heating and stirring at 120°C. Sonication, filtration, and centrifugation.	Peel yield 30.17%, bagasse 7.5%	(V. T. W. Thuppahige et al., 2023)
Jackfruit peel and seed (*Artocarpus heterophyllus*)	Pectin	Pulsed electric fields and microwave-assisted extraction	Yield 18.24 %	(Lal et al., 2021)
Melon peel (*Cucumis melo L.*)	Pectin	Extraction with $C_6H_8O_7$. Ethanol precipitation 96%	Yield 29.48	(Raji et al., 2017)

7.4.1.4 Pectin

Pectin is a natural substance found in plants, especially fruits, and is a form of soluble fiber. It is a complex substance composed mainly of polysaccharides. Its composition varies depending on the source and maturity of the plant (Popoola-Akinola et al., 2022) but generally includes galacturonic acid (which forms the backbone of the pectin molecule) and galactose. Pectin also may contain side branches containing other sugars, such as arabinose and rhamnose, methoxyl groups (methyl groups attached to galacturonic acid), and small amounts of neutral sugars, such as glucose and mannose, which may be present in the side branches. Pectin is widely known for its ability to form gels when combined with sugars and acids under suitable conditions. This makes it a valuable ingredient in the food industry, especially in the production of jellies, jams, baked goods, and dairy products (Popoola-Akinola et al., 2022).

The main ingredient for the production of pectin is the peel of citrus fruits, especially lemons, oranges, and tangerines (Suri et al., 2022). Although citrus fruit peels are the primary source of pectin in the industry, pectin can also be extracted from other sources such as watermelon peel (Liu et al., 2023), melon peels (Raji et al., 2017), and jackfruit peel (Lal et al., 2021). The pectin production process involves extracting the pectin from the peels, followed by a series of purification and processing steps to obtain the final product. The peel is ground and treated with acid and heated to release the pectin, which is then separated from the pulp and other impurities.

7.4.2 Polypeptides

Collagen is a fibrous protein found in the extracellular matrix of connective tissues, playing a fundamental role in providing strength, elasticity, and support to various organs and tissues in animals (Gautam et al., 2022). It is composed of three polypeptide chains – alpha chains – that are coiled together to form a triple helix, being rich in amino acids, particularly glycine (around 30%), proline, and hydroxyproline (Oliveira et al., 2021). Collagen molecules are cross-linked to enhance their stability and resistance to enzymatic degradation through covalent bonds between specific amino acids. This process involves the formation of lysine and hydroxylysine in adjacent collagen molecules (Oliveira et al., 2021). Cross-linking is responsible for collagen's ability to withstand mechanical stress and maintain its structural integrity, making it an attractive biomaterial for various industrial applications. In the food industry, it is commonly used as a gelling agent in processed meats, sausages, and jellied desserts, improving texture, increasing water retention, and enhancing the organoleptic attributes (Irastorza et al., 2021). Collagen-based films and coatings are also employed to extend the shelf life of perishable food items (Gautam et al., 2022). Additionally, collagen peptides derived from food waste have gained popularity as functional ingredients in dietary supplements and functional foods due to their potential health benefits (Irastorza et al., 2021).

In the pharmaceutical and biomedicine sectors, collagen serves as a crucial biomaterial widely used in the formulation of drug delivery systems, wound dressings, and tissue engineering scaffolds (Irastorza et al., 2021). Collagen's biocompatibility, biodegradability, and ability to support cell adhesion make it an ideal choice for tissue regeneration and repair (Nagarajan et al., 2019). Collagen has a long history in the cosmetic and dermatological industries due to its skin-enhancing properties. Collagen-based cosmetics and skincare products claim to promote skin hydration, elasticity, and firmness (Gupta et al., 2022). This particular demand has led to the development of innovative collagen formulations sourced from food waste, promoting sustainability in the cosmetics industry. Collagen is integral to various biotechnological processes, including cell culture and tissue engineering. Collagen-coated culture dishes facilitate cell adhesion and growth, essential for in vitro studies and biopharmaceutical production (Nagarajan et al., 2019). Additionally, collagen-based matrices serve as three-dimensional substrates for tissue engineering, enabling the development of functional tissue constructs (Grabska-Zielińska et al., 2021).

Collagen can be extracted from various food waste sources, including discarded fish scales and skin obtained during fish processing, poultry by-products such as chicken feet, cartilage, and bones, and bones from other animals, such as cattle or pigs, which are typically discarded as waste in the food industry. The extraction involves several steps, including cleaning, deproteinization, demineralization, and purification (Gautam et al., 2022). Different sources for collagen extraction are listed in Table 7.4.

Gelatin, in its most fundamental form, is a proteinaceous substance primarily composed of collagen. Gelatin's structure is very similar to its precursor, with some alterations resulting from the extraction process. During extraction, collagen undergoes hydrolysis, breaking down the triple helix into individual polypeptide chains (Alipal et al., 2021). These chains, also known as gelatin peptides, possess various functional groups, including amine, carboxyl, and hydroxyl groups, which confer distinctive properties to gelatin. Typically, gelatin comprises a mixture of amino acids, including glycine (23–27%), proline (12–15%), hydroxyproline (9–12%), and alanine (8–12%) (Tyuftin and Kerry 2021). This amino acid profile contributes to the gelling, emulsifying, and film-forming properties of gelatin. Furthermore, the molecular weight of gelatin peptides can vary, ranging from small peptides to larger molecules, influencing its rheological and functional characteristics (Irastorza et al., 2021).

Gelatin is commonly derived from animal connective tissues, which serve as the primary source for industrial production. These tissues include skin, bones, cartilage, and tendons, obtained from various animal species, such as cattle, pigs, and poultry, among others presented in Table 7.4. Gelatin has a crucial role in the food industry, where it serves as a gelling agent, stabilizer, thickener, and emulsifier in a wide array of products (Alipal et al., 2021). Common applications include the production of gummy candies, marshmallows, and jelly desserts, providing the desired texture

TABLE 7.4
Alternative Sources for Collagen and Gelatin

Biopolymer	Food Waste	Yield	Reference
Collagen	Bigeye tuna – skin	16.7%	(Ahmed, Haq, and Chun 2019)
	Tilapia – skin	27.2%	(Liao et al., 2018)
	Chicken – fat lungs	31.2%	(Zou et al., 2020)
	Squid – outer skin	56.8%	(Veeruraj et al., 2015)
	Pufferfish – skin	67.3%	(Chen et al., 2019)
	Sea cucumber – body wall	72.2%	(P.-H. Li et al., 2020)
	Bovine hides	73.4%	(Ocak 2020)
	Giant croaker – skin	84.8%	(Yu et al., 2018)
	Chicken – sternal cartilage	87.2%	(Akram and Zhang 2020)
	Golden carp – skin	94.9%	(Ali, Kishimura, and Benjakul 2018)
Gelatin	Turbot – skin	4.5–9.1%	(Valcarcel et al., 2021)
	Bovine bones	4.7%–11.7%	(Cao et al., 2020)
	Chicken bone by-products	6.6%–10.6%	(Saenmuang, Phothiset, and Chumnanka 2020)
	Porcine bones	13.5%	(Ma et al., 2019)
	Bovine skins	16.1%–23.7%	(Tanbir Ahmad et al., 2018)
	Salamander skin	17.6%	(Jin et al., 2019)
	Bovine skins	17.9%–23.5%	(Tanbir Ahmad et al., 2020)
	Tilapia – skin	18.4%–22%	(Zhang et al., 2020)
	Camel skin	36.8%–42.4%	(Al-Hassan 2020)
	Tilapia – bones and skin	60.1% and 74.4%	(Tinratat and Sila-asna 2017)

and chewiness. It is employed in yogurt, ice cream, and dairy-based desserts to enhance texture and mouthfeel. Gelatin can improve the quality of processed meats, sausages, and meat-based soups by binding water and improving juiciness. In baked goods like cakes and pastries, gelatin can be used to improve moisture retention and extend shelf life (Gautam et al., 2022). Gelatin also serves as a crucial ingredient in pharmaceutical formulations due to its biocompatibility and ease of incorporation, such as in capsules for oral drug delivery, providing a convenient and bioavailable dosage form; coatings for tablets and pills, facilitating drug release, taste masking, and protection against environmental factors; and microencapsulation for controlled drug release and targeted delivery systems (Irastorza et al., 2021). Gelatin is a common ingredient in facial masks, where it forms a flexible film to moisturize and rejuvenate the skin; it can also be found in hair and nail care products for conditioning and protection treatments (Gupta et al., 2022). Its biocompatibility, biodegradability, and ability to support cell growth make it a valuable biomaterial for tissue scaffolds and healing by creating a moist environment conducive to tissue repair (Grabska-Zielińska et al., 2021).

7.4.3 Microbial Production of Biopolymers

For the microbial production of biopolymers from food waste, it is necessary to choose the most suitable microorganism for the process since each species produces a specific type of biopolymer (Table 7.5). For example, *Cupriavidus necator*, *Pseudomonas* sp., *Azotobacter* sp., and *Bacillus* sp. can produce polyhydroxyalkanoates (PHAs) from waste streams (Pagliano et al., 2017). *Lactobacillus casei* and *Lactobacillus farciminis* can be used to produce lactic acid for the further obtainment of PLA (Gautam et al., 2022). *Xanthomonas campestris* and *Xanthomonas pelargonii* are used to produce xanthan gum (XG) (Ranganathan et al., 2020). *Acetobacter xylinum*, *A. hansenii*, and *A. pasteurianus* are used to produce bacterial cellulose (BC) (Wang et al., 2019b). BC can also be produced by a symbiotic culture of bacteria and yeast (SCOBY) present in fermented kombucha beverages (Ramírez- Tapias et al., 2022).

Each biopolymer has specific characteristics, such as their cellular localization: PHAs are accumulated inside the cells while the others (lactic acid, xanthan gum, and BC) are liberated into the media. Also, BC, PHAs, and XG are synthesized in their polymerized form, but lactic acid needs

TABLE 7.5
Microbial Biopolymers Production from Food Waste

Food Waste	Microorganism	Biopolymer	Production	Reference
Corn cob waste and dairy wastewater	*Lactobacillus plantarum*	LA	11.15 g/L	(Naomi David, Sewsynker-Sukai, and Gueguim Kana 2022)
Orange peel waste	*Lactobacillus delbrueckii* ssp. *delbrueckii*	LA	96.3 g/L	(de la Torre, Ladero, and Santos 2020)
Food waste oil	*Pseudomonas* sp. H3	PHA	8.64 g/L	(Pan et al., 2021)
Urban food waste	*Cupriavidus necator* ATCC 17699	PHA	5.32 g/L	(Hathi et al., 2022)
Waste bread	*Xanthomonas axonopodis vesicatoria*	XG	14.3 g/L	(Demirci et al., 2019)
Kitchen waste	*Xanthomonas campestris* LRELP-1	XG	11.73 g/L	(P. Li et al., 2016)
Sarchy kitchen waste	*Komagataeibacter hansenii* AS.5	BC	2.11 g/L	(A. K. Saleh et al., 2021)
Mango peel waste	*Achromobacter* S3	BC	1.22 g/L	(Hasanin et al., 2023)

to go through a polymerization phase to form PLA. Overall, microbial biopolymers share similar mechanical characteristics. Additionally, biopolymers can be combined with each other to produce a biomaterial with enhanced properties. For example, a BC-PLA composite was obtained by cultivating BC SCOBY bacteria in yerba mate waste, and the resulting BC was used to reinforce a PLA matrix (Agüero et al., 2023). The final biopolymer showed improved water barrier effects, mechanical properties, and antimicrobial activity, being suitable for food packaging or agricultural mulch films.

It is important to understand the biopolymer characteristics to select its field of application. For example, due to their mechanical strength, PHAs can be applied in bone tissue engineering and other medical applications like drug delivery carriers and implants (Pandey et al., 2022). BC can be used in food packaging, pharmaceuticals, and cosmetics industries (Wang et al., 2019). XG is mostly used as an additive in food industries but can also be applied in pharmaceutical, biomedical, and cosmetics fields (Petri 2015). Finally, PLA is suitable to be used in food packaging, tissue scaffolds, and medical applications (Sreekumar et al., 2021). It has been noticed that the use of nutrient-rich waste has enabled the use of biopolymers in a larger range of fields (Gautam et al., 2022).

7.5 NEW INNOVATIONS REGARDING BIOPOLYMER PRODUCTION FROM FOOD WASTE

One way to produce biopolymers in a sustainable manner is through their obtainment in biorefineries, which aim to replicate petroleum-based refineries but use biomass as a raw material rather than a non-renewable source. Therefore, multiple bioproducts can be obtained in the same platform, applying the residues obtained from other processes, ideally generating zero waste (Kumar et al., 2019). Biopolymers can benefit from these systems, as these molecules are of great importance nowadays and can bring a high added value to the system. In addition, using bio-based feedstocks, especially food waste, is an important step towards the sustainable development of biopolymers. For example, Patsalou and collaborators (2020) used citrus peel waste for the recovery of different biomolecules, among them pectin. The biopolymer was recovered through dilute acid hydrolysis with a recovery of 30%, while an essential oil containing limonene was recovered through distillation (0.43% yield). Additionally, the liquid fraction from the dilute acid hydrolysis was used for the production of succinic acid with *Actinobacillus succinogenes*, which produced up to 22.4 g/L of the organic acid, and the remaining solids of the overall process were evaluated as a fertilizer. Similarly, Talekar and colleagues (2018) used pomegranate peels for the recovery of pectin (up to 21%) and phenolics (12%) through hydrothermal treatment. The remaining solids were subjected to enzymatic hydrolysis, and the hydrolysate applied to ethanol production, which reached up to 34 g/L of the biofuel.

Genetic engineering can also be an innovative approach for the sustainable production of biopolymers, with the development of strains that can produce either the biomaterials or their precursors. Microorganisms that can natively produce the biopolymers could have their productivity enhanced by genetic improvement or could be engineered to consume a certain substrate that they do not usually consume. Also, PHAs-producing bacteria could be engineered, for example, for the production of co-polymers of PHAs or with other biopolymers (e.g., polyhydroxybutyrate-co-lactate, P(3HB-co-LA)) or could be enhanced to accumulate the PHAs in milder conditions that do not require considerable metabolic stress, which can further reduce costs of production. On the other hand, strains that do not naturally produce biopolymers or their precursors but are able to consume specific carbon sources derived from waste can also be engineered to produce the biomolecules of interest (Nielsen et al., 2017; Wu et al., 2021).

Rodríguez et al., (2021) developed an engineered *Cupriavidus necator* strain by introducing two lipolytic enzyme-coding genes from *Pseudomonas stutzeri* for the obtainment of PHAs from slaughterhouse waste. The engineered strain showed great lipolytic activity (121 mU/(min·mL)) compared to the native strain of *C. necator*. The best results of PHB production were still obtained with glucose; however, when using swine jowl as a substrate, the engineered strain was able to accumulate up to 65% of PHB while the native one accumulated 18%, showing the success of the bacterial transformation. The developed process can also be further optimized in terms of physico-chemical conditions for the obtainment of higher yield and productivity. In turn, Pais and colleagues (2014) generated a recombinant *Escherichia coli* strain by introducing a PHB-producing gene from *C. necator* and further conducting mutation on *E. coli* by proton suicide methodology. The recombinant strain could accumulate up to 21% of PHB using cheese whey as a substrate in a fed-batch operation mode, while the recombinant-mutant strain reached 38% of accumulation.

Another promising and innovative mode of operation is consolidated bioprocessing, in which enzyme production occurs simultaneously with biomass hydrolysis and product obtainment, in the same bioreactor. Usually, genetically engineered strains are applied in this case. This type of operation has the advantage of being less energy-intensive and requires low maintenance costs. It is mostly researched for the production of cellulose-based ethanol; however, other industries and researchers are adapting this mode of operation for different biomolecules (Ibrahim, Kim, and Abd-Aziz 2018). In this sense, Brojanigo et al., (2022) developed an engineered *Cupriavidus necator* that could produce extracellular amylases, which in turn converted starchy materials such as broken rice to glucose that was metabolized by the same *C. necator* strain for the production of PHB, reaching up to 5.78 g/L of the biopolymer. Consolidated bioprocessing is a highly innovative technology; therefore, literature and reports on this theme are still scarce, especially regarding the production of biopolymers. However, the development of new processes and strains can boost the implementation of this technology in the industry.

Processes for biopolymer extraction from food waste, as discussed throughout the chapter, are usually well-established methods regarding the type of polymer to be obtained and the raw material of origin. The best alternatives for extraction are the ones that do not apply severe conditions for the biopolymer recovery, either in terms of temperature, pressure, pH, or the use of hazardous solvents (Borić et al., 2018). Therefore, innovations regarding extraction are directed to methods that can be considered greener and more sustainable, using milder conditions for the processes, usually involving microorganisms. For example, Suresh and collaborators (2023) used the proteolytic bacteria *Stenotrophomonas koreensis* for the fermentation of fish scales and the further recovery of chitin. Extraction occurred in a biological manner, as the bacteria could degrade the protein and non-chitin part of the fish scales, leaving the desired biopolymer at the end. A final yield of 2.8g of chitin per 100g of fish scales was obtained. Despite being a sustainable method, biological methods for the recovery of biopolymers can lead to low yields. In order to counter this drawback, Borić and colleagues (2018) developed a process for the recovery of chitin through a plasma-based process. The chitin content increased from 25 to 52% from the untreated to the plasma-treated waste of crustacean biomass. Plasma is known to degrade proteins and is commonly used in the sterilization of medical devices. In addition, this method does not require any solvent to form plasma or to pretreat the substrate, which is a great alternative, as it can be considered greener.

7.6 CONCLUSIONS AND PERSPECTIVES

The production of biopolymers from food waste is a great alternative for the utilization of this undervalued by-product. The natural composition of some wastes enables the direct extraction of different biopolymers, such as polysaccharides and polypeptides. Also, through fermentation, there is the possibility of obtaining high-quality materials and/or their precursors. There have been research and technological efforts in order to properly implement this type of production; however, there is still the need to evaluate overall process viability and sustainability through techno-economic analysis and life-cycle assessments. Along with that, public policies on the sustainable production and use of biomaterials should boost the development of new biopolymers. In order to guarantee greener processes with high yields, recent developments show the use of biological and solventless extraction of biopolymers and their production in biorefineries and consolidated bioprocessing, broadening the frontiers for further evolution of these biomolecules.

REFERENCES

Adesra, Ankita, Vijay Kumar Srivastava, and Sunita Varjani. 2021. "Valorization of dairy wastes: Integrative approaches for value added products". *Indian Journal of Microbiology* 61 (3): 270–78. doi:10.1007/s12088-021-00943-5.

Agarwal, Shilpi, Shailey Singhal, Chirag B. Godiya, and Sanjay Kumar. 2021. "Prospects and applications of starch based biopolymers". *International Journal of Environmental Analytical Chemistry* 00 (00). Taylor & Francis: 1–20. doi:10.1080/03067319.2021.1963717.

Agüero, Ángel, Esther Corral Perianes, Sara Soledad Abarca de las Muelas, Diego Lascano, María del Mar de la Fuente García-Soto, Mercedes Ana Peltzer, Rafael Balart, and Marina Patricia Arrieta. 2023. "Plasticized mechanical recycled PLA films reinforced with microbial cellulose particles obtained from kombucha fermented in yerba mate waste". *Polymers* 15 (2). doi:10.3390/polym15020285.

Ahmad, Ashfaq, Israa Othman, Blaise Leopold Tardy, Shadi W. Hasan, and Fawzi Banat. 2022. "Enhanced lactic acid production with indigenous microbiota from date pulp waste and keratin protein hydrolysate from chicken feather waste". *Bioresource Technology Reports* 18 (June): 101089. doi:10.1016/j.biteb.2022.101089.

Ahmad, Imran, Natasha Nabila Binti Ibrahim, Norhayati Abdullah, Iwamoto Koji, Shaza Eva Mohamad, Kuan Shiong Khoo, Wai Yan Cheah, Tau Chuan Ling, and Pau Loke Show. 2023. "Bioremediation strategies of palm oil mill effluent and landfill leachate using microalgae cultivation: An approach contributing towards environmental sustainability". *Chinese Chemical Letters* 34 (5): 107854. doi:10.1016/j.cclet.2022.107854.

Ahmad, Talha, Rana Muhammad Aadil, Haassan Ahmed, Ubaid ur Rahman, Bruna C.V. Soares, Simone L.Q. Souza, Tatiana C. Pimentel, et al., 2019. "Treatment and utilization of dairy industrial waste: A review". *Trends in Food Science & Technology* 88 (June): 361–72. doi:10.1016/j.tifs.2019.04.003.

Ahmad, Tanbir, Amin Ismail, Siti Aqlima Ahmad, Khalilah Abdul Khalil, Leo Teik Kee, Elmutaz Atta Awad, and Awis Qurni Sazili. 2020. "Extraction, characterization and molecular structure of bovine skin gelatin extracted with plant enzymes bromelain and zingibain". *Journal of Food Science and Technology* 57 (10): 3772–81. doi:10.1007/s13197-020-04409-2.

Ahmad, Tanbir, Amin Ismail, Siti Aqlima Ahmad, Khalilah Abdul Khalil, Leo Teik Kee, Elmutaz Atta Awad, Kazeem Dauda Adeyemi, and Awis Qurni Sazili. 2018. "Autolysis of bovine skin, its endogenous proteases, protease inhibitors and their effects on quality characteristics of extracted gelatin". *Food Chemistry* 265 (November): 1–8. doi:10.1016/j.foodchem.2018.05.046.

Ahmed, Raju, Monjurul Haq, and Byung-Soo Chun. 2019. "Characterization of marine derived collagen extracted from the by-products of bigeye tuna (Thunnus obesus)". *International Journal of Biological Macromolecules* 135 (august): 668–76. doi:10.1016/j.ijbiomac.2019.05.213.

Akram, Ayesha Noreen, and Chunhui Zhang. 2020. "Extraction of collagen-II with pepsin and ultrasound treatment from chicken sternal cartilage; physicochemical and functional properties". *Ultrasonics Sonochemistry* 64 (June): 105053. doi:10.1016/j.ultsonch.2020.105053.

Al-Hassan, A.A. 2020. "Gelatin from camel skins: Extraction and characterizations". *Food Hydrocolloids* 101 (April): 105457. doi:10.1016/j.foodhyd.2019.105457.

Ali, Ali Muhammed Moula, Hideki Kishimura, and Soottawat Benjakul. 2018. "Extraction efficiency and characteristics of acid and pepsin soluble collagens from the skin of golden carp (Probarbus Jullieni) as affected by ultrasonication". *Process Biochemistry* 66 (March): 237–44. doi:10.1016/j.procbio.2018.01.003.

Alipal, J., N.A.S. Mohd Pu'ad, T.C. Lee, N.H.M Nayan, N. Sahari, H. Basri, M.I. Idris, and H.Z. Abdullah. 2021. "A review of gelatin: Properties, sources, process, applications, and commercialisation". *Materials Today: Proceedings* 42: 240–50. doi:10.1016/j.matpr.2020.12.922.

Amorim, Julia Didier Pedrosa de, Karina Carvalho de Souza, Cybelle Rodrigues Duarte, Izarelle da Silva Duarte, Francisco de Assis Ribeiro, Sales, Girlaine Santos Silva, et al., 2020. "Plant and bacterial nanocellulose: production, properties and applications in medicine, food, cosmetics, electronics and engineering. A review". *Environmental Chemistry Letters* 18: 851–69. doi:10.1007/s10311-020-00989-9.

Aremu, Mujidat O., Mofoluwake M. Ishola, and Mohammad J. Taherzadeh. 2021. "Polyhydroxyalkanoates (PHAs) production from Volatile Fatty Acids (VFAs) from organic wastes by pseudomonas oleovorans". *Fermentation* 7 (4): 287. doi:10.3390/fermentation7040287.

Arrutia, Fátima, Mohamed Adam, Miguel Ángel Calvo-Carrascal, Yujie Mao, and Eleanor Binner. 2020. "Development of a continuous-flow system for microwave-assisted extraction of pectin-derived oligosaccharides from food waste". *Chemical Engineering Journal* 395 (September): 125056. doi:10.1016/j.cej.2020.125056.

Atasoy, Merve, Isaac Owusu-Agyeman, Elzbieta Plaza, and Zeynep Cetecioglu. 2018. "Bio-based volatile fatty acid production and recovery from waste streams: Current status and future challenges". *Bioresource Technology* 268 (november): 773–86. doi:10.1016/j.biortech.2018.07.042.

Awogbemi, Omojola, Emmanuel Idoko Onuh, and Freddie L Inambao. 2019. "Comparative study of properties and fatty acid composition of some neat vegetable oils and waste cooking oils". *International Journal of Low-Carbon Technologies* 14 (3): 417–25. doi:10.1093/ijlct/ctz038.

Azelee, Nur Izyan Wan, D. Digvijay, Seenivasan Ayothiraman, Norhayati Mohamed Noor, Zaitul Iffa Abd Rasid, Aizi Nor Mazila Ramli, Balasubramani Ravindran, Felicitas U. Iwuchukwu, and Rangabhashiyam Selvasembian. 2023. "Sustainable valorization approaches on crustacean wastes for the extraction of chitin, bioactive compounds and their applications - A review". *International Journal of Biological Macromolecules*, August, 126492. doi:10.1016/j.ijbiomac.2023.126492.

Banu J, Rajesh, and Godvin Sharmila V. 2023. "Review on food waste valorisation for bioplastic production towards a circular economy: Sustainable approaches and biodegradability assessment". *Sustainable Energy & Fuels*. Royal Society of Chemistry, 3165–84. doi:10.1039/d3se00500c.

Baranwal, Jaya, Brajesh Barse, Antonella Fais, Giovanna Lucia Delogu, and Amit Kumar. 2022. "Biopolymer: A Sustainable Material for Food and Medical Applications". *Polymers* 14 (5): 1–22. doi:10.3390/polym14050983.

Berenguer, Cristina V., Carolina Andrade, Jorge A.M. Pereira, Rosa Perestrelo, and José S. Câmara. 2023. "Current challenges in the sustainable valorisation of agri-food wastes: A review". *Processes* 11 (1). doi:10.3390/pr11010020.

Bhuimbar, Madhuri V., Prashant K. Bhagwat, and Padma B. Dandge. 2019. "Extraction and characterization of acid soluble collagen from fish waste: Development of collagen-chitosan blend as food packaging film". *Journal of Environmental Chemical Engineering* 7 (2): 102983. doi:10.1016/j.jece.2019.102983.

Binma-ae, Hatsalinda, Poonsuk Prasertsan, and Wanna Choorit. 2020. "Preparation and characterization of biopolymers recovered from palm oil mill effluent and their complex hydrogels compared to commercial xylan". *Waste and Biomass Valorization* 11 (10): 5109–21. doi:10.1007/s12649-019-00823-6.

Borić, M., H. Puliyalil, U. Novak, and B. Likozar. 2018. "An intensified atmospheric plasma-based process for the isolation of the chitin biopolymer from waste crustacean biomass". *Green Chemistry* 20 (6): 1199–1204. doi:10.1039/c7gc03735j.

Bouiche, Cilia, Aicha, Asma Houfani, Zahra Azzouz, Lamia Taouzinet, Lamia Medouni-Haroune, Meriem Amrane-Abider, and Khodir Madani. 2023. "Exploring the potential of xylooligosaccharides as prebiotics: Insights from CAZymes families and their emerging health benefits". *Biologia* 78: 1789–1806. doi:10.1007/s11756-023-01440-4.

Brojanigo, Silvia, Nicoletta Gronchi, Tiziano Cazzorla, Tuck Seng Wong, Marina Basaglia, Lorenzo Favaro, and Sergio Casella. 2022. "Engineering cupriavidus necator DSM 545 for the one-step conversion of starchy waste into polyhydroxyalkanoates". *Bioresource Technology* 347 (November 2021). Elsevier Ltd: 126383. doi:10.1016/j.biortech.2021.126383.

Bühlmann, Christopher H., Bede S. Mickan, Stephan Tait, and Parisa A. Bahri. 2022. "Developing a food waste biorefinery: Lactic acid extraction using anionic resin and impacts on downstream biogas production". *Chemical Engineering Journal* 431 (March): 133243. doi:10.1016/j.cej.2021.133243.

Cao, Songmin, Yi Wang, Lujuan Xing, Wangang Zhang, and Guanghong Zhou. 2020. "Structure and physical properties of gelatin from bovine bone collagen influenced by acid pretreatment and pepsin". *Food and Bioproducts Processing* 121 (May): 213–23. doi:10.1016/j.fbp.2020.03.001.

Carter, B.G., N. Cheng, R. Kapoor, G.H. Meletharayil, and M.A. Drake. 2021. "Invited review: Microfiltration-derived casein and whey proteins from milk". *Journal of Dairy Science* 104 (3): 2465–79. doi:10.3168/jds.2020-18811.

Carus, Michael, and Lara Dammer. 2018. "The circular bioeconomy - Concepts, opportunities, and limitations". *Industrial Biotechnology* 14 (2): 83–91. doi:10.1089/ind.2018.29121.mca.

Cebin, Aleksandra Vojvodić, Marie Christine Ralet, Jacqueline Vigouroux, Sara Karača, Arijana Martinić, Draženka Komes, and Estelle Bonnin. 2021. "Valorisation of walnut shell and pea pod as novel sources for the production of xylooligosaccharides". *Carbohydrate Polymers* 263 (July). Elsevier: 117932. doi:10.1016/J.CARBPOL.2021.117932.

Chandrasekar, Chandra Mohan, Harini Krishnamachari, Stefano Farris, and Diego Romano. 2023. "Development and characterization of starch-based bioactive thermoplastic packaging films derived from banana peels". *Carbohydrate Polymer Technologies and Applications* 5 (June): 100328. doi:10.1016/j.carpta.2023.100328.

Chatzifragkou, Afroditi, and Seraphim Papanikolaou. 2012. "Effect of impurities in biodiesel-derived waste glycerol on the performance and feasibility of biotechnological processes". *Applied Microbiology and Biotechnology* 95 (1): 13–27. doi:10.1007/s00253-012-4111-3.

Chen, Junde, Min Li, Ruizao Yi, Kaikai Bai, Guangyu Wang, Ran Tan, Shanshan Sun, and Nuohua Xu. 2019. "Electrodialysis extraction of pufferfish skin (Takifugu flavidus): A promising source of collagen". *Marine Drugs* 17 (1): 25. doi:10.3390/md17010025.

Corrado, Sara, Carla Caldeira, Mattias Eriksson, Ole Jørgen Hanssen, Hans-Eduard Hauser, Freija van Holsteijn, Gang Liu, et al., 2019. "Food waste accounting methodologies: Challenges, opportunities, and further advancements". *Global Food Security* 20 (March): 93–100. doi:10.1016/j.gfs.2019.01.002.

Dan, Tingxing, Huiyan Jing, Tian Shen, Jia Zhu, and Yanping Liu. 2023. "Performance of production of polyhydroxyalkanoates from food waste fermentation with Rhodopseudomonas palustris". *Bioresource Technology* 385 (October): 129165. doi:10.1016/j.biortech.2023.129165.

Dang, Bao-Trong, Xuan-Thanh Bui, Duyen P.H. Tran, Huu Hao Ngo, Long D. Nghiem, Thi-Khanh-Dieu Hoang, Phuong-Thao Nguyen, et al., 2022. "Current application of algae derivatives for bioplastic production: A review". *Bioresource Technology* 347 (March): 126698. doi:10.1016/j.biortech.2022.126698.

Demirci, Ahmet Sukru, Ibrahim Palabiyik, Demet Apaydın, Mustafa Mirik, and Tuncay Gumus. 2019. "Xanthan gum biosynthesis using Xanthomonas isolates from waste bread: Process optimization and fermentation kinetics". *Lwt* 101 (June 2018): 40–47. doi:10.1016/j.lwt.2018.11.018.

Deshwal, Gaurav Kr, Tanweer Alam, Narender, Raju Panjagari, and Aastha Bhardwaj. 2021. "Utilization of cereal crop residues, cereal milling, sugarcane and dairy processing by-products for sustainable packaging solutions". *Journal of Polymers and the Environment* 29: 2046–61. doi:10.1007/s10924-020-02034-w.

Devadas, Vishno Vardhan, Kuan Shiong Khoo, Wen Yi Chia, Kit Wayne Chew, Heli Siti Halimatul Munawaroh, Man-Kee Lam, Jun-Wei Lim, Yeek-Chia Ho, Keat Teong Lee, and Pau Loke Show. 2021. "Algae biopolymer towards sustainable circular economy". *Bioresource Technology* 325 (April): 124702. doi:10.1016/j.biortech.2021.124702.

Edwiges, Thiago, Laercio Frare, Bruna Mayer, Leonardo Lins, Jin Mi Triolo, Xavier Flotats, and Mônica Sarolli Silva de Mendonça Costa. 2018. "Influence of chemical composition on biochemical methane potential of fruit and vegetable waste". *Waste Management* 71 (January): 618–25. doi:10.1016/j.wasman.2017.05.030.

Fu, Xuemei, Heng Xu, Qi Zhang, Jiaxing Xi, Huihui Zhang, Mingxia Zheng, Beidou Xi, and Li'an Hou. 2023. "A review on polyhydroxyalkanoates production from various organic waste streams: Feedstocks, strains, and production strategy". *Resources, Conservation and Recycling* 198 (November): 107166. doi:10.1016/j.resconrec.2023.107166.

Garcia-Garcia, Guillermo, Elliot Woolley, Shahin Rahimifard, James Colwill, Rod White, and Louise Needham. 2017. "A methodology for sustainable management of food waste". *Waste and Biomass Valorization* 8 (6): 2209–27. doi:10.1007/s12649-016-9720-0.

García-Moreno, Pedro J., Andres Pelayo, Sen Yu, María Busolo, Jose M. Lagaron, Ioannis S. Chronakis, and Charlotte Jacobsen. 2018. "Physicochemical characterization and oxidative stability of fish oil-loaded electrosprayed capsules: Combined use of whey protein and carbohydrates as wall materials". *Journal of Food Engineering* 231 (August): 42–53. doi:10.1016/j.jfoodeng.2018.03.005.

Gautam, Krishna, Reena Vishvakarma, Poonam Sharma, Amarnath Singh, Vivek Kumar Gaur, Sunita Varjani, and Janmejai Kumar Srivastava. 2022. "Production of biopolymers from food waste: Constrains and perspectives". *Bioresource Technology* 361 (July). Elsevier Ltd: 127650. doi:10.1016/j.biortech.2022.127650.

Gong, Xiaoqiang, Menghan Wu, Yong Jiang, and Hui Wang. 2021. "Effects of different temperatures and pH values on volatile fatty acids production during codigestion of food waste and thermal-hydrolysed sewage sludge and subsequent volatile fatty acids for polyhydroxyalkanoates production". *Bioresource Technology* 333 (August): 125149. doi:10.1016/j.biortech.2021.125149.

Grabska-Zielińska, Sylwia, Alina Sionkowska, Ângela Carvalho, and Fernando J. Monteiro. 2021. "Biomaterials with potential use in bone tissue regeneration—Collagen/Chitosan/Silk fibroin scaffolds cross-linked by EDC/NHS". *Materials* 14 (5): 1105. doi:10.3390/ma14051105.

Gupta, Shreya, Swati Sharma, Ashok Kumar Nadda, Mohamed Saad Bala Husain, and Arun Gupta. 2022. "Biopolymers from waste biomass and its applications in the cosmetic industry: A review". *Materials Today: Proceedings* 68: 873–79. doi:10.1016/j.matpr.2022.06.422.

Hasanin, Mohamed S., Mohamed Abdelraof, Amr H. Hashem, and Houssni El Saied. 2023. "Sustainable bacterial cellulose production by Achromobacter using mango peel waste". *Microbial Cell Factories* 22 (1). BioMed Central: 1–12. doi:10.1186/s12934-023-02031-3.

Hassan, Mohd Ali, Lian-Ngit Yee, Phang Lai Yee, Hidayah Ariffin, Abdul Rahim Raha, Yoshihito Shirai, and Kumar Sudesh. 2013. "Sustainable production of polyhydroxyalkanoates from renewable oil-palm biomass". *Biomass and Bioenergy* 50 (March): 1–9. doi:10.1016/j.biombioe.2012.10.014.

Hathi, Zubeen J., Md Ariful Haque, Anshu Priya, Zi hao Qin, Shuquan Huang, Chun Ho Lam, Dimitris Ladakis, et al., 2022. "Fermentative bioconversion of food waste into biopolymer poly(3-hydroxybutyrate-co-3-hydroxyvalerate) using Cupriavidus necator". *Environmental Research* 215 (P1). Elsevier Inc.: 114323. doi:10.1016/j.envres.2022.114323.

Ibrahim, Mohamad Faizal, Seung Wook Kim, and Suraini Abd-Aziz. 2018. "Advanced bioprocessing strategies for biobutanol production from biomass". *Renewable and Sustainable Energy Reviews* 91 (March 2017). Elsevier Ltd: 1192–1204. doi:10.1016/j.rser.2018.04.060.

Irastorza, Ainhoa, Iratxe Zarandona, Mireia Andonegi, Pedro Guerrero, and Koro de la Caba. 2021. "The versatility of collagen and chitosan: From food to biomedical applications". *Food Hydrocolloids* 116 (July): 106633. doi:10.1016/j.foodhyd.2021.106633.

Ji, Mengke, Taoran Zheng, Ziyu Wang, Weijian Lai, Lizhan Zhang, Qianyi Zhang, Hongyi Yang, et al., 2023. "PHB production from food waste hydrolysates by Halomonas bluephagenesis Harboring PHB operon linked with an essential gene". *Metabolic Engineering* 77 (May): 12–20. doi:10.1016/j.ymben.2023.03.003.

Jin, Wen-Gang, Jinjin Pei, Yi-Nan Du, Jinfeng Pan, Ruichang Gao, De-Jing Chen, Hai-Tao Wu, and Bei-Wei Zhu. 2019. "Characterization and functional properties of gelatin extracted from Chinese Giant Salamander (*Andrias Davidianus*) skin". *Journal of Aquatic Food Product Technology* 28 (8): 861–76. doi:10.1080/10498850.2019.1652716.

Kaur, Samandeep, Parmjit S. Panesar, and Harish K. Chopra. 2023. "Extraction of dietary fiber from kinnow (Citrus reticulata) peels using sequential ultrasonic and enzymatic treatments and its application in development of cookies". *Food Bioscience* 54 (August): 102891. doi:10.1016/j.fbio.2023.102891.

Khalil, Mohammed N. A., Hebatullah H. Farghal, and Mohamed A. Farag. 2022. "Outgoing and potential trends of composition, health benefits, juice production and waste management of the multi-faceted Grapefruit *Citrus X paradisi*: A comprehensive review for maximizing its value". *Critical Reviews in Food Science and Nutrition* 62 (4): 935–56. doi:10.1080/10408398.2020.1830364.

Khatami, Kasra, Mariel Perez-Zabaleta, and Zeynep Cetecioglu. 2022. "Pure cultures for synthetic culture development: Next level municipal waste treatment for polyhydroxyalkanoates production". *Journal of Environmental Management* 305 (March): 114337. doi:10.1016/j.jenvman.2021.114337.

Kongala, Sippi Issac, and Anil Kondreddy. 2023. "A review on plant and pathogen derived carbohydrates, oligosaccharides and their role in plant's immunity". *Carbohydrate Polymer Technologies and Applications* 6 (December). Elsevier: 100330. doi:10.1016/J.CARPTA.2023.100330.

Kumar, P. Senthil, A. Saravanan, R. Jayasree, and S. Jeevanantham. 2019. *Food industry waste biorefineries. Refining biomass residues for sustainable energy and bioproducts: Technology, advances, life cycle assessment, and economics*. Elsevier Inc. doi:10.1016/B978-0-12-818996-2.00018-1.

Kwan, Tsz Him, Yunzi Hu, and Carol Sze Ki Lin. 2016. "Valorisation of food waste via fungal hydrolysis and lactic acid fermentation with Lactobacillus casei Shirota". *Bioresource Technology* 217 (October): 129–36. doi:10.1016/j.biortech.2016.01.134.

la Torre, Isabel de, Miguel Ladero, and Victoria E. Santos. 2020. "D-lactic acid production from orange waste enzymatic hydrolysates with L. delbrueckii cells in growing and resting state". *Industrial Crops and Products* 146 (September 2019). Elsevier: 112176. doi:10.1016/j.indcrop.2020.112176.

Lal, A. M. Nandhu, M. V. Prince, Anjineyulu Kothakota, R. Pandiselvam, Rohit Thirumdas, Naveen Kumar Mahanti, and R. Sreeja. 2021. "Pulsed electric field combined with microwave-assisted extraction of pectin polysaccharide from jackfruit waste". *Innovative Food Science & Emerging Technologies* 74 (December). Elsevier: 102844. doi:10.1016/J.IFSET.2021.102844.

Li, Panyu, Ting Li, Yu Zeng, Xiang Li, Xiaolong Jiang, Yabo Wang, Tonghui Xie, and Yongkui Zhang. 2016. "Biosynthesis of xanthan gum by Xanthomonas campestris LRELP-1 using kitchen waste as the sole substrate". *Carbohydrate Polymers* 151. Elsevier Ltd.: 684–91. doi:10.1016/j.carbpol.2016.06.017.

Li, Po-Hsien, Wen-Chien Lu, Yung-Jia Chan, Wen-Ching Ko, Chao-Chuen Jung, Dung Thi Le Huynh, and Yu-Xiang Ji. 2020. "Extraction and characterization of collagen from sea cucumber (Holothuria cinerascens) and its potential application in moisturizing cosmetics". *Aquaculture* 515 (January): 734590. doi:10.1016/j.aquaculture.2019.734590.

Liao, Wei, Xia Guanghua, Yongcheng Li, Xuan Ri Shen, and Chuan Li. 2018. "Comparison of characteristics and fibril-forming ability of skin collagen from barramundi (Lates calcarifer) and tilapia (Oreochromis niloticus)". *International Journal of Biological Macromolecules* 107 (February): 549–59. doi:10.1016/j.ijbiomac.2017.09.022.

Liu, Zeqi, Shaozong Wu, Huixin Zuo, Jie Lin, Hua Zheng, Hongtao Lei, Qunli Yu, Xiangheng Wu, and Zonglin Guo. 2023. "Freeze-drying pretreatment of watermelon peel to improve the efficiency of pectin extraction: RSM optimization, extraction mechanism, and characterization". *International Journal of Biological Macromolecules* 249 (September). Elsevier: 125944. doi:10.1016/J.IJBIOMAC.2023.125944.

Ma, Yanli, Xiaofang Zeng, Xiaotian Ma, Ruijin Yang, and Wei Zhao. 2019. "A simple and eco-friendly method of gelatin production from bone: One-step biocatalysis". *Journal of Cleaner Production* 209 (February): 916–26. doi:10.1016/j.jclepro.2018.10.313.

Mandic, Mina, Jelena Spasic, Marijana Ponjavic, Marija S. Nikolic, Vladan R. Cosovic, Kevin E. O'Connor, Jasmina Nikodinovic-Runic, Lidija Djokic, and Sanja Jeremic. 2019. "Biodegradation of poly(ε-caprolactone) (PCL) and medium chain length polyhydroxyalkanoate (mcl-PHA) using whole cells and cell free protein preparations of Pseudomonas and Streptomyces strains grown on waste cooking oil". *Polymer Degradation and Stability* 162 (April): 160–68. doi:10.1016/j.polymdegradstab.2019.02.012.

Manzanares, P., I. Ballesteros, M.J. Negro, A. González, J.M. Oliva, and M. Ballesteros. 2020. "Processing of extracted olive oil pomace residue by hydrothermal or dilute acid pretreatment and enzymatic hydrolysis in a biorefinery context". *Renewable Energy* 145 (january): 1235–45. doi:10.1016/j.renene.2019.06.120.

Martins, S. H.F., K. V. Pontes, R. L. Fialho, and F. M. Fakhouri. 2022. "Extraction and characterization of the starch present in the avocado seed (Persea americana mill) for future applications". *Journal of Agriculture and Food Research* 8 (June). Elsevier: 100303. doi:10.1016/J.JAFR.2022.100303.

Melikoğlu, Arzu Yalçın, Seda Ersus Bilek, and Serap Cesur. 2019. "Optimum alkaline treatment parameters for the extraction of cellulose and production of cellulose nanocrystals from apple pomace". *Carbohydrate Polymers* 215 (July). Elsevier: 330–37. doi:10.1016/J.CARBPOL.2019.03.103.

Mello, Ariane Fátima Murawski de, Luciana Porto de Souza Vandenberghe, Clara Matte Borges Machado, Manuela Cardozo Ribeiro Mendonça Ravazoli, Gustavo Amaro Bittencourt, Leonardo Wedderhoff Herrmann, Priscilla Zwiercheczewski de Oliveira, and Carlos Ricardo Soccol. 2023. "Emerging technologies, recent developed processes, patents, and innovation about second- and third-generation bioplastics". Em *Second and Third Generation Bioplastics*, 191–206. Boca Raton: CRC Press. doi:10.1201/9781003344018-14.

Mishra, Prateeksha, and Bhabatarini Panda. 2023. "Polyhydroxybutyrate (PHB) accumulation by a mangrove isolated cyanobacteria Limnothrix planktonica using fruit waste". *International Journal of Biological Macromolecules* 252 (December): 126503. doi:10.1016/j.ijbiomac.2023.126503.

Moita, R., A. Freches, and P.C. Lemos. 2014. "Crude glycerol as feedstock for polyhydroxyalkanoates production by mixed microbial cultures". *Water Research* 58 (July): 9–20. doi:10.1016/j.watres.2014.03.066.

Moradi, Elnaz, and Milad Fathi. 2023. "Production of cellulose nanocrystals from tomato pomace as a food waste and their application for stabilizing of Pickering emulsions". *Bioactive Carbohydrates and Dietary Fibre* 30 (November). Elsevier: 100378. doi:10.1016/J.BCDF.2023.100378.

Muley, Abhijeet B., Sandeep A. Chaudhari, Ketan H. Mulchandani, and Rekha S. Singhal. 2018. "Extraction and characterization of chitosan from prawn shell waste and its conjugation with cutinase for enhanced thermo-stability". *International Journal of Biological Macromolecules* 111 (May): 1047–58. doi:10.1016/j.ijbiomac.2018.01.115.

Nagarajan, Sakthivel, Socrates Radhakrishnan, S. Narayana Kalkura, Sebastien Balme, Philippe Miele, and Mikhael Bechelany. 2019. "Overview of protein-based biopolymers for biomedical application". *Macromolecular Chemistry and Physics* 220 (14). doi:10.1002/macp.201900126.

Naomi David, Anthea, Y. Sewsynker-Sukai, and E. B. Gueguim Kana. 2022. "Co-valorization of corn cobs and dairy wastewater for simultaneous saccharification and lactic acid production: Process optimization and kinetic assessment". *Bioresource Technology* 348 (December 2021). Elsevier Ltd: 126815. doi:10.1016/j.biortech.2022.126815.

Nielsen, Chad, Asif Rahman, Asad Ur Rehman, Marie K. Walsh, and Charles D. Miller. 2017. "Food waste conversion to microbial polyhydroxyalkanoates". *Microbial Biotechnology* 10 (6): 1338–52. doi:10.1111/1751-7915.12776.

Ocak, Bugra. 2020. "Properties and characterization of thyme essential oil incorporated collagen hydrolysate films extracted from hide fleshing wastes for active packaging". *Environmental Science and Pollution Research* 27 (23): 29019–30. doi:10.1007/s11356-020-09259-1.

Oliveira, Vagne de Melo, Caio Rodrigo Dias Assis, Beatriz de Aquino Marques Costa, Robson Coelho de Araújo Neri, Flávia Thuane Duarte Monte, Helane Maria Silva da Costa Vasconcelos Freitas, Renata Cristina Penha França, Juliana Ferreira Santos, Ranilson de Souza Bezerra, and Ana Lúcia Figueiredo Porto. 2021. "Physical, biochemical, densitometric and spectroscopic techniques for characterization collagen from alternative sources: A review based on the sustainable valorization of aquatic by-products". *Journal of Molecular Structure* 1224 (January): 129023. doi:10.1016/j.molstruc.2020.129023.

Pagliano, Giorgia, Valeria Ventorino, Antonio Panico, and Olimpia Pepe. 2017. "Integrated systems for biopolymers and bioenergy production from organic waste and by-products: A review of microbial processes". *Biotechnology for Biofuels* 10 (1). BioMed Central: 1–24. doi:10.1186/s13068-017-0802-4.

Pais, Joana, Inês Farinha, Filomena Freitas, Luísa S. Serafim, Virginia Martínez, Juan Carlos Martínez, Miguel Arévalo-Rodríguez, M. Auxiliadora Prieto, and Maria A.M. Reis. 2014. "Improvement on the yield of polyhydroxyalkanotes production from cheese whey by a recombinant Escherichia coli strain using the proton suicide methodology". *Enzyme and Microbial Technology* 55. Elsevier Inc.: 151–58. doi:10.1016/j.enzmictec.2013.11.004.

Palaniappan, Ayyappan, Shanmuga Sundharam Yuvaraj, Saravanan Sonaimuthu, and Usha Antony. 2017. "Characterization of xylan from rice bran and finger millet seed coat for functional food applications". *Journal of Cereal Science* 75 (may). Academic Press: 296–305. doi:10.1016/J.JCS.2017.03.032.

Pan, Lanjia, Jie Li, Ruming Wang, Yu Wang, Qinghuai Lin, Chunxing Li, and Yin Wang. 2021. "Biosynthesis of polyhydroxyalkanoate from food waste oil by Pseudomonas alcaligenes with simultaneous energy recovery from fermentation wastewater". *Waste Management* 131 (May). Elsevier Ltd: 268–76. doi:10.1016/j.wasman.2021.06.008.

Pandey, Aishwarya, Ndao Adama, Kokou Adjallé, and Jean François Blais. 2022. "Sustainable applications of polyhydroxyalkanoates in various fields: A critical review". *International Journal of Biological Macromolecules* 221 (May): 1184–201. doi:10.1016/j.ijbiomac.2022.09.098.

Patidar, Mukesh Kumar, Sadhana Nighojkar, Anil Kumar, and Anand Nighojkar. 2018. "Pectinolytic enzymes-solid state fermentation, assay methods and applications in fruit juice industries: a review". *3 Biotech* 8 (4): 199. doi:10.1007/s13205-018-1220-4.

Patsalou, Maria, Antonios Chrysargyris, Nikolaos Tzortzakis, and Michalis Koutinas. 2020. "A biorefinery for conversion of citrus peel waste into essential oils, pectin, fertilizer and succinic acid via different fermentation strategies". *Waste Management* 113. Elsevier Ltd: 469–77. doi:10.1016/j.wasman.2020.06.020.

Peinemann, Jan Christoph, Francesca Demichelis, Silvia Fiore, and Daniel Pleissner. 2019. "Techno-economic assessment of non-sterile batch and continuous production of lactic acid from food waste". *Bioresource Technology* 289 (October): 121631. doi:10.1016/j.biortech.2019.121631.

Petri, Denise F.S. 2015. "Xanthan gum: A versatile biopolymer for biomedical and technological applications". *Journal of Applied Polymer Science* 132 (23). doi:10.1002/app.42035.

Popoola-Akinola, Oyekemi Olabisi, Temiloluwa Joy Raji, and Babatunde Olawoye. 2022. "Lignocellulose, dietary fibre, inulin and their potential application in food". *Heliyon* 8 (8). Elsevier: e10459. doi:10.1016/J.HELIYON.2022.E10459.

Prakash, Priyanka, Wing Hin Lee, Ching-Yeeloo, Hau Seung Jeremy Wong, and Thaigarajan Parumasivam. 2022. "Advances in polyhydroxyalkanoate nanocarriers for effective drug delivery: An overview and challenges". *Nanomaterials* 12 (1): 1–20. doi:10.3390/nano12010175.

Priya, Anshu, Zubeen Hathi, Md Ariful Haque, Sunil Kumar, Aman Kumar, Ekta Singh, and Carol S.K. Lin. 2022. "Effect of levulinic acid on production of polyhydroxyalkanoates from food waste by Haloferax mediterranei". *Environmental Research* 214 (November): 114001. doi:10.1016/j.envres.2022.114001.

Raji, Zarifeh, Faramarz Khodaiyan, Karamatollah Rezaei, Hossein Kiani, and Seyed Saeid Hosseini. 2017. "Extraction optimization and physicochemical properties of pectin from melon peel". *International Journal of Biological Macromolecules* 98 (May). Elsevier: 709–16. doi:10.1016/J.IJBIOMAC.2017.01.146.

Ramírez Tapias, Yuly A., M. Victoria Di Monte, Mercedes A. Peltzer, and Andrés G. Salvay. 2022. "Bacterial cellulose films production by Kombucha symbiotic community cultured on different herbal infusions". *Food Chemistry* 372 (October 2021). doi:10.1016/j.foodchem.2021.131346.

Ranganathan, Saranya, Sayantani Dutta, J. A. Moses, and C. Anandharamakrishnan. 2020. "Utilization of food waste streams for the production of biopolymers". *Heliyon* 6 (9). Elsevier Ltd: e04891. doi:10.1016/j.heliyon.2020.e04891.

Rodríguez G., Jesús E., Silvia Brojanigo, Marina Basaglia, Lorenzo Favaro, and Sergio Casella. 2021. "Efficient production of polyhydroxybutyrate from slaughterhouse waste using a recombinant strain of Cupriavidus necator DSM 545". *Science of the Total Environment* 794. doi:10.1016/j.scitotenv.2021.148754.

Saenmuang, Soraya, Suphatta Phothiset, and Chuleeporn Chumnanka. 2020. "Extraction and characterization of gelatin from black-bone chicken by-products". *Food Science and Biotechnology* 29 (4): 469–78. doi:10.1007/s10068-019-00696-4.

Saleh, Ahmed K., Hamada El-Gendi, Julie Basu Ray, and Tarek H. Taha. 2021. "A low-cost effective media from starch kitchen waste for bacterial cellulose production and its application as simultaneous absorbance for methylene blue dye removal". *Biomass Conversion and Biorefinery*. Springer Berlin Heidelberg, 12437–49. doi:10.1007/s13399-021-01973-1.

Saleh, Aiman, A Mohammed, Muhammad Naveed, and Norbert Jost. 2021. "Polysaccharides; Classification, chemical properties, and future perspective applications in fields of pharmacology and biological medicine (A review of current applications and upcoming potentialities)". *Journal of Polymers and the Environment* 29: 2359–71. doi:10.1007/s10924-021-02052-2.

Sar, Taner, Sharareh Harirchi, Mohaddaseh Ramezani, Gülru Bulkan, Meltem Yesilcimen Akbas, Ashok Pandey, and Mohammad J. Taherzadeh. 2022. "Potential utilization of dairy industries by-products and wastes through microbial processes: A critical review". *Science of The Total Environment* 810 (March): 152253. doi:10.1016/j.scitotenv.2021.152253.

Sariatli, Furkan. 2017. "Linear economy versus circular economy: A comparative and analyzer study for optimization of economy for sustainability". *Visegrad Journal on Bioeconomy and Sustainable Development* 6 (1): 31–34. doi:10.1515/vjbsd-2017-0005.

Sebastián-Nicolás, Jessica L., Luis G. González-Olivares, Gabriela A. Vázquez-Rodríguez, Carlos A. Lucho-Constatino, Araceli Castañeda-Ovando, and Alma E. Cruz-Guerrero. 2020. "Valorization of whey using a biorefinery". *Biofuels, Bioproducts and Biorefining* 14 (5): 1010–27. doi:10.1002/bbb.2100.

Seddiqi, Hadi, Erfan Oliaei, Hengameh Honarkar, Jianfeng Jin, Lester C. Geonzon, Rommel G. Bacabac, and Jenneke Klein-Nulend. 2021. "Cellulose and its derivatives: Towards biomedical applications". *Cellulose* 28. doi:10.1007/s10570-020-03674-w.

Sharma, Poonam, Vivek Kumar Gaur, Sang-Hyoun Kim, and Ashok Pandey. 2020. "Microbial strategies for bio-transforming food waste into resources". *Bioresource Technology* 299 (March): 122580. doi:10.1016/j.biortech.2019.122580.

Sharmila, G., C. Muthukumaran, N. Manoj Kumar, V.M. Sivakumar, and M. Thirumarimurugan. 2020. "Food waste valorization for biopolymer production". Em *Current Developments in Biotechnology and Bioengineering*, 233–49. Elsevier. doi:10.1016/B978-0-444-64321-6.00012-4.

Sixto-Berrocal, Ana María, Marlenne Vázquez-Aldana, Susana Patricia Miranda-Castro, M. Aurora Martínez-Trujillo, and Martín R. Cruz-Díaz. 2023. "Chitin/chitosan extraction from shrimp shell waste by a completely biotechnological process". *International Journal of Biological Macromolecules* 230 (March): 123204. doi:10.1016/j.ijbiomac.2023.123204.

Soni, Sweeta, Vinod Chhokar, Vikas Beniwal, Ravinder Kumar, Hemlata Badgujjar, Rohit Chauhan, Shruti Dudeja, and Anil Kumar. 2023. "Cost effective media optimization for PHB production by Bacillus badius MTCC 13004 using the statistical approach". *International Journal of Biological Macromolecules* 233 (April): 123575. doi:10.1016/j.ijbiomac.2023.123575.

Sreekumar, Kalyani, B. Bindhu, and K. Veluraja. 2021. "Perspectives of polylactic acid from structure to applications". *Polymers from Renewable Resources* 12 (1–2): 60–74. doi:10.1177/20412479211008773.

Suresh, Sreehari, Mridul Umesh, and Adhithya Sankar Santosh. 2023. "Biological extraction of chitin from fish scale waste using proteolytic bacteria Stenotrophomonas koreensis and its possible application as an active packaging material". *Biomass Conversion and Biorefinery*, nº 0123456789. Springer Berlin Heidelberg. doi:10.1007/s13399-023-03865-y.

Suri, Shweta, Anupama Singh, and Prabhat K. Nema. 2022. "Current applications of citrus fruit processing waste: A scientific outlook". *Applied Food Research* 2 (1). Elsevier B.V.: 100050. doi:10.1016/j.afres.2022.100050.

Talekar, Sachin, Antonio F. Patti, R. Vijayraghavan, and Amit Arora. 2018. "An integrated green biorefinery approach towards simultaneous recovery of pectin and polyphenols coupled with bioethanol production from waste pomegranate peels". *Bioresource Technology* 266 (June). Elsevier: 322–34. doi:10.1016/j.biortech.2018.06.072.

Thakare, Vishal B., Girirajsinh C. Jadeja, and Meghal. A. Desai. 2023. "Extraction of mangiferin and pectin from mango peels using process intensified tactic: A step towards waste valorization". *Chemical Engineering Research and Design* 192 (April): 280–88. doi:10.1016/j.cherd.2023.02.036.

Thi Thuy Van, Nguyen, Pag-asa Gaspillo, Ho Gia Thien Thanh, Nguyen Huynh Thao Nhi, Huynh Nhat Long, Nguyen Tri, Nguyen Thi Truc Van, Tien-Thanh Nguyen, and Huynh Ky Phuong Ha. 2022. "Cellulose from the banana stem: optimization of extraction by response surface methodology (RSM) and charaterization". *Heliyon* 8 (12): e11845. doi:10.1016/j.heliyon.2022.e11845.

Thuppahige, Vindya Thathsaranee, Lalehvash Moghaddam, Zachary G. Welsh, Tony Wang, Hong-Wei Xiao, and Azharul Karim. 2023. "Extraction and characterisation of starch from cassava (Manihot esculenta) agro-industrial wastes". *LWT* 182 (June): 114787. doi:10.1016/j.lwt.2023.114787.

Tinratat, Sirikhwan, and Monnipha Sila-asna. 2017. "Optimization of gelatin extraction and physico-chemical properties of fish skin and bone gelatin: Its application to panna cotta formulas". *Current Research in Nutrition and Food Science Journal* 5 (3): 263–73. doi:10.12944/CRNFSJ.5.3.11.

Treichel, Helen, Gislaine Fongaro, Thamarys Scapini, Aline Frumi Camargo, Fábio Spitza Stefanski, and Bruno Venturin. 2020. "Waste biomass pretreatment methods". Em *Utilising Biomass in Biotechnology*, organized by Helen Treichel, Gislaine Fongaro, Thamarys Scapini, Aline Frumi Camargo, Fabio Spitza Stefanski, and Bruno Venturin, 1º ed, 19–48. Cham: Springer International Publishing. doi:10.1007/978-3-030-22853-8_3.

Tsang, Yiu Fai, Vanish Kumar, Pallabi Samadar, Yi Yang, Jechan Lee, Yong Sik Ok, Hocheol Song, Ki-Hyun Kim, Eilhann E. Kwon, and Young Jae Jeon. 2019. "Production of bioplastic through food waste valorization". *Environment International* 127 (June): 625–44. doi:10.1016/j.envint.2019.03.076.

Tsermoula, Paraskevi, Bekzod Khakimov, Jacob Holm Nielsen, and Søren Balling Engelsen. 2021. "WHEY - The waste-stream that became more valuable than the food product". *Trends in Food Science & Technology* 118 (December): 230–41. doi:10.1016/j.tifs.2021.08.025.

Tyuftin, Andrey A., and Joe P. Kerry. 2021. "Gelatin films: Study review of barrier properties and implications for future studies employing biopolymer films". *Food Packaging and Shelf Life* 29 (September): 100688. doi:10.1016/j.fpsl.2021.100688.

Valcarcel, Jesus, Javier Fraguas, Carolina Hermida-Merino, Daniel Hermida-Merino, Manuel M. Piñeiro, and José Antonio Vázquez. 2021. "Production and physicochemical characterization of gelatin and collagen hydrolysates from turbot skin waste generated by aquaculture activities". *Marine Drugs* 19 (9): 491. doi:10.3390/md19090491.

Veeruraj, Anguchamy, Muthuvel Arumugam, Thangappan Ajithkumar, and Thangavel Balasubramanian. 2015. "Isolation and characterization of collagen from the outer skin of squid (Doryteuthis singhalensis)". *Food Hydrocolloids* 43 (January): 708–16. doi:10.1016/j.foodhyd.2014.07.025.

Vries, Lisanne de, Sydne Guevara-Rozo, MiJung Cho, Li-Yang Liu, Scott Renneckar, and Shawn D Mansfield. 2021. "Tailoring renewable materials via plant biotechnology". *Biotechnol Biofuels* 14: 167. doi:10.1186/s13068-021-02010-z.

Wang, Jing, Javad Tavakoli, and Youhong Tang. 2019a. "Bacterial cellulose production, properties and applications with different culture methods – A review". *Carbohydrate Polymers* 219 (April). Elsevier: 63–76. doi:10.1016/j.carbpol.2019.05.008.

Wang, Xiao-Li, Jin-Jie Zhou, Ya-Qin Sun, and Zhi-Long Xiu. 2019b. "Bioconversion of raw glycerol from waste cooking-oil-based biodiesel production to 1,3-Propanediol and lactate by a microbial consortium". *Frontiers in Bioengineering and Biotechnology* 7 (Febuary). doi:10.3389/fbioe.2019.00014.

Wu, Ju, Xiangju Wei, Pengye Guo, Aiyong He, Jiaxing Xu, Mingjie Jin, Yanjun Zhang, and Hui Wu. 2021. "Efficient poly(3-hydroxybutyrate-co-lactate) production from corn stover hydrolysate by metabolically engineered Escherichia coli". *Bioresource Technology* 341 (July). doi:10.1016/j.biortech.2021.125873.

Yadav, Bhoomika, Shraddha Chavan, Anusha Atmakuri, R.D. Tyagi, and Patrick Drogui. 2020. "A review on recovery of proteins from industrial wastewaters with special emphasis on PHA production process: Sustainable circular bioeconomy process development". *Bioresource Technology* 317 (December): 124006. doi:10.1016/j.biortech.2020.124006.

Yousuf, Ahasa, Juan-Rodrigo Bastidas-Oyanedel, and Jens Ejbye Schmidt. 2018. "Effect of total solid content and pretreatment on the production of lactic acid from mixed culture dark fermentation of food waste". *Waste Management* 77 (July): 516–21. doi:10.1016/j.wasman.2018.04.035.

Yu, Fangmiao, Chuhong Zong, Shujie Jin, Jiawen Zheng, Nan Chen, Ju Huang, Yan Chen, et al., 2018. "Optimization of extraction conditions and characterization of pepsin-solubilised collagen from skin of giant croaker (Nibea japonica)". *Marine Drugs* 16 (1): 29. doi:10.3390/md16010029.

Zema, D.A., P.S. Calabrò, A. Folino, V. Tamburino, G. Zappia, and S.M. Zimbone. 2018. "Valorisation of citrus processing waste: A review". *Waste Management* 80 (October): 252–73. doi:10.1016/j.wasman.2018.09.024.

Zhang, Ting, Rui Sun, Mengzhen Ding, Lina Tao, Lijie Liu, Ningping Tao, Xichang Wang, and Jian Zhong. 2020. "Effect of extraction methods on the structural characteristics, functional properties, and emulsion stabilization ability of Tilapia skin gelatins". *Food Chemistry* 328 (October): 127114. doi:10.1016/j.foodchem.2020.127114.

Zhang, Wenting, Wei Duan, Gangliang Huang, and Hualiang Huang. 2023. "Ultrasonic-assisted extraction, analysis and properties of mung bean peel polysaccharide". *Ultrasonics Sonochemistry* 98 (August): 106487. doi:10.1016/j.ultsonch.2023.106487.

Zheng, Dingyuan, Yangyang Zhang, Yunfeng Guo, and Jinquan Yue. [s.d.]. "Isolation and characterization of nanocellulose with a novel shape from walnut (Juglans Regia L.) shell agricultural waste". doi:10.3390/polym11071130.

Zou, Ye, Heng Yang, Xinxiao Zhang, Pingping Xu, Di Jiang, Muhan Zhang, Weimin Xu, and Daoying Wang. 2020. "Effect of ultrasound power on extraction kinetic model, and physicochemical and structural characteristics of collagen from chicken lung". *Food Production, Processing and Nutrition* 2 (1): 3. doi:10.1186/s43014-019-0016-1.

8 Production of Organic Acids from Food Waste

Walter José Martínez-Burgos,
Priscilla Zwiercheczewski de Oliveira, Juliana de Oliveira,
Rafael Novaes Barros, and Cristine Rodrigues

8.1 INTRODUCTION

Organic acids are composed of carbon chains linked to an acidic functional group; among these are, for example, lactic, citric, propionic, malic, itaconic, butyric, and acetic acids, among others. These compounds are widely used as additives in different industries, among which the food industry, the chemical industry, the pharmaceutical industry, and the cosmetics industry stand out (Li et al. 2023).

In the food industry, organic acids can be used as preservatives, pH regulators, emulsifiers, and flavorings. Some of these are also used to manufacture animal feed, as they have a positive impact on the gastrointestinal health of cattle (Melaku et al. 2021). In the chemical industry, organic acids can be used as chelating agents in paints. They also have great applications in the manufacture of polymers for the production of synthetic fibers, biodegradable plastics, and synthetic latex, among others (Kuenz et al. 2012; Sándor et al. 2021). In the cosmetics sector, organic acids are used as additives to improve the notes of fragrances and perfumes (Jiang et al. 2018). The wide application of organic acids in high consumption products influences their market. The market for these reached USD 11.33 billion in 2022, and an annual growth of 5.3% is expected so that by 2032 this market will grow to USD 18.8 billion (Future-Market-Insights, 2023; Industry-ARC, 2023).

The chemical synthesis of organic acids is generally carried out under extreme conditions, which requires large amounts of energy and consequently releases significant amounts of pollutants into the atmosphere. Therefore, sustainable technologies are necessary for the production of these metabolites. Organic acids can be produced microbially as these are generally intermediate metabolites of the tricarboxylic acid cycle (Vion et al. 2023). Although fermentative processes have been developed using pure substrates, there is great interest in the use of waste, such as food waste, as around 1.3 billion tons is generated worldwide (Teleky and Vodnar, 2021). The objective of this chapter is to present the potential of food waste as a substrate for the production of organic acids.

8.2 CITRIC ACID

Citric acid (CA) is a weak and naturally occurring organic acid found primarily in citrus fruits, such as lemons, limes, and oranges; however, it is also present in varying quantities in other fruits and vegetables. Its chemical structure, a tricarboxylic acid ($C_6H_8O_7$), grants a versatile range of properties that make it invaluable in various sectors. The unique combination of its properties, such as acidity, chelating ability, and water solubility, contributes to its widespread applications (Reena et al. 2022). CA has a long history of human use, dating back to ancient times when it was used as a preservative and flavoring agent. Its isolation and identification from citrus fruits occurred in the late 18th century, marking the beginning of its industrial production and utilization. Over the years, it has grown to become a vital ingredient in numerous industries, fueling its economic importance (Wang et al. 2020).

DOI: 10.1201/9781032706030-8

One of the primary sectors where CA finds extensive use is human consumption. Its role as a food additive serves as an acidity regulator, flavor enhancer, and preservative in the food and beverage industry. It imparts tartness to soft drinks, and enhances the taste of candies and fruit-flavored snacks (Coban 2020). Additionally, it acts as a preservative, preventing spoilage and extending the shelf life of various food products (Coban 2020). In the pharmaceutical sector, it is used as an excipient in drug formulations to adjust pH levels, enhance solubility, and improve the stability of certain medications (Nangare et al. 2021). Furthermore, CA plays a critical role in the production of effervescent tablets and syrups, providing a pleasant effervescence and taste, which enhances patient compliance. CA also finds a niche in the cosmetics and personal care industry. Due to its gentle exfoliating properties and ability to adjust the pH of formulations, it is used in skincare products, shampoos, and conditioners (Nangare et al. 2021). Due to its chelating abilities, CA is a key ingredient in cleaning products (Coban 2020). It efficiently binds to mineral ions, preventing the formation of scale and soap scum in household cleaners. Its eco-friendly nature has led to its increasing use in environmentally conscious cleaning solutions (Ibrahim et al. 2022).

CA's market is characterized by steady demand across various industries (Wang et al. 2020). Its stability as a food additive and the growth of the processed food industry have ensured a consistent market. Additionally, the global trend toward natural and organic products has led to increased demand for CA as a safe and eco-friendly ingredient. Its economic significance is notable in both developed and emerging economies. The global trade generates substantial revenue, and the global market of CA is expected to reach USD 3.9 Billion by 2024, with major producers including China, the United States, and India (Reena et al. 2022). CA is a valuable commodity in international trade. Countries with surplus production capacity export CA to meet the demand of importing countries, contributing to foreign exchange earnings, trade balance, and economic growth (Wang et al. 2020).

8.2.1 Production of Citric Acid

As a commercial product, CA was traditionally extracted from citrus fruits, but today, it is predominantly produced through microbial fermentation (Hu et al. 2019). Strains of *Aspergillus niger* are commonly used in the fermentation process to convert carbohydrates into CA under controlled conditions, ensuring consistent and efficient production for industrial purposes. *A. niger* is a robust fermentor capable of thriving in a wide range of environmental conditions, including temperature and pH. It exhibits a high yield of CA during fermentation, making it economically attractive for large-scale production and allowing for genetic modifications to enhance production. *A. niger* dominates the CA production industry; however, other fungal species have also been explored (Tong et al. 2019). Some *Penicillium* strains, such as *P. restrictum* and *P. camemberti*, exhibit CA-producing capabilities and have been investigated for their potential in specific niche applications. Certain *Candida* yeast strains have been studied for their CA production potential, particularly in the context of biofuel production. Although less common, some bacterial strains have been investigated for CA production, such as *Bacillus* strains, but their CA-producing yields are typically lower than those of fungi. Engineered *Escherichia coli* strains have been developed for CA production as part of metabolic engineering efforts, where this process involves complex metabolic pathways within microorganisms (Hu et al. 2019). Understanding these pathways is essential for optimizing CA yields and production efficiency.

To achieve high yields of CA accumulation, it is essential to optimize several factors during the fermentation process. The optimal temperature and pH conditions vary among microorganisms; for *A. niger,* the typical temperature range is 25–30 °C, with an acidic pH of approximately 2.5–3.5. Oxygen supply is essential for the growth and metabolism of microorganisms. Initial aerobic conditions are required for biomass production, while oxygen limitations in later stages promote CA biosynthesis. The nutrient composition of the fermentation medium may include carbon sources (e.g., glucose), nitrogen sources (e.g., ammonium sulfate), and trace elements (e.g., iron and manganese) (Behera et al. 2021). When employing food residues as substrates, it is essential to consider their composition, potential inhibitors, and pre-treatment requirements. Some food residues may contain inhibitors or contaminants that hinder CA fermentation, such as preservatives, pesticides, or contaminants that affect the growth and activity of microorganisms. Recent advancements in genetic engineering have enabled the modification of microorganisms to enhance CA production (Hu et al. 2019). Strategies include overexpressing key enzymes in the CA biosynthesis pathway, optimizing transporters for citrate accumulation, and improving metabolic flux (Tong et al. 2019). Co-cultivation of microorganisms with complementary metabolic pathways can also enhance substrate utilization and CA yields.

Challenges in using food residues for CA production could involve substrate variability (quality, composition), inhibitors and contaminants, and nutrient imbalance. Usual substrates are carbohydrate-rich content (Sweta et al. 2021), such as citrus peel (CP) residues, which are a prominent source, as they have a high natural CA content. These residues are generated during citrus fruit processing, where the outer peels are removed to extract juice. CP is rich in carbohydrates, primarily pectin and cellulose, making them suitable substrates for CA production (Mahato et al. 2021). Pectin, a complex polysaccharide, can be broken down into simpler sugars through pectinase enzymes (Mahato et al. 2021). Sugarcane bagasse (SB) is a fibrous residue left after sugarcane processing to extract sugar. It is composed primarily of cellulose, hemicellulose, and lignin (Özüdoğru et al. 2019). SB has gained attention as a potential feedstock for CA production due to cellulose being the dominant component, consisting of glucose molecules linked together, which can be hydrolyzed into glucose. Hemicellulose is a more complex carbohydrate, comprising various sugar monomers, and it can also be hydrolyzed into fermentable sugars. Lignin, a complex and rigid polymer, is less suitable for CA production but can be converted into value-added products. SB typically requires pretreatment to break down cellulose and hemicellulose into fermentable sugars. Common pretreatment methods include steam explosion, acid hydrolysis, or enzymatic treatment (Özüdoğru et al. 2019). Molasses is a by-product of sugar refining from sugar beet or SB (Ozdal and Kurbanoglu 2019). It is rich in sucrose and other fermentable sugars, making it a readily available and suitable substrate for CA production. Besides sucrose, molasses can contain glucose, fructose, and other sugars, providing diverse carbon sources for fermentation (Ozdal and Kurbanoglu 2019). Different carbon sources and different microorganisms found in the literature for CA obtention are listed in Table 8.1.

Solid-state fermentation (SSF) is a preferred method for CA production from food residues. In this process, the food residues are inoculated with CA-producing microorganisms, typically strains of *Aspergillus niger* or other filamentous fungi. The microorganisms break down the carbohydrates into CA through a series of enzymatic reactions (Behera et al. 2021). SSF offers several advantages, including reduced water usage, lower energy requirements, and minimal effluent production. Submerged fermentation (SmF) involves the immersion of food residues in a liquid medium, where CA-producing microorganisms are cultivated, more commonly using liquid-rich residues

TABLE 8.1
Different Microorganisms and Carbon Sources Used in Citric Acid Fermentation

Microorganism	Carbon Source	CA Production	Reference
Yarrowia lipolytica K-168	Carrot juice	80.53 g/L	(Urak et al. 2015)
Aspergillus niger	Corn flour	173.2 g/L	(Wang et al. 2017)
A. niger 831	Wheat bran extract	162.7 g/L	(Yu et al. 2018)
A. niger SIIM M288	Corn stover	136.3 g/L	(Hou and Bao 2018)
A. niger chsC	Liquefied corn	180.3 g/L	(Sun et al. 2018)
Y. lipolytica AWG7 INU 8	Inulin	203 g/L	(Rakicka et al. 2019)
A. niger B60	Pomegranate peel	351.5 g/L	(Roukas and Kotzekidou 2020)
A. niger B6 – CCT 7717	Cocoa pod husks	978.52 g/kg	(de Oliveira et al. 2022)

like molasses (Sweta et al. 2021). Although SmF requires more water and energy compared to SSF, it can be employed effectively with certain food residues (Behera, Mishra, and Mohapatra 2021). The fermentation technique may enable the downstream processing phase, where CA recovery typically involves several steps (Sawant et al. 2018): filtration (separation of fermentation broth and solid biomass), precipitation (CA is precipitated from the broth by adjusting pH or temperature), crystallization (CA is further purified through crystallization), and drying (CA crystals are dried to obtain the final product). Scaling up these processes requires larger equipment, efficient filtration systems, and precise control of precipitation and crystallization conditions. Purification techniques such as chromatography or ion-exchange are essential for producing high-purity CA (Sawant et al. 2018).

CA production using food residues as substrate contributes to waste reduction by repurposing residues that would otherwise be discarded. This approach minimizes landfill waste and reduces the environmental burden associated with waste disposal (Reena et al. 2022). Utilizing food residues as feedstock enhances resource efficiency; instead of relying solely on dedicated crops or sugarcane, for example, this method maximizes the value of existing agricultural and food processing operations (Özüdoğru et al. 2019). CA production from food residues often results in a reduced carbon footprint compared to traditional methods. SSF, in particular, is less energy-intensive and emits fewer greenhouse gases, contributing to a lower environmental impact. By integrating CA production with food residue utilization, the circular economy concept is realized, and valuable compounds are recovered and repurposed, promoting a sustainable and closed-loop system (Reena et al. 2022).

8.3 LACTIC ACID

Lactic acid (LA) is an important organic acid due to its versatility as a basic input and the large amount of applications found in the food, cosmetic, chemical, pharmaceutical industries, among others. LA is a chiral molecule, which exists as enantiomers L- and D-lactic acid. Because of carboxylic and hydroxyl groups in its structure, LA can be applied as a flavoring agent, inhibitor of bacteria, acidulant, pH regulator, humectant, and antimicrobial compound. In addition, LA can be converted into different potentially useful chemicals such as pyruvic acid, acrylic acid, and esters, among others (Sharma et al. 2022).

Lactic acid polymerization is also gaining importance due to the polymer characteristics and applications in the packaging, textile, biomedical, and pharmaceutical industries, among others. Poly(lactic acid) (PLA) has properties comparable to existing petroleum-based polymers, such as its high strength, modulus, and transparency, etc. (Liu et al. 2011). However, properties of PLA such as its thermal properties, toughness, and water vapor and gas barrier are inferior to those of conventional petroleum-based polymers. Combining the polymer with reinforcing elements, thus making blends and composites, is a strategy to improve the properties of PLA (Fortunati et al. 2012).

The global LA market size was valued at USD 3.1 billion in 2022 and is projected to grow at a compound annual growth rate of 8.0% from 2023 to 2030 (Lactic Acid Market Size, 2023).

8.3.1 Production of Lactic Acid

LA can be obtained by chemical synthesis or fermentation processes. Chemical synthesis is mainly based on the hydrolysis of lactonitrile with the use of strong acids. This process produces a racemic mixture of the isomers L- and D-lactic acid. However, fermentation processes lead to the production of optically pure L- or D-lactic acid, depending on the selected strain. Furthermore, biotechnological processes offer advantages such as renewable substrates, low energy requirements, mild production temperatures, and high purity of final products (Cubas et al. 2018).

Refined carbohydrates provide high amounts of LA, but the cost makes production economically unfavorable. Several materials have been considered as attractive alternative substrates and renewable resources, including non-food substrates, byproducts of agricultural industries, food industries, and natural unutilized biomasses such as starchy biomass, lignocellulosic biomass, food waste, and microalgae (Abdel-Rahman et al. 2013; Ahmad et al. 2020).

An obstacle to the use of waste is the need for pretreatment due to its intimate association with lignin and the lack of hydrolytic enzyme production by lactic acid-producing strains (Abdel-Rahman et al. 2011). The conversion of starch or cellulose to sugar consumes energy during saccharification and increases production costs. Furthermore, the efficiency of lactic acid fermentation processes mainly depends on the microorganism, substrate used, and operational modes.

In this way, different studies have been conducted using various strategies of fermentation, such as batch, fed-batch, semi-continuous/repeated batch, and continuous fermentation. Simultaneous saccharification and fermentation (SSF) and separate hydrolysis and fermentation (SHF) have also been investigated. Some processes and strategies for lactic acid production are shown in Table 8.2.

TABLE 8.2
Different Substrates and Microorganisms Used in the Fermentation Process for Lactic Acid Production

Substrate	Microorganism	Fermentation Mode	Lactic Acid Production	Reference
Corn stover hydrolysate	*Lactobacillus pentosus* LB-1	Batch	210.1 g/l in 100-mL flasks 157.4 g/l in 3-L bioreactor	(Zhang, et al. 2023)
Brewer's spent grain	*Enterococcus mundtii CGMCC 22227*	Batch simultaneous saccharification and fermentation	53.1 g/L	(Wang and Gao 2023)
Household food waste	*Lactobacillus rhamnosus ATCC 7469*	Repeated batch through simultaneous saccharification and fermentation	47.7 g/L	(Song, et al. 2023)
Rice straw	*Lactobacillus lactis 2369*	Batch	82.2 g/L	(Sivagurunathan et al. 2022)
Oil palm trees	*Lactobacillus rhamnosus* ATCC 10863	Open repeated batch; r epeated fed-batch	91.30 g/L 95.94 g/L	(Saelee 2022)
Pasta waste hydrolysate	*Bacillus coagulans* A559	Batch	52 g/L	(Marzo-Gago et al. 2022)
Potato processing waste	*Lactobacillus pentosus*	Batch	150 g/L	(De Oliveira, et al. 2021)
Inedible starchy biomass	*Lactobacillus rhamnosus DUT1908*	One step simultaneous liquefaction, saccharification, and fermentation	107.8 g/L	(Sun et al. 2021)
Cassava bagasse	*Lactobacillus rhamnosus and Bacillus coagulans*	Simultaneous saccharification and co-fermentation	112.5 g/L	(Chen et al. (2020)
Glucose and hydrolysate of microalga *Chlorella vulgaris* ESP-31	*L. plantarum* 23	Continuous	14.22 g/L/h	(Chen et al. 2020)

8.3.2 Downstream Process for Recovering and Purification of Lactic Acid

One of the most challenging aspects of the LA process is recovering and purifying LA from the fermentation broth. The fermentation broth contains a large amount of impurities, for example residual sugar compounds, color, and other organic acids. Fermentation conditions also determine the quality of the broth and suitability of downstream processing options (Ghaffar et al. 2014; Li et al. 2021).

Downstream processes for purification of LA can reach more than 50% of overall operational costs depending on the case; high purities of the product are desired. The conventional downstream steps involve precipitation, with the generation of a large amount of solid waste, gypsum (calcium sulfate). This process involves many steps, including precipitation, filtration, acidification, carbon adsorption, evaporation, and crystallization (Li et al. 2021).

Several alternatives have been proposed for lactic acid separation and purification, and depending on the objectives, each can present advantages and disadvantages. The most important ones are distillation, esterification, nano and ultrafiltration, liquid-liquid extraction, adsorption, ion exchange chromatography, and electrodialysis. Kumar et al. (2020) proposed the purification of LA using the phase partitioning method. In this method, LA is subjected to a extraction from the aqueous phase to the organic phase in the presence of ammonium sulfate and n-butanol. Under the LA downstream process's optimized conditions, a yield of 86% and 93% purity was obtained.

Among the alternative technologies that have been proposed in the literature for the downstream separation and purification of lactic acid from complex fermentation broths, electrodialysis was performed followed by decolorization and chromatography techniques in order to remove remaining ions (Olszewska-Widdrat et al. 2019). The effective lactic acid downstream enabled reaching a solution containing 905.8 g/L of LA, with an optical purity of 98.9%, indicating that the purification process was effective. Guo et al. (2023) studied a microfiltration membrane separation process system to separate and extract the LA in the fermentation broth of kitchen waste. The separation performance of the membrane on the LA fermentation broth was studied,as well as the turbidity removal efficiency, color removal efficiency, total sugar removal efficiency, protein removal efficiency, and LA recovery efficiency. The study concluded that the microfiltration process is promising as an effective separation method for the preliminary purification of kitchen waste fermentation broth.

Depending on the application, different degrees of LA purity are required. However, the cost associated with downstream processing has limited the economic viability of LA. Therefore, the development of innovative, economical, and sustainable technologies for the production of LA with high purity is still needed.

8.4 PROPIONIC ACID

Propionic acid (PA) is a short-chain carboxylic acid, consisting of three carbon atoms, six hydrogen atoms, and two oxygen atoms. PA is used as a food additive, in animal feed, and as a food preservative, especially in bakery products, helping to preserve freshness and inhibit the growth of fungi and bacteria. It is also commonly used in the production of plastics, where it is used as an esterifying agent, and in pharmaceuticals, as an intermediate for the production of several medicines, including ibuprofen and naproxen. It can be found naturally in small quantities in various foods, such as cheeses, breads, and dairy products, due to microbial activity during fermentation.

The global PA market was valued at USD 1.57 billion in 2022. It is estimated that this market will register an annual growth rate (CAGR) of 7.3% between 2023 and 2032, expected to reach USD 3.12 billion by 2032 (Market.us 2023). The growing demand for PA is related to a combination of factors, including population growth, growth in the food, chemical, pharmaceutical, and feed industries, as well as technological innovations that expand its applications.

PA is typically produced from petrochemical raw materials via Reppe synthesis. Several routes can be used for production, one of the most common being the carbonylation of ethylene with carbon monoxide and water, and the oxidation of propanol (Es et al. 2017). The production of PA, like any industrial process, presents specific challenges that may vary depending on the production method, the technology used, and market conditions. Current methods of producing PA are energy-consuming and involve the handling of potentially dangerous chemicals and petroleum derivatives. Finding ways to make the process more energy-efficient is a challenge. Furthermore, the search for more sustainable methods and new sources of carbon that do not use petroleum derivatives is also an important consideration.

8.4.1 Production of Propionic Acid from Food Waste

The use of food waste in the production of PA has several important advantages and benefits. One of the main benefits is the reduction of food waste and the added value to it. It can also help meet the ever-increasing market demand. Organic wastes such as sludge and food waste are rich in complex organic compounds such as polysaccharides, proteins, and lipids, which provide the carbon and nitrogen sources necessary for microbial growth (Chen et al. 2021). Several cheap carbon sources can be used, such as glucose, lactose, and lactic acid, as well as residues generated in food processing, residues rich in hemicelluloses (plant peels and straw), whey, and soy molasses. sugarcane. Table 8.3 presents recent developments in the microbial production of PA.

The production of PA from food waste can be carried out through microbial fermentation processes. Several microorganisms, such as the genera *Propionibacteria*, *Clostridium*, *Fusobacterium*, *Veillonella, and Selenomonas*, are capable of fermentative production of PA, but *Propionibacteria* is the genus most commonly used. The biological fermentation process is not yet economically viable when compared to chemical synthesis. The main obstacle is the feedback inhibition of the microorganism by PA and other acids, such as acetate and succinate, which reduce the final yield, productivity, and concentration (Gu et al. 1998).

The fermentation of food waste, which in most cases is cheap or even at no cost, can be considered an economic advantage of biological production over chemical production. However, it is still necessary to develop and genetically improve strains that are more tolerant to propionic acid and secondary metabolites. The expression of different genes involved in the propionic acid production pathway has been successfully used; the great challenge for metabolic engineering is to increase the carbon flow, directing it towards the biosynthesis of PA for a more efficient conversion of the substrate.

8.5 MALIC ACID

Malic acid (MA) is an organic acid with a four-carbon chain. This compound has antioxidant properties. MA is applied in different industries in which it stands out: beverages, confectionery and food in general, personal care, and the pharmaceutical sector. However, around 90% of MA is used in the food and beverage industry. MA, along with citric acid, is one of the predominant organic acids in fruits, such as apples, lychees, and plums (Walker and Famiani 2018).

The MA market is constantly growing. According to Kövilein et al. (2020), between 80,000 and 100,000 tons are produced each year with an annual increase of approximately 5.1%. In 2022, the MA market reached USD 212 million, and it is expected that by the year 2028, the MA market will reach USD 284 million (Imarc, 2023).

TABLE 8.3
Recent Advances in the Production of Propionic Acid from Food Waste

Food Waste	Strain	Fermentation Conditions	Yield (g g^{-1})	Product Concentration (g L^{-1})	Productivity (g/L/h)	References
Cocoa pod husk	*P. jensenii*	Fermentation carried out in a 100 ml Schott bottle	0.59	10.28	0.08	(Sarmiento-Vásquez et al. 2021)
Hydrolyzed soy molasses	*P. acidipropionici*	Fermentation in a 1 L bioreactor at 32 °C	0.39	n.a.	0.35	(Yang et al. 2018)
Corn stover hydrolysate	*P. acidipropionici*	Batch fermentation was fed with high cell density (HCD) in a 500 mL bioreactor.	0.5	64.00	0.77	(Wang et al. 2017)
Whey lactose	*P. acidipropionici*	Fed-batch fermentation in a 5 L stirred tank fermenter	0.67	135.00	0.61	(Jiang et al. 2015)
Potato juice	*P. acidipropionici*	Sequential batch with cell recycle	0.6	50.8	1.42	(Dishisha et al. 2013)
Hydrolyzed sugarcane molasses	*P. freudenreichii*	Fibrous-bed bioreactor (Fed-batch)	0.46	91.89	0.36	(Feng et al. 2011)

n.a = Not available

8.5.1 Production of Malic Acid

Currently, MA is produced based on petroleum derivatives (West, 2017). According to Chi et al. (2016), the industrial production of MA occurs through the hydration of maleic anhydride from the oxidation of benzene or butane. The process takes place at high pressures and temperatures where a racemic acid mix of D- and L-malic with chiral result is synthesized. MA can also be produced enzymatically, where fumaric acid is transformed to L-malate using purified fumarase (Chi et al. 2016). However, these processes have many disadvantages, including the fact that they have to use petroleum derivatives.

On the other hand, MA can also be produced from renewable sources. According to Vion et al. (2023), it can be produced microbially, as it is an intermediate product of the tricarboxylic acid cycle. Some native strains and others genetically modified have the ability to produce large amounts of L-malic acid. Among the producing strains are, for example, *Aspergillus flavus*, *Aspergillus niger* ATCC 9029, *Aspergillus oryzae* NRRL 3488, *Penicillium* spp., and *Escherichia coli* WGS-10. Among the renewable substrates that have been used for MA production are glucose, food waste, and agro-industrial waste. These last two are the most promising as they are much cheaper. According to Kövilein et al. (2020), the cost of producing a ton of L-malic acid from by-products such as molasses is approximately USD 52, while with glucose the cost is approximately six times higher.

In the production of MA using microorganisms, the nitrogen source must be limited, and carbonate salts must be used. Furthermore, these processes require small amounts of oxygen, generally 0.5 vvm, and when the substrate to be used is lignocellulosic biomass, it must be previously hydrolyzed. Zou et al. (2016) obtained a maximum MA concentration of 32.4 g/L using the strain *Aureobasidium pullulans* and corncob hydrolysate as the substrate. The process was developed at 25 °C with stirring at 220 rpm. The medium contained 90 g/L of the carbon source (xylose, glucose, fructose, maltose, and sucrose). A 500 mL Erlenmeyer flask was used as the bioreactor, and the working volume was 50 mL, while the inoculum was 10% (v/v) of the working volume. Leathers and Manitchotpisit (2013) produced MA using the strain *Aureobasidium pullulans* and, as a carbon source, they used corn fiber hydrolysate; the maximum concentration of MA was 11.6 g/L. According to the authors, $CaCO_3$ improves MA yield. Other examples are presented in Table 8.4.

8.6 ITACONIC ACID

Itaconic acid (IA) is a carboxylic acid, crystalline in appearance, biodegradable, and non-toxic to the environment (Data-Bridge-Market-Research, 2021). Chemically, IA is known as (1-propene-2,3-dicarboxylic acid, or 2-methylenesuccinic acid) and is an unsaturated organic acid considered a suitable chemical platform for monomers or their synthesis (Teleky and Vodnar, 2021). According to Kuenz et al. (2012) and Sándor et al. (2021), important chemical properties can be indexed in the conjugated double bonds of methylene groups, which allow polymerization through condensation and esterification with monomers different from carboxylic groups.

IA can be applied to different products, including synthetic latex, unbalanced polyester resin, detergents, different types of polymers (including unsaturated polyester resins), acrylic replacement fibers, superabsorbent polymer, polyitaconic acid, methacrylate, and some chemical intermediates such as 2-methyl-1,4-butanediol and 3-methyltetrahydrofuran (Data-Bridge-Market-Research, 2021; Kuenz et al. 2012; Sándor et al. 2021). Due to the various applications of IA, the market for this raw material is constantly increasing. According to Zhao et al. (2018), around 40,000 tons are produced each year, which is equivalent to approximately USD 165 million, and it is expected that by the year 2028 the IA market will reach USD 176.55 million (Data-Bridge-Market-Research, 2021), mainly because this raw material can be used to manufacture food packaging.

TABLE 8.4
Production of Malic Acid Using Different Food Waste

Microrganism	Substrates	Bioreactor	Conditions	L-Malic Acid (g/L)	References
Aureobasidium pullulans	Soy molasses (60 g/L), c orn steep liquor (2 g/L), KH_2PO_4 (0.1 g/L), $MgSO_4$+$7H_2O$ (0.1 g/L), $CaCO_3$ (30 g/L)	Bioreactor volume: 5 L. Work volume: 2 L	Operation mode: fed-batch. Temperature: 25 °C. Fermentation time: 70 h. Agitation: 600 rpm. Aeration: 1 vvm.	71. 9	(Cheng et al. 2017)
Aureobasidium pullulans	Soybean hull hydrolysate (60 g/L), corn steep liquor (2 g/L), KH_2PO_4 (0.1 g/L), $MgSO_4$+$7H_2O$ (0.1 g/L), $CaCO_3$ (30 g/L)	Bioreactor volume: 5 L. Work volume: 2 L	Operation mode: fed-batch. Temperature: 25 °C. Fermentation time: 70 h. Agitation: 600 rpm. Aeration: 1 vvm.	31.3	(Cheng et al. 2017)
Aureobasidium pullulans	Sugarcane molasses (35.8 % w/w), y east extract 1 g/L, $CaCO_3$ (20 g/L), KH_2PO_4 (0.1 g/L), $MgSO_4$+$7H_2O$ (0.1 g/L)	Bioreactor volume: 5 L. Work volume: 2.7 L	Operation mode: fed-batch. Temperature: 25 °C. Fermentation time: 168 h. Agitation: 300–500 rpm. Aeration: 1 vvm.	95.4	(Xia et al. 2016)
Aureobasidium pullulans	Sugarcane molasses (60 g/L), NH_4NO_3 (2 g/L), KH_2PO_4 (0.1 g/L), $ZnSO_4$ (0.1 -g/L), CaCO3 (20 -g/L)	Bioreactor volume: 5 L. Work volume:	Operation mode: fed-batch. Temperature: 25 °C. Fermentation time: 140 h. Agitation: 180 rpm. Aeration: 0.8 vvm.	94.2	(Feng et al. 2018)
Aspergillus wentii	Cassava peel (glucose: 100 g/L) $(NH_4)_2SO_4$ (1.2 g/L), KH_2PO_4 (0.75 g/L), K_2HPO_4 (0.1 g/L), $MgSO_4$+$7H_2O$ (0.1 g/L)	Bioreactor volume: 0.25 L. Work volume: 0.05 L	Operation mode: batch. Temperature: 30 °C. Fermentation time: 48 h. Agitation: 150 rpm.	37.9	(Gopaliya et al. 2023)

8.6.1 Production of Itaconic Acid

IA has been a metabolite traditionally produced microbially using the fungus *Aspergillus terreus* and renewable substrates, including glucose. This can be produced by solid fermentation and submerged fermentation. Although IA can be produced from complex sugars such as disaccharides and polysaccharides, the highest yields are obtained with monosaccharides (Krull et al. 2020; Teleky and Vodnar, 2021). Other microorganisms, such as Candida and Pseudozyma, also produce this metabolite. However, the highest yields have been achieved with *Aspergillus terreus* (Kerssemakers et al. 2020; Teleky and Vodnar, 2021). While with *Aspergillus terreus*, concentrations of up to 150 g/L have been achieved, with *Candida* and *Pseudozyma* the maximum concentrations achieved are between 30 g/L and 75 g/L (Krull et al. 2020; Teleky and Vodnar, 2021). IA is produced in the form of cis-aconitate during the tricarboxylic acid cycle in the mitochondria with the help of some genes, such as cadA, mfsA, and mttA, first found in *A. terreus*.

Currently, the interest in producing IA is mainly focused on the use of alternative substrates to reduce the costs of producing the metabolite. Among the alternative substrates, the use of agro-industrial waste and food waste stands out, as around 1.3 billion tons of the latter are generated each year (Teleky and Vodnar, 2021). Reddy and Singh (2002) obtained 35–37 g/L IA using food waste extract as the main substrate and *A. terreus* m.N45 as inoculum. The process was developed in a 100 mL Erlenmeyer flask at 120 rpm, a pH of 4.0, and a temperature of 34 °C. Narisetty et al. (2021) obtained a maximum IA concentration of 41.1 g/L using starchy food waste carbon as a source, which was composed of potatoes, rice, and noodles in a 1:1:1 ratio. Before being used, the residue was hydrolyzed to simpler sugars (mainly monosaccharides) using the enzyme amyloglucosidase. The production of IA was carried out with the *Aspergillus terreus* strain.

Narisetty et al. (2021) obtained a maximum concentration of 41.1 g/L of IA using starchy food waste carbon as a source, which was composed of potatoes, rice, and noodles in a 1:1:1 ratio. Before being used, the residue was hydrolyzed to simpler sugars (mainly monosaccharides) using the enzyme amyloglucosidase. The production of IA was carried out with the *Aspergillus terreus* strain. The process was conducted in a 500 mL Erlenmeyer flask with a working volume of 100 mL, stirring at 250 rpm and a temperature of 43 °C. Gnanasekaran et al. (2018) used three types of food waste to produce IA: corn starch, wheat flour, and sweet potatoes. Using *Aspergillus niveus,* the maximum concentrations obtained from IA were 15.65 g/L, 9.25 g/L, and 7.45 g/L, for each residue, respectively. The fermentation pH was maintained at 6.0 and temperature at 30 °C, and fermentation time was between 5 and 7 days.

Currently, work is being done on the genetic modification of microorganisms to improve production and increase the speed of metabolite production. Among the strains that have been modified are: *Corynebacterium glutamicum*, *Escherichia coli*, *Pseudomonas putida*, *Ustilago maydis*, *Neurospora crassa*, *Shewanella livingstonensis*, and *Candida lignohabitans*. Among the genes that have been expressed in each of these microorganisms are: CAD1 – *A. terreus*, cadA – *A. terreus, Del – ras2, fuz7, ubc3*, among others (Teleky and Vodnar, 2021).

8.7 BUTYRIC ACID

Butyric acid (BA) is a short-chain organic acid widely used in the chemical, food, pharmaceutical, and cosmetic industries. For example, in the chemical industry, BA is used as a precursor in the production of thermoplastic cellulose acetate butyrate, as it has good characteristics in terms of solubility with organic solvents and high resistance to cold and light (Zhang et al. 2009). In the food industry, BA is used to improve flavor and aroma. It is also used in the manufacture of some ingredients, such as thickeners, emulsifiers, and stabilizers. BA also has great application in the production of animal feed, due to the great benefits it brings to animals. According to Melaku et al. (2021), butyric and citric acid positively impact the well-being and intestinal health of cattle by reducing pathogenic bacteria and maintaining the gastrointestinal tract, which reflects on the animal's growth and performance.

In the cosmetic sector, BA esters such as amyl, methyl, and ethyl butyrate are used as additives to improve the fruity notes of fragrances and perfumes (Jiang et al. 2018). The applications of BA in high-consumption products mean that the market for this acid grows significantly each year. In the year 2021, the BA market was around USD 183 million, and it is projected that by the year 2028 the market for this will be USD 484.55 million (Globel-Newswire, 2022).

8.7.1 Production of Butyric Acid

Currently, the industrial production of BA faces a situation similar to that of fuels, as it also comes from the chemosynthesis of raw materials derived from petroleum, which are more economically viable processes (Jiang et al. 2018). Therefore, processes are needed where sustainable raw materials are used to produce this acid. BA can be produced biologically using different types of microorganisms, including bacteria from the genera *Clostridium*, *Butyrivibrio*, and *Butyribacterium*. Some bacteria from the genera *Fusobacterium*, *Sarcina*, *Eubacterium*, *Roseburia*, *Coprococcus*, and *Megasphaera* are also potential producers of this acid (Jiang et al. 2018). BA can be produced by pure bacteria as well as by bacterial consortia; the process can be continuous or in batch.

In general, BA is an intermediate metabolite and is therefore produced together with other organic acids such as acetic acid, succinic acid, and lactic acid, which makes downstream processes difficult. The main metabolic route for BA production is the Embden-Meyerhof-Parnas pathway. BA is an intermediate product generally produced in acetogenesis in anaerobic digestion or dark fermentation (De Carvalho et al. 2023). According to Sydney et al. (2020), in the digestion process, for each mole of glucose, two moles of BA and two moles of hydrogen are produced.

BA can be produced using different types of substrates such as glucose, fructose, sucrose, and different polysaccharides, such as starch, cellulose, and hemicellulose among others. Therefore, food waste is a promising source for BA production. For example, Do Nascimento et al. (2021) in a dark fermentation process using soft drink wastewater and corn steep liquor and a microbial consortium as substrate, in addition to producing significant amounts of biohydrogen (506 mL H_2/g $COD_{removed}$), also produced significant amounts of BA (8.95±0.256 g/L), propionic acid (1.29±0.01 g/L), and acetic acid (0.498±0.018 g/L). The process was developed under mesophilic conditions (37 °C) in batch. Martinez-Burgos et al. (2021) in a batch-operated biohydrogen production process, reported significant amounts of BA and acetic acid, approximately 12 g/L and 1.5 g/L, respectively, using *Clostridium beijerinckii* as inoculum and soft drink wastewater, corn steep liquor, and cheese whey as substrates. Fermentation time was 40 hours. Other examples are presented in Table 8.5.

8.8 CONCLUSIONS AND PERSPECTIVES

The increase in the use and global market of organic acids, coupled with the growing concern about environmental problems, has encouraged research and development of technologies for cheaper and less polluting processes. The use of food residues as a substrate is an alternative that has been widely studied, but some bottlenecks of the processes still need further study. Studies on the optimization of variables in production processes, such as pH, temperature, and composition of the medium, are already, in many cases, such as the production of citric acid, well advanced. However, downstream stages in other cases, such as the production of lactic acid, still need more studies. In addition, prospects of increasing the production of organic acids using alternative residues, such as food wastes, should consider the genetic improvement of strains.

TABLE 8.5
Butyric Acid Production from Different Food Waste

Microorganism	Substrates	Bioreactor and Operation Mode	Conditions	Butyric Acid (g/L)	References
Treated sludge	Food waste	Batch	T: 35°±1 °C; pH: 7.9–8.5	~4.5	(Cao and Zhao, 2009)
Treated sludge	Food waste; k itchen wastewater	Anaerobic baffled reactor; HRT: 1.5 d	T: 35 °C	1.8±0.3	(Tawfik et al. 2015)
Microbial consortia	Food waste (carbohydrate-rich); b iochar (pyrolysis of pine sawdust)	Batch	T: 35 °C; pH: 7.0	13.3±0.7	(Sunyoto et al. 2016)
Treated sludge	Food waste and o live husks	Batch	T: 37 °C; pH: 7.0–8.0	~6.0	(Pagliaccia et al. 2016)
Microbial consortia (*Clostridium, Sporonaerobacter, Coprococcus*): Vac, Esg, Lod	Cassava processing wastewater; y east extract	Batch	T: 37 °C; pH: 5.5–7.0	~0.25, ~0.30, ~0.30	(Martinez-Burgos et al. 2019)
Microbial consortia (Clostridiaceae, Porphyromonadaceae, Ruminococcaceae, Enterococcaceae): Vir	Cassava processing wastewater; c orn steep liquor	Batch	T: 37 °C; pH: 7.0	~8.5	(Martinez-Burgos et al. 2021)

REFERENCES

Abdel-Rahman, M. A., Tashiro, Y., & Sonomoto, K. (2011). Lactic acid production from lignocellulose-derived sugars using lactic acid bacteria: Overview and limits. *Journal of Biotechnology*, 156(4), 286–301. ISSN 0168-1656. Available at: http://www.sciencedirect.com/science/article/pii/S0168165611003166.

Abdel-Rahman, M. A., Tashiro, Y., & Sonomoto, K. (2013). Recent advances in lactic acid production by microbial fermentation processes. *Biotechnology Advances*, 31(6), 877–902.

Ahmad, A., Banat, F., & Taher, H. (2020). A review on the lactic acid fermentation from low-cost renewable materials: Recent developments and challenges. *Environmental Technology & Innovation*, 20, 101138.

Behera, Bikash Chandra, Mishra, Rashmiranjan, and Mohapatra, Sonali. (2021). Microbial citric acid: Production, properties, application, and future perspectives. *Food Frontiers,* 2(1), 62–76. https://doi.org/10.1002/fft2.66.

Cao, X., & Zhao, Y. (2009). The influence of sodium on biohydrogen production from food waste by anaerobic fermentation. *Journal of Material Cycles and Waste Management*, 11(3), 244–250. https://doi.org/10.1007/s10163-009-0237-5.

Chen, P.-T., Hong, Z.-S., Cheng, C.-L., Ng, I.-S., Lo, Y.-C., Nagarajan, D., & Chang, J.-S. (2020). Exploring fermentation strategies for enhanced lactic acid production with polyvinyl alcohol-immobilized Lactobacillus plantarum 23 using microalgae as feedstock. *Bioresource Technology*, 308, 2020.

Chen, Y., Zhang, X., & Chen, Y. (2021). Propionic Acid-Rich Fermentation (PARF) production from organic wastes: A review. *Bioresource Technology,* 339, 125569. https://doi.org/10.1016/j.biortech.2021.125569.

Cheng, C., Zhou, Y., Lin, M., Wei, P., & Yang, S. T. (2017). Polymalic acid fermentation by Aureobasidium pullulans for malic acid production from soybean hull and soy molasses: Fermentation kinetics and economic analysis. *Bioresource Technology*, 223, 166–174. https://doi.org/10.1016/j.biortech.2016.10.042.

Chi, Z., Wang, Z. P., Wang, G. Y., Khan, I., & Chi, Z. M. (2016). Microbial biosynthesis and secretion of l-malic acid and its applications. *Critical Reviews in Biotechnology*, 36(1), 99–107. https://doi.org/10.3109/07388551.2014.924474.

Coban, H. B. (2020). Organic acids as antimicrobial food agents: Applications and microbial productions. *Bioprocess and Biosystems Engineering,* 43(4), 569–591. https://doi.org/10.1007/s00449-019-02256-w.

Cubas-Cano, E., Gonzalez-Fernandez, C., Ballesteros, M., & Tomas-Pejo, E. (2018). Biotechnological advances in lactic acid production by lactic acid bacteria: Lignocellulose as novel substrate. *Biofuels, Bioproducts and Biorefining*, 12, 290–303.

Data-Bridge-Market-Research. (2021). Global Itaconic Acid Market - Industry Trends and Forecast to 2028. Available at:: https://www.databridgemarketresearch.com/reports/global-itaconic-acid-market. (Accessed September 26, 2023).

De Carvalho, J. C., De Souza Vandenberghe, L. P., Sydney, E. B., Karp, S. G., Magalhães, A. I., Martinez-Burgos, W. J., & Soccol, C. R. (2023). Biomethane production from sugarcane vinasse in a circular economy: Developments and innovations. *Fermentation*, 9(4), 1–27. https://doi.org/10.3390/fermentation9040349.

De Carvalho, J. C., Molina-Aulestia, D. T., Martinez-Burgos, W. J., Karp, S. G., Manzoki, M. C., Medeiros, A. B. P., & Soccol, C. R. (2022). Agro-industrial wastewaters for algal biomass production, bio-based products, and biofuels in a circular bioeconomy. *Fermentation*, 8(12). https://doi.org/10.3390/fermentation8120728.

De Oliveira, J., de Souza Vandenberghe, L. P., de Oliveira, P. Z., de Mello, A. F. M., Rodrigues, C., Nigam, P. S., & Soccol, C. R. (2021). Bioconversion of potato-processing wastes into an industrially-important chemical lactic acid. *Bioresource Technology Reports*, 15, 100698.

De Oliveira, P. Z., Vandenberghe, L. P. S., Rodrigues, C., Pereira, G. V. M., & Soccol, C. R. (2022). Exploring cocoa pod husks as a potential substrate for citric acid production by solid-state fermentation using aspergillus niger mutant strain. *Process Biochemistry,* 113(February), 107–112. https://doi.org/10.1016/j.procbio.2021.12.020.

Dishisha, Tarek, Ståhl, Åke, Lundmark, Stefan, and Hatti-Kaul, Rajni. (2013). An economical biorefinery process for propionic acid production from glycerol and potato juice using high cell density fermentation. *Bioresource Technology,* 135, 504–512. https://doi.org/10.1016/j.biortech.2012.08.098.

Do Nascimento Junior, J. R., Zevallos Torres, L. A., Medeiros, A. B. P., Woiciechowski, A. L., Martinez-Burgos, W. J., & Soccol, C. R. (2021). Enhancement of biohydrogen production in industrial wastewaters with vinasse pond consortium using lignin-mediated iron nanoparticles. *International Journal of Hydrogen Energy*, 46(54), 27431–27443. https://doi.org/10.1016/j.ijhydene.2021.06.009.

Es, I., Khaneghah, A. M., Hashemi, S. M. B., & Koubaa, M. (2017). Current advances in biological production of propionic acid. *Biotechnology Letters,* 39, 635–645. https://doi.org/10.1007/s10529-017-2293-6.

Feng, J., Yang, J., Yang, W., Chen, J., Jiang, M., & Zou, X. (2018). Metabolome- and genome-scale model analyses for engineering of *Aureobasidium pullulans* to enhance polymalic acid and malic acid production from sugarcane molasses. *Biotechnology for Biofuels*, 11(1), 1–13. https://doi.org/10.1186/s13068-018-1099-7.

Feng, X., Chen, F., Xu, H., Wu, B., Li, H., Li, S., & Ouyang, P. (2011). Green and economical production of propionic acid by propionibacterium freudenreichii CCTCC M207015 in plant fibrous-bed bioreactor. *Bioresource Technology Journal,* 102, 6141–6146. https://doi.org/10.1016/j.biortech.2011.02.087.

Fortunati, E. et al. (2012). Microstructure and nonisothermal cold crystallization of PLA composites based on silver nanoparticles and nanocrystalline cellulose. *Polymer Degradation and Stability*, 97(10), 2027–2036. ISSN 0141-3910. Available at: http://www.sciencedirect.com/science/article/pii/S014139101200105X.

Future-Market-Insights. (2023). Organic acid market outlook (2022 to 2032). Available at: https://www.futuremarketinsights.com/reports/global-organic-acids-market. (Accessed September 25, 2023).

Ghaffar, T. et al. (2014). Recent trends in lactic acid biotechnology: A brief review on production to purification. *Journal of Radiation Research and Applied Sciences*, 7(2), 222–229. http://dx.doi.org/10.1016/j.jrras.2014.03.002.

Globel-Newswire. (2022). Globel-Newswire. 2022. Butyric Acid Market Size Worth $484.55 Million, Globally, by 2028 at 14.9% CAGR. Available at: https://www.globenewswire.com/en/news-release/2022/03/24/2409834/0/en/Butyric-Acid-Market-Size-Worth-484-55-Million-Gl.

Gnanasekaran, R., Dhandapani, B., Gopinath, K. P., & Iyyappan, J. (2018). Synthesis of itaconic acid from agricultural waste using novel Aspergillus niveus. *Preparative Biochemistry and Biotechnology*, 48(7), 605–609. https://doi.org/10.1080/10826068.2018.1476884.

Gopaliya, D., Zaidi, S., Srivastava, N., Rani, B., Kumar, V., & Khare, S. (2023). Integrated fermentative production and downstream processing of L-malic acid by Aspergillus wentii using cassava peel waste. *Bioresource Technology*, 377, 128946. https://doi.org/10.1016/j.biortech.2023.128946.

Gu, Zhong, Glatz, Bonita A., & Glatz, Charles E. (1998). Effects of propionic acid on propionibacteria eermentation. *Enzyme and Microbial Technology,* 22, 13–18. https://doi.org/10.1016/S0141-0229(97)00094-X.

Guo, Y., Li, C., Zhao, H., Gao, M., & Wang, Q. (2023). The performance of microfiltration process for purifying lactic acid in the fermented broth of kitchen waste. *Membranes,* 13, 280. https://doi.org/10.3390/membranes13030280.

Hou, W., & Bao, J. (2018). Simultaneous saccharification and aerobic fermentation of high titer cellulosic citric acid by filamentous fungus aspergillus niger. *Bioresource Technology,* 253(April), 72–78. https://doi.org/10.1016/j.biortech.2018.01.011.

Hu, Wei, Li, , Wen-jian, Hai-quan, Yang, & Chen, Ji-hong. (2019). Current strategies and future prospects for enhancing microbial production of citric acid. *Applied Microbiology and Biotechnology,* 103(1), 201–209. https://doi.org/10.1007/s00253-018-9491-6.

Ibrahim, Nabil A., Eid, Basma M., & Emam, El-Amir M. (2022). An eco-friendly facile approach for imparting multifunctional protection properties to cellulose/wool blends. *Polymer Bulletin,* 79(11), 10313–10331. https://doi.org/10.1007/s00289-021-04072-3.

Imarc. (2023). Global industry trends, share, size, growth, opportunity and forecast 2023–2028, 87(1,2), 149–200. Available at: https://repositorio.ufsc.br/xmlui/bitstream/handle/123456789/167638/341506.pdf?sequence=1&isAllowed=y%0Ahttps://repositorio.ufsm.br/bitstream/handle/1/8314/LOEBLEIN%2CLUCINEIA CARLA.pdf?sequence=1&isAllowed=y%0Ahttps://antigo.mdr.gov.br/saneamento/proper.

Industry-ARC. (2023). Organic acids market-forecast (2023–2028). Available at: https://www.industryarc.com/Report/16128/organic-acids-market.html?gclid=Cj0KCQjwpc- oBhCGARIsAH6ote-exuRFZ_4NidISILCXbwNx5F1GelbP6Kjt-DKmLzgh4i4UU1n7bNUaAg- ZEALw_wcB. (Accessed September 25)

Jiang, L., Cui, H., Zhu, L., Hu, Y., Xu, X., Li, S., & Huang, H. (2015). Enhanced propionic acid production from whey lactose with immobilized propionibacterium acidipropionici and the role of trehalose synthesis in acid tolerance. *Green Chemistry,* 17, 250. https://doi.org/10.1039/c4gc01256a.

Jiang, L., Fu, H., Yang, H. K., Xu, W., Wang, J., & Yang, S. T. (2018). Butyric acid: Applications and recent advances in its bioproduction. *Biotechnology Advances*, 36(8), 2101–2117. https://doi.org/10.1016/j.biotechadv.2018.09.005.

Kerssemakers, A. A. J., Doménech, P., Cassano, M., Yamakawa, C. K., Dragone, G., & Mussatto, S. I. (2020). Production of itaconic acid from cellulose pulp: Feedstock feasibility and process strategies for an efficient microbial performance. *Energies*, 13(7), 1–12. https://doi.org/10.3390/en13071654.

Kövilein, A., Kubisch, C., Cai, L., & Ochsenreither, K. (2020). Malic acid production from renewables: A review. *Journal of Chemical Technology and Biotechnology*, 95(3), 513–526. https://doi.org/10.1002/jctb.6269.

Krull, S., Lünsmann, M., Prüße, U., & Kuenz, A. (2020). Ustilago rabenhorstiana—An alternative natural itaconic acid producer. *Fermentation*, 6(1), 1–15. https://doi.org/10.3390/fermentation6010004.

Kuenz, A., Gallenmüller, Y., & Willke, T. (2012). Microbial production of itaconic acid : Developing a stable platform for high product concentrations. *Applied Microbiology and Biotechnology*, 96, 1209–1216. https://doi.org/10.1007/s00253-012-4221-y.

Kumar, S., Yadav, N., Nain, L., & Khare, S. K. (2020). A simple downstream processing protocol for the recovery of lactic acid from the fermentation broth. *Bioresource Technology*, 318, 124260.

Leathers, T. D., & Manitchotpisit, P. (2013). Production of poly(β-L-malic acid) (PMA) from agricultural biomass substrates by Aureobasidium pullulans. *Biotechnology Letters*, 35(1), 83–89. https://doi.org/10.1007/s10529-012-1045-x

Li, C., Gao, M., Zhu, W., Wang, N., Ma, X., Wu, C., & Wang, Q. (2021). Recent advances in the separation and purification of lactic acid from fermentation broth. *Process Biochemistry*, 104, 142–151.

Li, J., Ma, D., Tian, J., Sun, T., Meng, Q., Li, J., & Shan, A. (2023). The responses of organic acid production and microbial community to different carbon source additions during the anaerobic fermentation of Chinese cabbage waste. *Bioresource Technology*, 371(600), 128624. https://doi.org/10.1016/j.biortech.2023.128624.

Liu, H. et al. (2011). Interaction of microstructure and interfacial adhesion on impact performance of polylactide (PLA) ternary blends. *Macromolecules*, 44(6), 1513–1522. ISSN 0024-9297. http://dx.doi.org/10.1021/ma1026934.

Mahato, N., Sharma, K., Sinha, M., Dhyani, A., Pathak, B., Jang, H., Park, S., Pashikanti, S., & Cho, S. (2021). Biotransformation of citrus waste-I: Production of biofuel and valuable compounds by fermentation. *Processes*, 9(2), 220. https://doi.org/10.3390/pr9020220.

Market Size – Lactic Acid, Share & Trends Analysis Report By Raw Material (Sugarcane, Corn, Cassava), By Application (Industrial, Food & Beverages, Pharmaceuticals, Personal Care, Polylactic Acid), By Region, And Segment Forecasts, 2023–2030. Available at: https://www.grandviewresearch.com/industry-analysis/lactic-acid-and-poly-lactic-acid-market.

Martinez-Burgos, Walter J., Junior, J. R., Medeiros, A. B. P., Herrmann, L. W., Sydney, E. B., & Soccol, C. R. (2021). Biohydrogen production from agro-industrial wastes using clostridium beijerinckii and isolated bacteria as inoculum. *Bioenergy Research*, 15, 987–997. https://doi.org/10.1007/s12155-021-10358-1.

Martinez-Burgos, Walter J., Sydney, E. B., Brar, S. K., Tanobe, V. O. de A., Pedroni Medeiros, A. B., Cesar de Carvalho, J., & Soccol, C. R. (2019). The effect of hydrolysis and sterilization in biohydrogen production from cassava processing wastewater medium using anaerobic bacterial consortia. *International Journal of Hydrogen Energy*, 44(47), 25551–25564. https://doi.org/10.1016/j.ijhydene.2019.08.085.

Martinez-Burgos, Walter J., Sydney, E. B., de Paula, D. R., Medeiros, A. B. P., De Carvalho, J. C., Molina, D., & Soccol, C. R. (2021). Hydrogen production by dark fermentation using a new low-cost culture medium composed of corn steep liquor and cassava processing water: Process optimization and scale-up. *Bioresource Technology*, 320, 124370. https://doi.org/10.1016/j.biortech.2020.124370.

Marzo-Gago, C., Venus, J., & López-Gómez, J. P. (2022). Production of lactic acid from pasta wastes using a biorefinery approach. *Biotechnol Biofuels*, 15, 128. https://doi.org/10.1186/s13068-022-02222-x.

Melaku, M., Zhong, R., Han, H., Wan, F., Yi, B., & Zhang, H. (2021). Butyric and citric acids and their salts in poultry nutrition: Effects on gut health and intestinal microbiota. *International Journal of Molecular Sciences*, 22(19). https://doi.org/10.3390/ijms221910392.

Nangare, S., Vispute, Y., Tade, R., Dugam, S., & Patil, P. (2021). Pharmaceutical āpplications of citric acid. *Future Journal of Pharmaceutical Sciences*, 7(1), 54. https://doi.org/10.1186/s43094-021-00203-9.

Narisetty, V., Prabhu, A. A., Al-Jaradah, K., Gopaliya, D., Hossain, A. H., Kumar Khare, S., & Kumar, V. (2021). Microbial itaconic acid production from starchy food waste by newly isolated thermotolerant Aspergillus terreus strain. *Bioresource Technology*, 337, 125426. https://doi.org/10.1016/j.biortech.2021.125426.

Olszewska-Widdrat, A., Alexandri, M., López-Gómez, J. P., Schneider, R., Mandl, M., & Venus, J. (2019). Production and purification of L-lactic acid in lab and pilot scales using sweet sorghum juice. *Fermentation*, 5(2), 36.

Ozdal, M., & Kurbanoglu, E. B. (2019). Citric acid production by aspergillus niger from agro-industrial by-products: Molasses and chicken feather peptone. *Waste and Biomass Valorization*, 10(3), 631–640. https://doi.org/10.1007/s12649-018-0240-y.

Özüdoğru, H. M. R., Nieder-Heitmann, M., Haigh, K. F., & Görgens, J. F. (2019). Techno-economic analysis of product biorefineries utilizing sugarcane lignocelluloses: Xylitol, citric acid and glutamic acid scenarios annexed to sugar mills with electricity co-production. *Industrial Crops and Products*, 133(July), 259–268. https://doi.org/10.1016/j.indcrop.2019.03.015.

Pagliaccia, P., Gallipoli, A., Gianico, A., Montecchio, D., & Braguglia, C. M. (2016). Single stage anaerobic bioconversion of food waste in mono and co-digestion with olive husks: Impact of thermal pretreatment on hydrogen and methane production. *International Journal of Hydrogen Energy*, 41(2), 905–915. https://doi.org/10.1016/j.ijhydene.2015.10.061.

Sivagurunathan, P., Raj, T., Chauhan, P.S., Kumari, P., Satlewal, A., Gupta, R. P., & Kumar, R. (2022). High-titer lactic acid production from pilot-scale pretreated non-detoxified rice straw hydrolysate at high-solid loading. *Biochemical Engineering Journal*, 187. https://doi.org/10.1016/j.bej.2022.108668.

Rakicka, M., Wolniak, J., Lazar, Z., & Rymowicz, W. (2019). Production of high titer of citric acid from inulin. *BMC Biotechnology,* 19(1), 11. https://doi.org/10.1186/s12896-019-0503-0.

Reddy, C. S. K., & Singh, R. P. (2002). Enhanced production of itaconic acid from corn starch and market refuse fruits by genetically manipulated *Aspergillus terreus* SKR10. *Bioresource Technology*, 85(1), 69–71. https://doi.org/10.1016/S0960-8524(02)00075-5.

Reena, R., Sindhu, R., Balakumaran, P. A., Pandey, A., Awasthi, M. K., & Binod, P. (2022). Insight into citric acid: A versatile organic acid. *Fuel,* 327(November), 125181. https://doi.org/10.1016/J.FUEL.2022.125181.

Roukas, Triantafyllos, & Kotzekidou, Parthena. (2020). Pomegranate peel waste: A new substrate for citric acid production by aspergillus niger in solid-state fermentation under non-aseptic conditions. *Environmental Science and Pollution Research,* 27(12), 13105–13113. https://doi.org/10.1007/s11356-020-07928-9.

Saelee, N. (2022). Lactic acid production from old oil palm trunk sap in the open batch, open repeated batch, fed-batch, and repeated fed-batch fermentation by lactobacillus rhamnosus ATCC 10863. *Fermentation*, 8, 430. https://doi.org/10.3390/fermentation8090430.

Sándor, E., Kolláth, I. S., Fekete, E., Bíró, V., Flipphi, M., Kovács, B., & Karaffa, L. (2021). Carbon-source dependent interplay of copper and manganese ions modulates the morphology and itaconic acid production in *Aspergillus terreus. Frontiers in Microbiology*, 12, 1–14. https://doi.org/10.3389/fmicb.2021.680420.

Sarmiento-Vásquez, Z., Vandenberghe, L. P. S., Rodrigues, C., Tanobe, V. O. A., Marín, O., Pereira, G. V. M., Rogez, H. L. G., Góes-Neto, A., & Soccol, C. R. (2021). Cocoa pod husk valorization: Alkaline-enzymatic pre-treatment for propionic acid production. *Cellulose,* 28, 4009–4024. https://doi.org/10.1007/s10570-021-03770-5.

Sawant, O., Mahale, S., Ramchandran, V., Nagaraj, G., & Bankar, A. (2018). Fungal citric acid production using waste materials: A mini-review. *Journal of Microbiology, Biotechnology and Food Sciences,* 8(2), 821–28. https://doi.org/10.15414/jmbfs.2018.8.2.821-828.

Sharma, A., Singh, S., Khare, S.K., Sharma, A., Tiwari, R., & Nain, L. (2022). Green lactic acid production using low-cost renewable sources and potential applications. In Anuj K. Chandel, & Fernando Segato (Eds.), *Production of Top 12 Biochemicals Selected by USDOE from Renewable Resources*, Elsevier, pp. 345–365, ISBN 9780128235317. https://doi.org/10.1016/B978-0-12-823531-7.00006-8.

Song, L., Liu, S., Yang, D., Liu, R., & Dai, X. (2023). Enhanced lactic acid production by Lactobacillus rhamnosus ATCC 7469 through simultaneous saccharification and fermentation from household food waste. *Clean - Soil, Air, Water*, 51, 2200329. https://doi.org/10.1002/clen.202200329.

Sun, X., Wu, H., Zhao, G., Li, Z., Wu, X., Liu, H., & Zheng, Z. (2018). Morphological regulation of *Aspergillus Niger* to improve citric acid production by ChsC gene silencing. *Bioprocess and Biosystems Engineering,* 41(7), 1029–1038. https://doi.org/10.1007/s00449-018-1932-1.

Sun, Y., Liu, H., Yang, Y. et al. (2021). High-efficient L-lactic acid production from inedible starchy biomass by one-step open fermentation using thermotolerant Lactobacillus rhamnosus DUT1908. *Bioprocess and Biosystems Engineering*, 44, 1935–1941. https://doi.org/10.1007/s00449-021-02573-z.

Sunyoto, N. M. S., Zhu, M., Zhang, Z., & Zhang, D. (2016). Effect of biochar addition on hydrogen and methane production in two-phase anaerobic digestion of aqueous carbohydrates food waste. *Bioresource Technology*, 219, 29–36. https://doi.org/10.1016/j.biortech.2016.07.089.

Sweta, V. L., Karemore, H., & Umekar, M. J. (2021). Review on production of citric acid by fermentation technology. *GSC Biological and Pharmaceutical Sciences,* 17(3), 85–93. https://doi.org/10.30574/gscbps.2021.17.3.0313.

Sydney, E. B., Duarte, E. R., Martinez Burgos, W. J., de Carvalho, J. C., Larroche, C., & Soccol, C. R. (2020). Development of short chain fatty acid-based artificial neuron network tools applied to biohydrogen production. *International Journal of Hydrogen Energy*, 45(8), 5175–5181. https://doi.org/10.1016/j.ijhydene.2019.10.128.

Tawfik, A., El-Qelish, M., & Salem, A. (2015). Efficient anaerobic co-digestion of municipal food waste and kitchen wastewater for bio-hydrogen production. *International Journal of Green Energy*, 12(12), 1301–1308. https://doi.org/10.1080/15435075.2014.909357.

Teleky, B.-E., & Vodnar, D. C. (2021). Recent advances in biotechnological itaconic acid production, and application for a sustainable approach. *Polymers*, 1–22. https://doi.org/10.3390/polym13203574.

Tong, Z., Zheng, X., Tong, Y., Shi, Y.-C., & Sun, J. (2019). Systems metabolic engineering for citric acid production by aspergillus niger in the post-genomic era. *Microbial Cell Factories,* 18(1), 28. https://doi.org/10.1186/s12934-019-1064-6.

Urak, S., Yeniay, O., & Karasu-Yalcin, S. (2015). Optimization of citric acid production from a carrot juice-based medium by yarrowia lipolytica using response surface methodology. *Annals of Microbiology,* 65(2), 639–649. https://doi.org/10.1007/s13213-014-0900-5.

Vion, C., Muro, M., Bernard, M., Richard, B., Valentine, F., Yeramian, N., … Marullo, P. (2023). New malic acid producer strains of Saccharomyces cerevisiae for preserving wine acidity during alcoholic fermentation. *Food Microbiology*, 112(December 2022). https://doi.org/10.1016/j.fm.2022.104209.

Wang, B., Li, H., Zhu, L., Tan, F., Li, Y., Zhang, L., Ding, Z., & Shi, G. (2017). High-efficient production of citric acid by aspergillus niger from high concentration of substrate based on the staged-addition glucoamylase strategy. *Bioprocess and Biosystems Engineering,* 40(6), 891–899. https://doi.org/10.1007/s00449-017-1753-7.

Wang, J., Cui, Z., Li, Y., Cao, L., & Lu, Z. (2020). Techno-economic analysis and environmental impact assessment of citric acid production through different recovery methods. *Journal of Cleaner Production,* 249(March), 119315. https://doi.org/10.1016/J.JCLEPRO.2019.119315.

Wang, Y., & Gao, M. (2023). Efficient biorefinery based on designed lignocellulosic substrate for lactic acid production. *Fermentation,* 9(8), 744. https://doi.org/10.3390/fermentation9080744.

Walker, R. P., & Famiani, F. (2018). Organic acids in fruits: Metabolism, functions and contents. In *Horticultural Reviews* (pp. 371–430). https://doi.org/10.1002/9781119431077.ch8.

West, T. P. (2017). Microbial production of malic acid from biofuel-related coproducts and biomass. *Fermentation*, 3(2), 1–10. https://doi.org/10.3390/fermentation3020014.

Xia, J., Xu, J., Hu, L., & Liu, X. (2016). Enhanced poly(L-malic acid) production from pretreated cane molasses by Aureobasidium pullulans in fed-batch fermentation. *Preparative Biochemistry and Biotechnology*, 46(8), 798–802. https://doi.org/10.1080/10826068.2015.1135464.

Yang, H., Wang, Z., Lin, M., & Yang, S.-T. (2018). Propionic acid production from soy molasses by propionibacterium acidipropionici: Fermentation kinetics and economic analysis. *Bioresource Technology,* 250, 1–9. https://doi.org/10.1016/j.biortech.2017.11.016.

Yu, B., et al. (2018). Continuous citric acid production in repeated-fed batch fermentation by aspergillus niger immobilized on a new porous foam. *Journal of Biotechnology,* 276–277(June), 1–9. https://doi.org/10.1016/j.jbiotec.2018.03.015.

Zhang, C., Yang, H., Yang, F., & Ma, Y. (2009). Current progress on butyric acid production by fermentation. *Current Microbiology*, 59(6), 656–663. https://doi.org/10.1007/s00284-009-9491-y.

Zhang, Y., et al. (2023). High titer (>200 g/L) lactic acid production from undetoxified pretreated corn stover. *Bioresource Technology*, 388. https://doi.org/10.1016/j.biortech.2023.129729

Zhao, M., Lu, X., Zong, H., Li, J., & Zhuge, B. (2018). Itaconic acid production in microorganisms. *Biotechnology Letters*, 40(3), 455–464. https://doi.org/10.1007/s10529-017-2500-5.

Zou, X., Yang, J., Tian, X., Guo, M., Li, Z., & Li, Y. (2016). Production of polymalic acid and malic acid from xylose and corncob hydrolysate by a novel Aureobasidium pullulans YJ 6-11 strain. *Process Biochemistry*, 51(1), 16–23. https://doi.org/10.1016/j.procbio.2015.11.018.

9 Production of Compost and Vermicompost from Food Waste

Shiqing Chen, Haoyuan Zhang, Yuzhe Chen, Suyue Hao, and Mukesh Kumar Awasthi

With the development of society, having sufficient food and clothing is no longer a problem in most countries, but in contrast, food waste is widespread. Statistics show that a huge amount of food is wasted around the world every day. Food waste is a good raw material that can yield high value products. However, in many countries and regions, food waste cannot be properly disposed of, which is not only a waste of resources but also causes great environmental pressure and increases the economic cost of disposal. When it comes to disposing of food waste, composting has great advantages over many disposal methods. Therefore, the composting technology in recent years is reviewed in this chapter, and the development of earthworm composting technology is prospected. Based on the principle of continuous composting, the importance of precursors is explored. How to improve the composting efficiency by mechanical modification is also discussed, as well as the efficient recycling methods of subsequent high-value products. In addition, the advantages of earthworm compost compared with traditional compost are analyzed, the key points of development of earthworm compost technology are clarified, and suggestions on how to improve the efficiency of earthworm compost are put forward.

9.1 INTRODUCTION

A huge amount of food is wasted around the world every day. According to the Food and Agriculture Organization of the United Nations (FAO), the world's countries and regions waste an average of about 1.3 billion tons of food each year, accounting for about a third of the annual global food production. It puts tremendous pressure on the environment and increases financial costs globally due to mismanagement. The FAO estimates that the additional economic burden of food waste in developed and developing countries is US \$680 billion and US \$310 billion, respectively. At the same time, huge waste endangers environmental health, such as polluting soil, water, and affecting human health. But food waste is a valuable organic resource that can be recycled through a variety of biotechnological methods, including incineration, bio-refining, anaerobic digestion, and composting. Incineration can recover the calorific value of food waste, but it will produce toxic and harmful gases. Research shows that biorefining cannot be universally applied to all food waste (Weber et al., 2020). According to the research (Wainaina et al., 2019), anaerobic digestion (AD) for food waste treatment has the problems of immature sustainable technology development and low product purity. Following the principles of circular economy, composting is currently a commonly used environmentally friendly disposal method.

Composting is a biological process that converts organic waste into fertilizer with economic value that can be applied to agriculture. In the traditional composting process, the microorganisms will first metabolize the easily degradable substrate, and the metabolic process is accompanied by the temperature increase of the compost. Later, when the compost temperature drops close to the ambient temperature, secondary composting begins, degrading previously unmetabolized organic

DOI: 10.1201/9781032706030-9

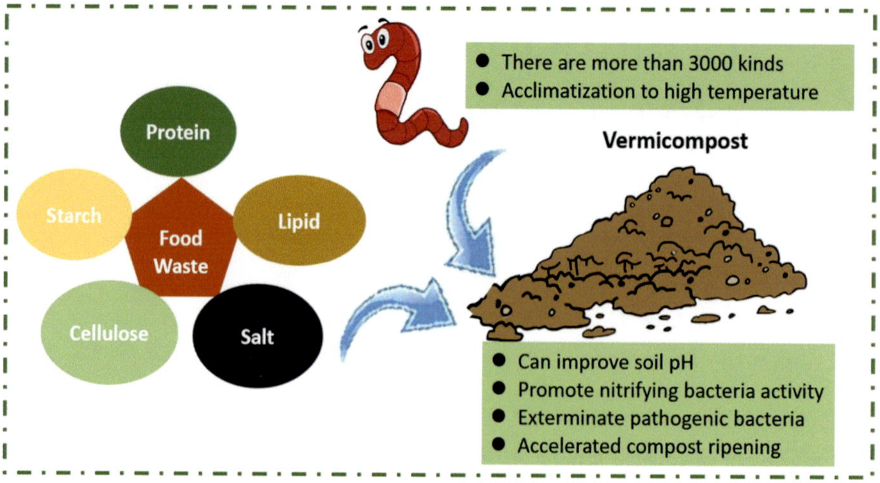

FIGURE 9.1 Earthworm composting diagram.

matter (Amuah et al., 2022). However, due to the decline in the temperature of secondary composting, some pathogenic bacteria and pathogens are difficult to kill due to insufficient temperature, and earthworm composting can effectively solve this problem.

Earthworm composting is a process in which earthworms and microorganisms work together to promote the decomposition of organic matter into humus and produce compost products. Earthworms are widely distributed and have many species. There are more than 3,000 known species of earthworms, and more than 170 species of earthworms live in China. Earthworms are saporous invertebrates, and their excrement is mainly ammonia and urea. Earthworms often inhabit soil surfaces that have higher temperatures and more humus, and can loosen the soil, improve soil acid-alkalinity and fertility, as well as promote the activity of nitrifying bacteria in the soil. Adding earthworms to compost can help eliminate pathogenic bacteria caused by insufficient compost temperature and uneven compost, reduce heavy metal toxicity, and accelerate compost maturation (Figure 9.1).

The purpose of this chapter is to review the technology of earthworm composting in recent years and to forecast the development of composting in the future. In addition, this chapter analyzes the advantages of earthworm compost compared with traditional compost, clarifies the key points of development of earthworm compost technology, and further puts forward some suggestions to improve the efficiency of earthworm compost.

9.2 NECESSITY AND GLOBAL SCENARIO OF FW CO-COMPOSTING

According to the latest Global Report on Food Crises, 258 million people in 58 countries and territories will be food insecure in 2022, up from 193 million people in 53 countries and territories in 2021. Food waste is common in rich countries. In the United States, about 40 million tons of food is wasted each year. Every year in Europe, more than 88 million tons of food is wasted. Ten million tons of food is wasted in Australia every year. Moreover, in developing countries, food is more vulnerable to loss and deterioration due to inadequate infrastructure and technology. India wastes about 21 million tons of food a year, while Africa wastes 54 million tons. As we all know, food waste deteriorates very quickly, and especially at the right temperature. This shows that the disposal of food waste is urgent, otherwise it will cause great environmental pressure. At present, the commonly used treatment method is incineration, but it does not conform to the principle of circular economy.

Co-composting is a method of degrading multiple organic waste compounds into useful organic products (Awasthi et al., 2022). Some organic wastes can only be co-composted due to their complex physical and chemical properties (Chi et al., 2020). It has bee shown that co-composting has good effects on regulating the ratio of carbon to nitrogen and initial water content of food and kitchen waste, increasing the treatment efficiency and improving the quality of compost. Therefore, in recent years, people have paid more and more attention to the co-composting treatment of different organic wastes. For example, the livestock industry, as a significant source of global methane emissions, should follow the principles of sustainable development and achieve the goal of global carbon neutrality. Manure produced by animal husbandry should be treated and disposed of accordingly, and co-composting cattle manure and biochar can play a good role in this process (Harrison et al., 2022). Commonly used organic materials added to compost as fillers include biochar, bark, and leaves (Huang et al., 2022; Schroeter-Zakrzewska et al., 2021), and inorganic materials such as zeolite, lime, and minerals (Awasthi et al., 2017; Zhang et al., 2023).

9.3 PRINCIPLES OF SUSTAINABLE FW COMPOSTING

9.3.1 Role of Precursor for FW Composting

Temperature, moisture content, C/N ratio, PH, aeration, and material particle size are all factors affecting the composting process. To optimize the composting process and improve the quality of composting products, one or several additives are usually selected, which are generally inorganic additives, organic additives, and microbial agents. Biochar has the characteristics of high carbon content, large surface, developed pore structure, and area abundant surface functional groups. The research showed that Bch treatment could maintain the thermophilic temperature range for a long time, promote organic degradation, reduce pile density, adjust the C/N ratio, relieve pH, prevent water evaporation, improve the humification process, and make humification more mature (Hoang et al., 2022; Nguyen et al., 2023; Parra-Orobio et al., 2023; Piao et al., 2023; Ravindran et al., 2022), thus improving the composting process. Meanwhile, biochar could regulate the bacterial dynamics in food waste compost and improve FW composting. Verma et al., (2023) found in an experiment that the proportion of wheat straw biochar (WSB) in treatment can change the overall composition of bacteria at a medium temperature, and at thermophilic and mature stages, indicating the role of WSB in composting. In addition, adding 15% of biochar improved FW composting by affecting bacterial dynamics.

Microbes play a huge role in the composting process. The dominant microbial population also changes at different stages. However, in the process of succession, due to the insufficient number of primary microbial communities and the disappearance of some functional microorganisms, the organic matter degradation efficiency is not ideal and nitrogen loss may be caused (Zhang et al., 2023), Therefore, microorganisms have been paid more and more attention as efficient additives to promote the composting process. Wang et al., (2022a, 2022b) found that through *Aneurinibacillus* sp. LD3, the degradation rate of organic matter and lignin was greatly increased, and at the same time, it significantly adjusted the composition and function of fungi, effectively improving the composting process. Meanwhile, He et al., (2022) found that inoculation of multifunctional microbial agents in FW compost could increase the degradation rate of organic matter. At the same time, microbial community structure changed, tricarboxylic acid cycle and amino acid metabolism were enhanced, the relationship between microorganisms became closer, and the refractory organic matter such as cellulose provided more carbon sources for microorganisms (Table 9.1).

Lime, as a common soil conditioner, is more suitable for regulating acidification of food waste (Jin et al., 2021). Because food waste contains a large amount of organic matter, it is easy to produce a large amount of organic acid in the composting process, and if there is no appropriate intervention, over-acidified waste is not easy to compost. Since lime is highly alkaline and soluble in water and can generate a lot of heat, the addition of lime can regulate the pH of the waste. The addition of lime enriched the biomarkers of bacteria (such as Tyberia and sphingococcus) and enhanced the organic biodegradation of humification. In addition, the addition of lime promotes the proliferation

TABLE 9.1
Major Microorganisms at Different Stages of Food Waste Composting

Phase	Dominant Population	Reference
Mesophilic,	*Stenotrophomonas*, *Xanthomonas*, *Pseudomonas*, *Klebsiella Alcaligenes*, *Achromobacter*, *Caulobacter*, *Luteimonas* sp., and *Achromobacter* sp.	(Pepe et al., 2013)
Thermophilic	*Desulfomicrobium*, *Desulfovibrio*, *Methanocella*, *Mageeibacillus*, *Acetobacterium*, *Desulfovibrio*, *Klebsiella pneumoniae*, *Pseudomonas aeruginosa* *Desulfobulbus*, *N itrosomonas* sp., *Nitrosospira* sp., *Thermoclostridium stercorarium Luteimonas* sp., *Achromobacter* sp., *Alcaligenes faecalis*, *Pseudomonas stutzeri*, *Rhodobacteraceae*, *Pseudomonas*, *Pseudomonadales*, *Bacillus*, *Pseudomonas*, *Luteimonas*, *Unidentified_ bacteria*, *Pseudomonas*, *Alcaligenes*, *Corynebacterium*, *Brevundimonas*, *Tamlana*, and *Thauera*	(Zhong et al., 2020; Yin et al., 2018; Yang et al., 2019; Sun et al., 2016; Yu et al., 2020; Liu et al., 2019; Guo et al., 2020; Tang et al., 2020)
Maturation	*Methanocella*, *Mageeibacillus*, *Acetobacterium*, *Desulfovibrio*, *Desulfobulbus*, *Bacillus cereus*, *Pseudomonas donghuensis*, *Bacillus Licheniformis*, *Nitrosomonas* sp., *Nitrosospira* sp., *Luteimonas* sp., *Achromobacter* sp., *Alcaligenes faecalis*, *Pseudomonas stutzeri*, *Rhodobacteraceae*, *Pseudomonas*, and *Unidentified_bacteria*	(Zhong et al., 2020; Yang et al., 2019; Yu et al., 2020; Guo et al., 2020)
Cooling	*Bacillus cereus*, *Pseudomonas donghuensis*, and *Bacillus licheniformis*	(Yang et al., 2019)

of thermophilic bacteria (such as bacillus and symbiotic bacteria) to undergo aerobic chemical heterotrophy, resulting in enhanced organic decomposition (Qi et al., 2022).

At present, new additives have received more attention. He et al., (2022) found that the degradation of organic matter by Fe-C particles increased by 12.3%, especially lignocellulose, resulting in an enhanced humification process (increased by 15.5%). At the same time, Fe-C significantly increased and decreased the relative abundance of *Bacillus* and *Aspergillus*, and greatly promoted the decomposition of organic substrates.

9.3.2 Improvement of Composting Efficacy by Machinery Modification

In the composting process of FW, temperature, PH, oxygen, moisture, particle size, and so on will affect the quality of compost. In order to improve composting efficiency and reduce pollution effects, additives (biochar, activated carbon, and ball milling) can affect the physical and chemical properties of the compost mixture, the community and activity of microorganisms, and the composting process, so it is widely used (Zhou et al., 2022; Ajibade et al., 2022). Nguyen et al., (2023) used FW (47% of vegetables and fruits, 16% of tea leaves, and 29% of red meat) and sawdust (particle size < 1cm) as raw materials, and biochar as additive (wood biochar obtained by pyrolysis at 550 °C, average particle size < 4 mm) for a 30-day composting experiment (Figure 9.2). The results show that the use of biochar can effectively reduce the emissions of NH_3 and H_2S, and has higher pore structure and surface functional groups. Among them, 20% of the biochar had the lowest emissions of ammonia (2 ppm) and hydrogen sulfide (3 ppm) on day 16 and stopped emissions on day 17.

He et al., (2022) investigated the effect of iron carbon particles on food waste composting. The composting material was made of food waste (cooked rice, vegetables, organic skin) and filler (2–3 cm of sawdust) as raw materials, with the addition of iron carbon particles (hyperactivated FeC particles with 1% hydrochloric acid for 10 minutes, then washed with deionized water). The composting time was 32 days. The results showed that the interpretation effect of Fe-C particles on organic matter increased by 12.3%, especially the humification process caused by lignin (increased by

FIGURE 9.2 Reducinggas emissions from food waste composting.

15.3%). Nitration increased NO_3-N production during cooling and maturation (15.9%). Meanwhile, Fe-C particles not only significantly increased the relative abundances of *Bacillus* and *Aspergillus* for organic matter decomposition but also decreased the relative abundances of acid-producing bacteria.

9.3.3 Nutrient Conservation and Recovery of Value-Added Products from Compost

Composting is the process of using organisms to stabilize organic waste. Good composting products not only contain a lot of organic matter and various elements needed by crops but also kill pathogenic microorganisms in the high temperature stage, so can be used as a soil conditioner or fertilizer. Composting has made outstanding contributions in maintaining soil fertility, increasing soil and water conservation capacity, saving irrigation water, and so on. However, how to reduce the loss of nutrients has become a problem that must be paid attention to in composting. In composting, the improper adjustment of temperature, humidity, pH, C/N ratio, particle size, and other factors may lead to the production of a large amount of gas. N, S, and other elements in compost products mostly overflow into the external environment as NH_3, H_2S, and VOCs. The N loss from ammonia volatilization accounts for about 79–94% of the total nitrogen loss; similarly, H_2S accounts for 39–43% of the total sulfur emissions. The loss of a large number of nutrients leads to a decline in the quality of compost products. Studies have shown that ammonia volatilization can be reduced by adjusting PH, adjusting initial C/N ratio, co-composting, adding adsorbent, struvite precipitation, promoting digestion process, and adjusting aeration rate.

Due to the different food cultures in different regions, the characteristics of the food waste generated are also different, so different composting methods are needed. Liming (2023) collected and analyzed the kitchen waste of university restaurants in western China, and composted the local kitchen waste to produce organic fertilizer. The experimental study found that the content of C, N, P, and other nutrients in the food waste of restaurants in western universities ensured the quality of compost. Before aerobic composting, three material ratios of structural materials, finished materials of secondary compost, and inoculated materials were used. Aerobic composting was carried out through two stages of high temperature fermentation and low temperature decomposition, and the maximum temperature of compost was maintained at 60~65 °C. The composting was completed within 36 days, and the quality of the resource products accounted for about 50% of the total material. The composting products could meet the technical requirements of NY525-2012 (Table 1) organic fertilizer in terms of appearance, nutrients, and heavy metal content, and could be promoted and applied in the region as an effective organic fertilizer for soil fertilizer distribution and crop yield improvement.

FW can also serve as a substrate for ryegrass growth. Taking ryegrass as an example, Chen et al., (2015) mixed FW compost, perlite, and peat in proportion to make a substrate, observed and tested biological indicators such as germination rate, plant height and single-photon avalanche diode (SPAD) value through cultivation experiments, screened the substrate suitable for seedling cultivation and growth, and obtained the best formula. The biological effects of different proportioned substrates were tested by using cooked food waste compost as raw material and ryegrass as test crop. The optimum matrix formula for ryegrass growth was: peat: compost: perlite (v/v)=0.6:0.4:1.4).

9.3.4 Policies and Regulations for Compost Production and Application

Developed countries mainly promote the standardization of the composting industry through the establishment of specialized agencies, the establishment of quality systems, and the unification of testing methods. The United Kingdom and Italy promoted standardization through the establishment of manure associations. The Organic Waste Composting Association (OWCA) was established in 1992 to standardize composting products. Since 2005, the Composting Association and the British Standards Institution have jointly established PAS 100:2005, PAS 100:2011, and PAS 100:2018, which mainly include compost production and monitoring procedures, sampling and analysis methods, and quality standards. In Italy, the Italian Consortium for Composting (CIC), as the certification system for composting products, unified certification of composing products. During the certification process, samples are taken by certified personnel of the CIC and analyzed in a certified laboratory.

In 2011, the U.S. Department of Agriculture (USDA), the Environmental Protection Agency (EPA), and the Food and Drug Administration (FDA) launched a joint effort to reduce food waste. This includes innovation research, community investment, educational exhibitions, voluntary projects, corporate collaboration, technical assistance, tool development, event participation, and policy discussion. The concept of sustainable use of food was put forward, and the "inverted pyramid" shape, which divides food recycling into six levels, was used for the classification of food and to promote its recycling (Figure 9.3).

China is the largest developing country in the world with a huge population. Every year, a large amount of food waste is generated, and in 2021, China implemented the "Anti-Food Waste

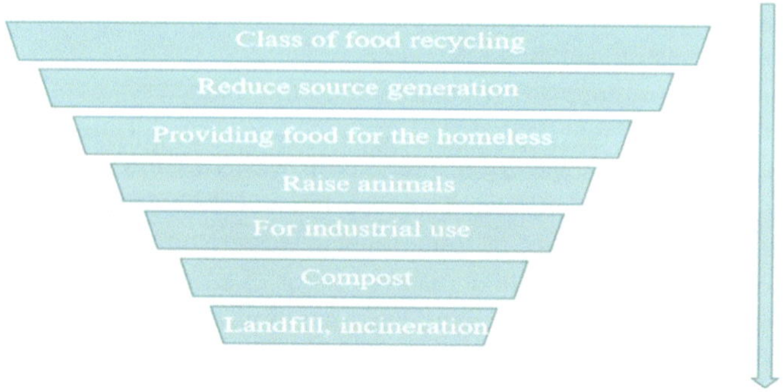

FIGURE 9.3 American levels of food recycling.

Law of the People's Republic of China" to prevent the production of food waste at the source. On October 1, 2022, the "Technical Specification for Pollution Control of Biomass Waste Compost" (HJ 1266-2022) was implemented for the composting treatment of food waste that has been generated. These include technical requirements for pollution control in the collection, storage, transportation, pretreatment, and fermentation processes of biomass waste, as well as monitoring and environmental management requirements. It is suitable for pollution control of biomass waste composting treatment, collection, transportation, storage, pretreatment, fermentation, and other processes. It can be used as the technical basis for environmental impact assessment, environmental protection facility design, environmental protection acceptance, emission permit management, cleaner production audit, and other aspects of biomass waste composting-related construction projects.

9.4 RESEARCH NEEDS AND CHALLENGES FOR FOOD WASTE COMPOSTING

Composting is a biological process that converts complex organic waste into relatively stable high-value humus compost, which has the advantages of having low costs, easy implementation, simple operation, and high-value products, and is a mature and efficient technology for food waste treatment (Awasthi et al., 2023; Wang et al., 2021). However, because of its complex physical and chemical structure, food waste still poses a huge challenge to composting. Another challenge with food waste composting is that low pH reduces the efficiency of food waste decomposition. This is because, compared with other compost substrates, food waste contains a large number of carbohydrates and fats, which produce high concentrations of organic acids during rapid fermentation in the early stage of compost, reducing the pH value of the compost and resulting in the inhibition of the growth of microorganisms conducive to the production of humus. As a result, the decomposition efficiency of food waste slows down (Cheung et al., 2010). For pH adjustment, people usually choose to use alkaline additives, such as fly ash and lime (Wong et al., 2009). However, the addition of alkaline substances such as lime enhanced the conversion of nitrogen to ammonia in the compost substrate. Wang et al., (2021) found that excessive aeration three days before composting can promote the decomposition of organic acids, so that the pH is above 8. However, due to excessive aeration, more ammonium ions are produced, which accelerates the volatilization of ammonia from the compost. All of these methods reinforce another serious problem in the composting process, which is the loss of nitrogen during composting.

In the process of food waste composting, there is a loss of ammonium nitrogen due to volatilization, microbial growth, and nitrification. The adjustment of pH at the beginning of composting exacerbates the problem. In general, nitrogen loss in compost occurs through gaseous volatilization, liquid leaching, and incomplete denitrification (Hoang et al., 2022). The loss of nitrogen reduces the quality and application value of compost, and the emission of nitrogen intensifies the pollution of atmospheric environment (Steiner et al., 2010). Therefore, fixing the release of ammonia during composting can not only reduce nitrogen loss but also reduce odor production, which is necessary. For nitrogen fixation, a variety of methods can be used, including C/N ratio regulation and the addition of physical, chemical, and microbial amendments (Manu et al., 2020; Ren et al., 2010; Steiner et al., 2010; Wang et al., 2018; Xiong et al., 2023). When magnesium hydroxide and phosphoric acid were added to pig manure and straw compost in different proportions, it was found that nitrogen loss was reduced from the original 35% to 1%. Wang et al., (2013) applied this method to the composting process of food waste to control pH and reduce nitrogen loss, and achieved good results. However, due to the addition of compounds containing magnesium and phosphorus, inorganic salt is introduced into the compost, which increases the electrical conductivity of the compost and then affects the quality of the compost.

When composting food waste as a substrate, we have to face the problem of high salt content. Salt ions can cause an increase in the electrical conductivity of compost, and the limit of electrical conductivity in general compost is 4 m/s, depending on the salt tolerance of crops (Waqas et al.,

2018). The formation of struvite to control pH and protect nitrogen has a good effect on nature, but it causes an increase in EC, which directly affects the quality of compost and the survival rate of crops. It may be possible to improve the salt content of food waste through pretreatment of the substrate and reducing the initial salt content; this, however, needs further research.

9.5 CONCLUSIONS AND PERSPECTIVES

Food waste composting is an efficient and environmentally friendly technology whose technical route includes the collection, treatment, composting, and subsequent application of food waste, which can convert organic waste into stable/mature compost for agricultural production. At present, many research results and methods have been used to enhance food waste composting, such as various physical, chemical, physico-chemical, and biological pretreatments of FW, adding different types of additives in the composting stage, filling materials, and microbial inoculation. However, food waste composting still has some shortcomings, such as the production of unpleasant odors, composting inefficiency, and stability problems. Using food waste for earthworm composting is a very good way to stabilize composting and improve composting efficiency and maturity. Further research is needed to continuously optimize the earthworm composting system, improve the safety and stability of earthworm composting, and improve the value of composting products applied to agriculture.

ACKNOWLEDGMENTS

This work was supported by a research fund from Shaanxi Introduced Talent Research Funding (A279021901 and F1020221012), China, Shaanxi Provincial Key R&D Plan Project (2022NY-052) and College of Natural Resources and Environment, Northwest A&F University, Yangling, Shaanxi Province 712100, China.

REFERENCES

Ajibade S, Nnadozie E C, Iwai C B, et al., Biochar-based compost: a bibliometric and visualization analysis. *Bioengineered*, 2022, 13(7–12): 15013–15032.

Amuah E Y, Fei-Baffoe B, Sackey L N A, et al., A review of the principles of composting: understanding the processes, methods, merits, and demerits. *Organic Agriculture*, 2022, 12(4): 547–562.

Awasthi M K, Harirchi S, Sar T, et al., Myco-biorefinery approaches for food waste valorization: present status and future prospects. *Bioresource Technology*, 2022, 360: 127592.

Awasthi M K, Kumar V, Hellwig C, et al., Filamentous fungi for sustainable vegan food production systems within a circular economy: present status and future prospects. *Food Research International*, 2023, 164: 112318.

Awasthi M K, Selvam A, Lai K M, et al., Critical evaluation of post-consumption food waste composting employing thermophilic bacterial consortium. *Bioresource Technology*, 2017, 245: 665–672.

Chen et al., Study on compost of kitchen waste as growth substrate of ryegrass. *Environment*, 2015, S1: 47–50.

Cheung H N B, Huang G H, Yu H. Microbial-growth inhibition during composting of food waste: effects of organic acids. *Bioresource Technology*, 2010, 101(15): 5925–5934.

Chi C P, Chu S, Wang B, et al., Dynamic bacterial assembly driven by Streptomyces griseorubens JSD-1 inoculants correspond to composting performance in swine manure and rice straw co-composting. *Bioresource Technology*, 2020, 313: 123692.

Guo H, Gu J, Wang X, et al., Beneficial effects of bacterial agent/bentonite on nitrogen transformation and microbial community dynamics during aerobic composting of pig manure. *Bioresource Technology*, 2020, 298: 122384.

Harrison B P, Gao S, Gonzales M, et al., Dairy manure co-composting with wood biochar plays a critical role in meeting global methane goals. *Environmental Science & Technology*, 2022, 56(15): 10987–10996.

He Y, Huang X, Zhang H, et al., Insights into the effect of iron-carbon particle amendment on food waste composting: physicochemical properties and the microbial community. *Bioresource technology*, 2022, 351. https://doi.org/10.1016/j.biortech.2022.126939

Hoang H G, Thuy B T P, Lin C, et al., The nitrogen cycle and mitigation strategies for nitrogen loss during organic waste composting: a review. *Chemosphere*, 2022, 300: 134514.

Huang X, He Y, Zhang Y, et al., Independent and combined effects of biochar and microbial agents on physico-chemical parameters and microbial community succession during food waste composting. *Bioresource Technology*, 2022, 366: 128023.

Jin C, Sun, S, Yang, et al., Anaerobic digestion: An alternative resource treatment option for food waste in China. *Science of the Total Environment*, 2021, 779: 146397.

Liming, Y. Treatment status of kitchen waste and its research progress of pyrolysis technology. *Chemical Industry and Engineering Progress*, 2023, 42(7): 3791–3801.

Liu N, Hou T, Yin H, et al., Effects of amoxicillin on nitrogen transformation and bacterial community succession during aerobic composting. *Journal of Hazardous Materials*, 2019, 362: 258–265.

Manu M K, Wang C, Li D, et al., Impact of zeolite amendment on composting of food waste digestate. *Journal of Cleaner Production*, 2022, 371: 133408.

Mingyue P, Ang L, Huishi D, et al., A review of additives use in straw composting. *Environmental Science and Pollution Research International*, 2023, 30(20): 18–20.

Nguyen M K, Lin C, Hoang H G, et al., Investigation of biochar amendments on odor reduction and their characteristics during food waste co-composting. *Science of The Total Environment*, 2023, 865: 161128.

Orobio P A B, Paz S J, Cruz H A J, et al., Evaluation of biochar as an additive in the co-composting of green waste and food waste. *Sustainability*, 2023, 15(9). https://doi.org/10.3390/su15097437

Pepe O, Ventorino V, Blaiotta G. Dynamic of functional microbial groups during mesophilic composting of agro-industrial wastes and free-living (N2)-fixing bacteria application. *Waste Management*, 2013, 33(7): 1616–1625.

Piao M, Li A, Du H, et al., A review of additives use in straw composting. 2023, 30(20): 57253–57270.

Proietti P, Federici E, Fidati L., et al., Effects of amendment with oil mill waste and its derived-compost on soil chemical and microbiological characteristics and olive (Olea europaea L.) productivity. 2015, 207: 51–60.

Qi C, Yin R, Cheng J, et al., Bacterial dynamics for gaseous emission and humification during bio-augmented composting of kitchen waste with lime addition for acidity regulation. *Science of the Total Environment*, 2022, 848. https://doi.org/10.1016/j.scitotenv.2022.157653

Qiaoli Zheng, Guofang Wang, Tingting Liu, et al., Research progress on resource treatment technology of kitchen waste. *Science and Technology Innovation Herald*, 2018, 15(7)143-144.

Ravindran B, Awasthi M K, Karmegam N, et al., Co-composting of food waste and swine manure augmenting biochar and salts: Nutrient dynamics, gaseous emissions and microbial activity. *Bioresource Technology*, 2022, 344: 126300.

Ren L, Schuchardt F, Shen Y, et al., Impact of struvite crystallization on nitrogen losses during composting of pig manure and cornstalk. *Waste Management*, 2010, 30(5): 885–892.

Schroeter-Zakrzewska A, Wolna-Maruwka A, Kleiber T, et al., Influence of compost from post-consumer wood on development, nutrition state of plants, microbiological and biochemical parameters of substrates in zonal Pelargonium (Pelargonium zonale). *Agronomy*, 2021, 11(5): 994.

Steiner C, Das K C, Melear N, et al., Reducing nitrogen loss during poultry litter composting using biochar. *Journal of Environmental Quality*, 2010, 39(4): 1236–1242.

Sun Y, Men M, Xu B, et al., Assessing key microbial communities determining nitrogen transformation in composting of cow manure using illumina high-throughput sequencing. *Waste Management*, 2019, 92: 59–67.

Tang J, Li X, Cui P, et al., Nitrification plays a key role in N_2O emission in electric-field assisted aerobic composting. *Bioresource Technology*, 2020, 297: 122470.

Technology of organic fertilizer from kitchen waste. Guangdong Chemical Industry, 2022, 49(04):126–128.

Verma S, Awasthi M K, Liu T, et al., Biochar as smart organic catalyst to regulate bacterial dynamics during food waste composting. *Bioresource Technology*, 2023, 373: 128745.

Wainaina S, Lukitawesa, Kumar Awasthi M, et al., Bioengineering of anaerobic digestion for volatile fatty acids, hydrogen or methane production: a critical review. *Bioengineered*, 2019, 10(1): 437–458.

Wang K, Mao H, and Li X. Functional characteristics and influence factors of microbial community in sewage sludge composting with inorganic bulking agent. *Bioresource Technology*, 2018, 249: 527–535. https://doi.org/10.1016/j.biortech.2017.10.034.

Wang N, Huang D, Bai X, et al., Mechanism of digestate-derived biochar on odorous gas emissions and humification in composting of digestate from food waste. *Journal of Hazardous Materials*, 2022, 434. https://doi.org/10.1016/j.jhazmat.2022.128878.

Wang Q, Awasthi M K, Zhang Z, et al., Sustainable composting and its environmental implications[M]// Sustainable resource recovery and zero waste approaches. *Elsevier*, 2019, 115–132.

Wang X, Selvam A, Chan M, et al., Nitrogen conservation and acidity control during food wastes composting through struvite formation. *Bioresource Technology*, 2013, 147: 17–22.

Wang Y, Tang Y, Li M, et al., Aeration rate improves the compost quality of food waste and promotes the decomposition of toxic materials in leachate by changing the bacterial community. *Bioresource Technology*, 2021, 340: 125716.

Wang Y, Tang Y, and Yuan Z. Improving food waste composting efficiency with mature compost addition. *Bioresource Technology*, 2022, 349. https://doi.org/10.1016/j.biortech.2022.126830.

Wang Y, Yuan Z, Tang Y. Enhancing food security and environmental sustainability: a critical review of food loss and waste management. *Resources, Environment and Sustainability*, 2021, 4:100023.

Waqas M, Nizami A S, Aburiazaiza A S, et al., Optimization of food waste compost with the use of biochar. *Journal of Environmental Management*, 2018, 216: 70–81.

Weber C T, Trierweiler L F, Trierweiler J O. Food waste biorefinery advocating circular economy: bioethanol and distilled beverage from sweet potato. *Journal of Cleaner Production*, 2020, 268: 121788.

Wong J W C, Fung S O, Selvam A. Coal fly ash and lime addition enhances the rate and efficiency of decomposition of food waste during composting. *Bioresource Technology*, 2009, 100(13): 3324–3331.

Wu J, Zhao Y, Qi H, et al., Identifying the key factors that affect the formation of humic substance during different materials composting. *Bioresource Technology*, 2017, 244: 1193–1196.

Xiong S, Liu Y, Zhang H, et al., Effects of chemical additives and mature compost on reducing nitrogen loss during food waste composting. *Environmental Science and Pollution Research*, 2023. https://doi.org/10.1007/s11356-022-24752-5

Yang X C, Han Z Z, Ruan X Y, et al., Composting swine carcasses with nitrogen transformation microbial strains: succession of microbial community and nitrogen functional genes. *Science of The Total Environment*, 2019. https://doi.org/10.1016/j.scitotenv.2019.06.283

Yin Y, Gu J, Wang X, et al., Impact of copper on the diazotroph abundance and community composition during swine manure composting. *Bioresource Technology*, 2018, 255, 257–265 https://.doi.org/10.1016/j.biortech.2018.01.120

Yingying H, Xia H, Hongnig Z, et al., Insights into the effect of iron-carbon particle amendment on food waste composting: physicochemical properties and the microbial community. *Bioresource Technology*, 2022, 351,126393.

Yu J, Gu J, Wang X, et al., Effects of inoculation with lignocellulose-degrading microorganisms on nitrogen conversion and denitrifying bacterial community during aerobic composting. *Bioresource Technology*, 2020, 313: 123664.

Zhang L, Shi T, Xu Z, et al., Effect of lime and calcium magnesium phosphate on gaseous emissions, maturity, and bacterial dynamics during food waste composting. *Environmental Technology & Innovation*, 2023, 32: 103306.

Zhong X-Z, Zeng Y, Wang S-P, et al., Insight into the microbiology of nitrogen cycle in the dairy manure composting process revealed by combining high-throughput sequencing and quantitative PCR. *Bioresource Technology*, 2020, 301, 122760.

Zhou Y, Xiao R, Klammsteiner T, et al., Recent trends and advances in composting and vermicomposting technologies: a review. *Bioresource Technology*, 2022, 360: 127591.

10 Production of Biochar from Food Waste

Payal Sahu, Mehak Bhardwaj, Varaprasad Kolla, Arjun Suresh, and Amita Shakya

10.1 INTRODUCTION

The waste generated through any anthropogenic activity is a subject of grave concern that not only burdens the economic structure of the country but also stresses the environment. Food waste (FW) is such an anthropogenic waste that not only strains production processes but also compromises the environment and economy.

> Food waste is defined as food (including a drink) and associated inedible parts removed from the human food supply chain in the following sectors: manufacturing of food products (under certain circumstances); food/grocery retail; food service; and households. "Removed from the human food supply" chain means one of the following end destinations: landfill, controlled combustion, sewer, litter/discards/ refuse, co/anaerobic digestion, compost / aerobic digestion, land application. (Forbes 2021)

In literature, food loss, and unavoidable/avoidable/potentially avoidable food waste are alternative terms used to define food waste. As per the UNEP Food Waste Index Report (2021), 8–10% of global greenhouse gas emissions are associated with food that is not consumed, and around one-third (≈1.3 billion tons/year) of the world's food is wasted (Forbes 2021). Yet, firm and accurate evidence-based assessment methods are lacking, that assess global food waste, its types, and loopholes in the supply chain system (Parfitt et al., 2010).

Rapid population growth and industrial development have raised the demand for products of basic necessity, including food. The inclusion of new cutting-edge technologies in the food processing industry and associated agricultural sectors, including poultry, meat, and dairy, which are highly unorganized sectors specifically in the developing world, is critical. FW includes mixed domestic food waste, catering wastes, animal byproducts, and vegetable residues. The wastage of food happens on three levels: (a) primary level (on the land, i.e., agricultural fields), (b) secondary level (industries), and (c) tertiary level (food consumption) (Morone et al., 2019) (Figure 10.1).

Dumping of FW to landfills causes odor as it decomposes, leading to public annoyance in addition to forming germs, attracting flies and vermin, and generating landfill leachate that further can add biological oxygen demand (BOD) to the groundwater (Ahmed and Gupta 2010). Suppression of FW production and its valorization are the only ways out to limit this present-day pressing issue. FW biorefinery is a viable and integrated approach for valorization that creates products/byproducts with additional value, reducing the amount of FW dumped in landfills and reducing GHG emissions (O'Connor et al., 2021).

10.2 PRODUCTION OF BIOCHAR FROM FOOD WASTE

Consequential FW and food loss in pre-harvest to post-harvest stages, during the supply chain and during consumption, have forced many countries to address the issue with impactful upcycling solutions (Figure 10.2). The opportunities provided by FW reduction have remained largely untapped and are under-exploited. An analysis of food waste approximates 80% as volatile matter, 15% as

DOI: 10.1201/9781032706030-10

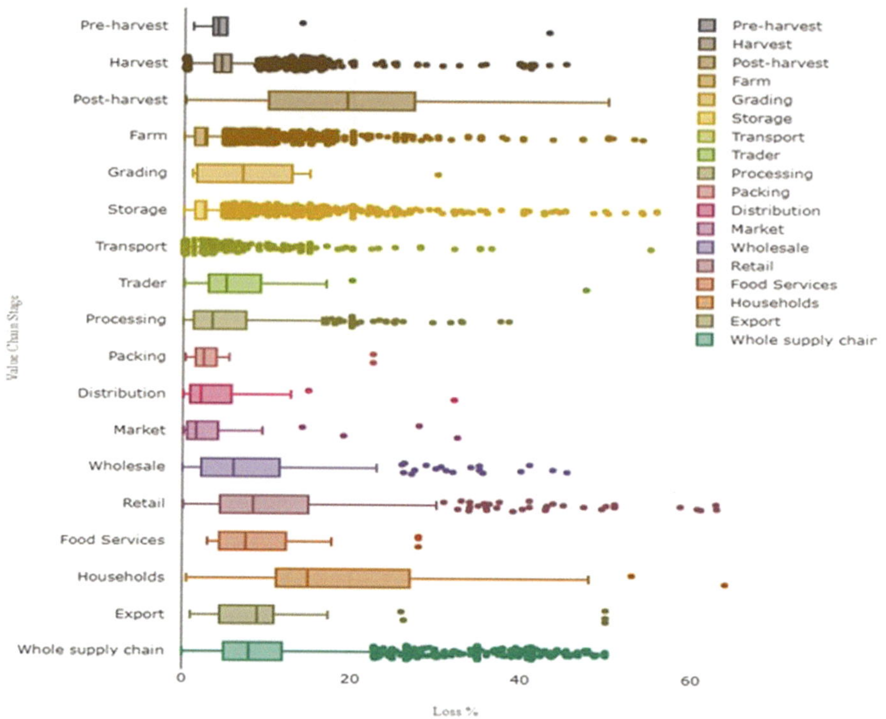

FIGURE 10.1 Food waste in different operations of the value chain stage (Source: Food and Agriculture Organization of the United States https://www.fao.org/platform-food-loss-waste/flw-data/en/).

fixed carbon, and 5% as ash (Ahmed and Gupta 2010). To utilize this huge amount of energy-intensive material, the concept of biorefinery has emerged and become a potentially viable approach for harnessing energy and valuable and value-added products from biomass. Specialized/commodity/consumer and niche chemicals such as bioplastics, biofertilizers, nutraceuticals, and biofuels can be produced through a biorefinery, using a variety of processes including fermentation, distillation, pyrolysis, hydrolysis, microwave irradiation, and pressing (O'Connor et al., 2021).

Conversion of FW to biochar through thermal treatment is a viable and sustainable approach to combat the environmental issues of waste management and climate change and achieve the SDGs. For FW-based biorefineries, biochar is one of the value-added end products that can be used as a soil amendment, adsorbent, catalyst, energy storage material, alternative fuel, and many more. Energy recovery from volatile components in FW can be possible through thermochemical processes as an alternative to traditional food disposal techniques that have recently attracted attention.

Biochar is no longer a material that requires an introduction. The thermochemical conversion of biomass at elevated temperatures (300–900°C) under limited or no oxygen conditions produces biochar, and the process is well known as "pyrolysis." The thermochemical conversion of FW biomass provides a quick and easy solution to reduce trash volume (by >80%), speed up response times, improve energy recovery efficiency, and broaden the range of garbage that can be disposed of. Thus, thermochemical processes offer a useful method for converting FW into heat or energy and may help lessen the negative effects on the environment. The diverse applications of biochar make it a material of the future, and the literature itself evidences the usefulness of biochar in, but not limited to, soil applications, crop yield improvement, enhancing soil microbial activity, denitrification (Cayuela et al., 2013; Guo et al., 2023), waste management (Kumar et al., 2023; Qambrani et al., 2017), organic (VOCs, gases, dyes, PPCPs, PAHs) and inorganic pollutant remediation from soil (Shakya and Agarwal 2020), water (Lee et al., 2019; Shakya et al., 2022) and air (Wen et al., 2023a),

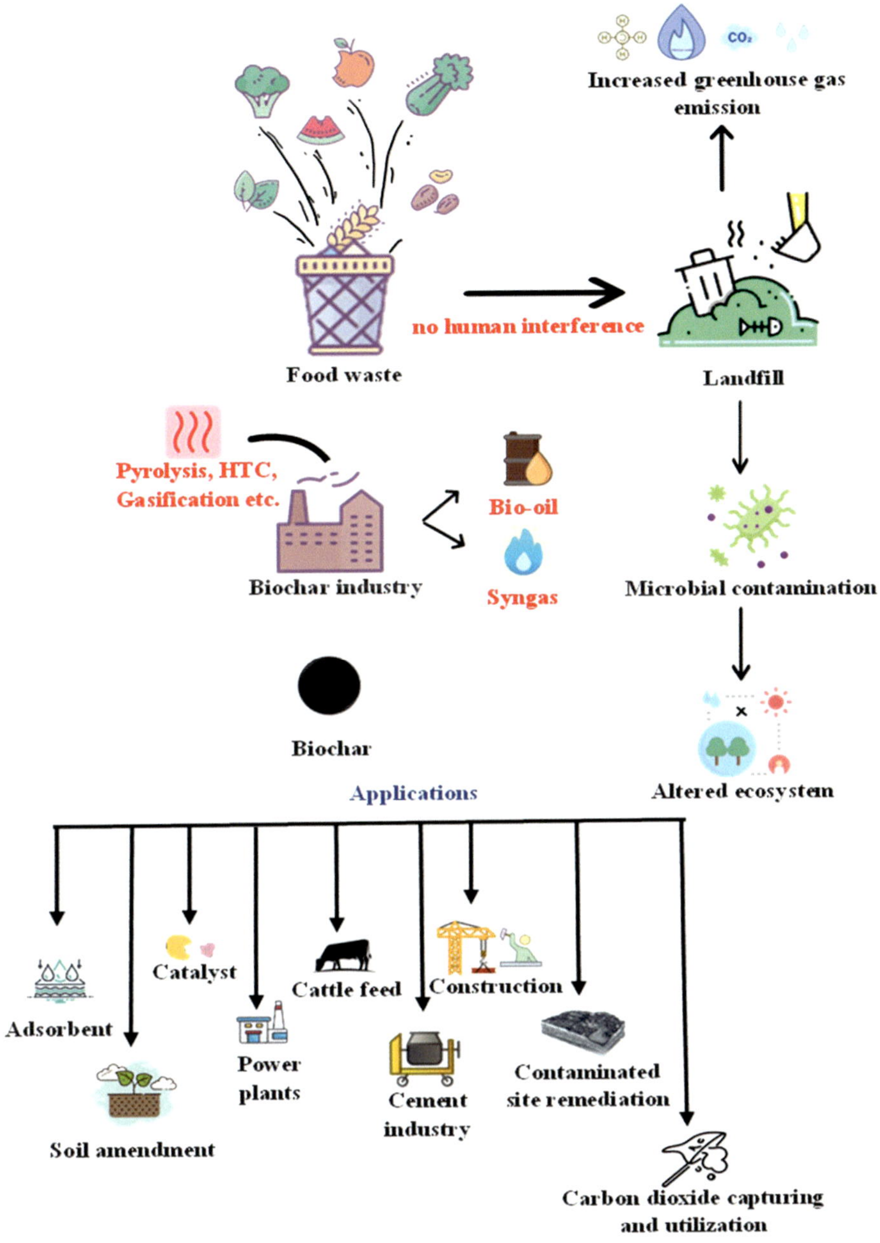

FIGURE 10.2 Biochar as an opportunity for environmental management of food waste.

applications in the construction industry (Gupta et al., 2018b; Senadheera et al., 2023; Wen et al., 2023b), as a catalyst (Yuan et al., 2023; Zhao et al., 2023a), as fuel and in fuel cells (Chakraborty et al., 2020; Dhanda et al., 2023; Safarian 2023), CO_2 capturing (Yuan et al., 2022; Zhang et al., 2022), and many other integrated approaches (Lin et al., 2023).

Seeds, peels, hard shells, agri-residues, aquaculture waste, and organic municipal (food) waste, which are UFW from food processing industries, are the most suitable raw materials or feedstocks for biochar production. However, continuous research on biochar has proved that any organic material can serve as the feedstock for biochar production, yet lignocellulosic materials with low moisture content (to make the process energy efficient) are considered as the most suitable ones for biochar production. With a plethora of variable physicochemical, morphological, and surface characteristics, biochar has huge potential to manage the waste of both animal and plant-based origins (Sirohi et al., 2023). The physicochemical properties of the generated biochar are significantly influenced by the type and composition of the feedstock, pyrolysis temperature (PyrT), and process parameters and types.

Interestingly, biochar is never just pure carbon but a maximum carbon material with hydrogen, oxygen, nitrogen, and minerals all together. The percentage of elements in biochar depends on the type of feedstock and process parameters though. According to the International Biochar Initiative (IBI), biochar has been categorized as Class I >60% C content, Class II (with C content ≥30% and <60%, and Class 3: ≥10% and <30% (IBI 2014). The stability of biochar (half-life >100 years) in the environment has made it the most reliable tool for direct carbon capture. Functionalization, activation, and modification of biochar through chemical, mechanical, or biological methods, or composite products, can enhance the properties of biochar in terms of high surface area (SA), porous structure, and abundant functional groups.

10.2.1 Types and Characteristic of the Food Waste for Biochar Production

As residue from the food supply chain, food wastes have high organic matter and water content, and could potentially trigger the development of pathogenic microorganisms, producing mycotoxins and other harmful substances. The FW can be of many types, including avoidable food waste such as (i) avoidable FW such as spoiled or discarded fruits, vegetables, grains, over-produced harvest, and cooked food; (ii) unavoidable FW such as peels, seeds, eggshells, vegetable skins, crop residues, residues of animal and poultry farming (meat and poultry) (bird feathers, manure, etc.); (iii) kitchen waste; (iv) catering waste; (v) commercial food outlet waste (a largely unorganized sector); and (vi) supply chain discards. All FW categories, except (i), equally contain non-edible parts. Lipids, cellulose, hemicellulose, starch, lignin, and protein are common components of food waste; together, they make up 82%–96% of its total volatile percentage. FW with high lipid content (vegetable oil and animal fat) is a potential substrate to be transformed into chemicals and biofuels through transesterification compared to other wastes having carbon content (Ahmed and Gupta 2010). Kitchen waste, municipal (food) waste, catering, and commercial FW can be considered as processed feedstock for biochar production, as they largely contain a mixture of more than one type of biomass/compounds in undefined ratios, likewise cattle manure and litter. Ben Hassen-Trabelsi et al., (2014) used animal fats for pyrolysis and reported that the produced biochar had low carbon content and high ash content, an unattractive renewable energy source. However, the bio-oil obtained from the same waste animal fats showed good fuel properties. Table 10.1 summarizes the studies that used FW as the potential feedstocks for biochar production.

As per the process feasibility, the FW biomass for biochar can be categorized as low MC biomass such as groundnut shells, eggshells, various nutshells, coconut husk and shell, and rice husk; moderate MC biomass such as jackfruit peels, potato peels, and corncobs; and high MC biomass such as pineapple peels, cooked food like rice, and most citrus peels. High MC biomass is preferably treated under high pressure and relatively low temperature through HTC, and when produced is termed "hydrochar." As per the IBI guidelines, the MC of the feedstocks should be reduced to <10%. To produce biochar, the MC, AC, calorific value, the percentage of lignin, cellulose, and hemicellulose percentage, fixed carbon, and volatile components are the properties that should be considered while selecting the feedstock (Kumar and Bhattacharya 2021).

10.2.2 Suitability/Criteria of Food Waste as the Potential Feedstock for Biochar

As discussed above, FW can be mixed with discarded organic waste having a high content of various organic and inorganic nutrients with usually high moisture. Being rich in hydrocarbons,

TABLE 10.1
Food Waste to Produce Biochar

S. No.	Food Waste	Thermochemical Process	Process Conditions	Yield (%)	Biochar Properties			Biochar Usage	References
					C Content (%)	Ash Content (%)	Surface Area (m^2/g)		
1	Food waste (mixed)	Pyrolysis	T= 700°C; RT= 30 min	NA	57.11	NA	110.98	Phenol adsorption	(Lee et al., 2019)
2	Leftover rice	Torrefaction	T= 240°C; RT= 30 min	90	46.26	0.36	NA	Energy	(Huang et al., 2019)
			T= 260°C; RT= 30 min	85	48.35	0.40			
			T= 270°C; RT= 30 min	65	55.12	0.82			
			T= 280°C; RT= 30 min	47	64.81	0.87			
			T= 300°C; RT= 30 min	43	69.50	0.87			
3	Pineapple peels	Pyrolysis	T= 350°C; RT= 60 min; HR= 5°C/min	37.17	62.54	7.93	5.548	Hexavalent chromium adsorption	(Shakya and Agarwal 2019)
			T= 450°C; RT= 60 min; HR= 5°C/min	32.99	64.87	10.73	3.372		
			T= 550°C; RT= 60 min; HR= 5°C/min	40.32	68.87	11.61	11.112		
			T= 650°C; RT= 60 min; HR= 5°C/min	54.80	73.55	12.08	6.391		
4	Food waste (mixed)	Torrefaction	T= 225°C; HR= 15°C/min; RT= 1h	85.16	41.93	13.34	NA	Alternate fuel	(Rago et al., 2018)
			T= 250°C; HR= 15°C/min; RT= 1h	66.58	48.95	17.40			
			T=275°C; HR= 15°C/min; RT= 1h	63.67	52.42	18.91			
			T=300°C; HR= 15°C/min; RT= 1h	56.27	55.86	24.91			
5	Food waste	Torrefaction	T= 225°C; HR= 10°C/min; RT= 40 min	87%	51.01	NA	NA	Energy	(Pahla et al., 2018)
			T= 275°C; HR= 10°C/min; RT= 40 min	78%	61.22				
			T= 300°C; HR= 10°C/min; RT= 40 min	63%	62.2				

(Continued)

TABLE 10.1 (CONTINUED)
Food Waste to Produce Biochar

S. No.	Food Waste	Thermochemical Process	Process Conditions	Yield (%)	Biochar Properties			Biochar Usage	References
					C Content (%)	Ash Content (%)	Surface Area (m^2/g)		
6	Food wastes (rice, noodle and pasta, meat, vegetable matters, bones)	Slow pyrolysis	T= 500°C; HR= 10°C/min; RT= 45–60 min	NA	70.90%	NA	9.70	Admixture for cement mortar	(Gupta et al., 2018a)
7	Food waste (mixed)	Pyrolysis	T= 700; HR= 10°C/min	NA	57.11	NA	110.98	Characterization	(Opatokun et al., 2016)
			T= 300; HR= 10°C/min		66.7	12.6	1.91		
			T= 400; HR= 10°C/min		69.6	15.2	0.79		
			PyrT= 500; HR= 10°C/min		71.3	17.4	1.90		
			PyrT= 600; HR= 10°C/min		74.6	18.1	63.83		
8	Digestate from waste food			42.50%	45.4 to 36.7%	35.7 to 60.2%	0.666 to 26.95		
9	Potato peel waste	Pyrolysis	450°C	30.50		25.4	NA	Characterization	(Liang et al., 2015)
10	Coconut pith	Pyrolysis	T= 300°C; HR= 10°C/min; RT= 60 min	61.17	59.74	3.71	3.98	For mercury removal	(Johari et al., 2016)
			T= 500°C; HR= 10°C/min; RT= 60 min	36.16	63.70	4.00	16.31		
			T= 700°C; HR= 10°C/min; RT= 60 min	31.42	76.24	4.42	315.06		
			T= 900°C; HR= 10°C/min; RT= 60 min	27.76	79.95	6.34	393.46		

proteins, and nutrients, the use of FW for biofuel production is the first choice. However, often, due to process feasibility and economic limitations, this is not possible. In such circumstances and others, pyrolyzing FW for biochar production, an emerging frontier of novel technologies for recycling FW, is an option (Tan et al., 2023). High moisture content, mixed components (in most KW, processing, and municipal waste), salinity, and the type of feedstock significantly affect the properties of what is produced (biochar, bio-oil, bio-vinegar, syngas). This affects the properties of the final yielded product, such as poor heat transfer and low surface area (Elkhalifa et al., 2019; Zhao et al., 2017). It is noticed that co-pyrolysis of FW with other feedstock (especially woody biomass) reduces the concentration of phenol and polycyclic aromatic hydrocarbons (PAHs) (Park et al., 2021). Pineapple peels and groundnut shells (both are UFW) were used to produce biochar at different PyrT of 350°C to 650°C, and significant differences in their characteristic properties were noticed with respect to PyrT; however, a close similarity in the physicochemical properties was noticed (Shakya and Agarwal 2019; Shakya et al., 2022). High MC (sometimes >80%) is the upper limitation of FW; however, it can be overcome by pre-treatment using low-grade heat from flue gas at temperatures in the range of 150 to 180°C or lower temperature solar heating systems (Bian et al., 2022); however, this is very energy consuming. The inhomogeneity of post-consumer FW such as remaining catering food includes many inorganic components like bone or table salt (NaCl). This inhomogeneity poses an entirely different challenge that may affect the chemical structures and morphological properties of produced biochar (Kane and Ryan 2022). Mixed FW (biodegradable and non-biodegradable) is another challenge for the thermal treatment of FW, as it will change the density, nature, and properties of the FW. FW is a mixed and compound type of feedstock that often comes from postharvest products, processing facilities, food services, and households. It affects the physical, chemical, and biological properties of the biochar (O'Connor et al., 2021).

10.3 PRODUCTION TECHNOLOGIES FOR BIOCHAR

The properties and end usages of any material are determined by its structural and physicochemical attributes (Chen et al., 2019). Besides feedstock type, the production technology (methods and parameter variations) is the most critical parameter that regulates the properties of the biochar. Several thermochemical conversion technologies can be utilized to produce biochar, and research is ongoing to make the available technologies and methods more efficient and multipurpose. The thermochemical conversion technologies can be subdivided into conventional (age-old methods) and advanced (improvised, efficient models of conventional technologies) methods (Figure 10.3). Depending on the type of thermochemical technique used, the possible products are (i) solid material (biochar), (ii) non-condensable gases, and (iii) condensable liquid (bio-oil) (Shakya and Ahmad 2020).

10.3.1 Conventional Methods

The conventional methods used for biochar production are pyrolysis (slow/flash/fast), HTC, gasification, and torrefaction. These processes are discussed briefly in the following sections.

10.3.1.1 Pyrolysis

Dry pyrolysis is the most common and widely accepted method for biochar production. Thermal treatment of biomass at elevated temperatures (350–1000°C) in the absence of oxygen is termed pyrolysis. Based on of HR and residence time (RT), pyrolysis is categorized as slow, flash, and fast pyrolysis. Slow pyrolysis is the conventional carbonization of biomass to produce biochar at slow and low HT with relatively longer RT (Qian et al., 2015). Fast pyrolysis requires shorter residence time (in seconds) and a high heating rate. Generally, fast pyrolysis gives a high yield of bio-oil while slow pyrolysis favors high biochar production (Qian et al., 2015). Peak temperature, heat transfer rate, and vapor-solid interactions are also important factors, along with heating rate and

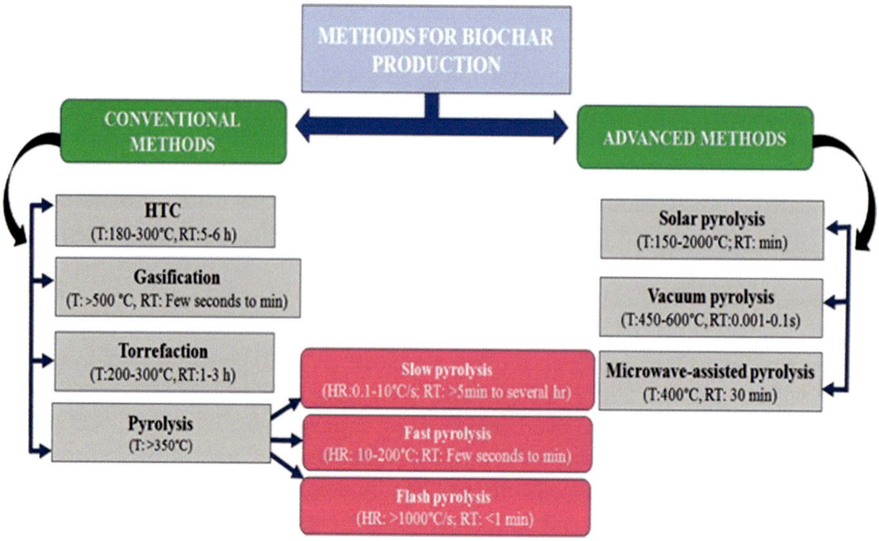

FIGURE 10.3 Biochar production methods.

residence time, that affect the pyrolysis process. During pyrolysis, smaller biomass particles favor the high reactivity of biochar. The mechanism of biochar formation is a complicated multistep process (Demirbas 2004):

Step 1:
Biomass $\longrightarrow$water + unreacted residue
Step 2:
Unreacted residue $\longrightarrow$ (volatile + gases)^1 + (char)^1
Step 3:
(Char)^1 $\longrightarrow$ (volatile + gases)^2 + (char)^2

In the first step, bound moisture and some volatile substances evaporate, and in the second step, primary char formation occurs. The third and final step is initially fast, followed by slow reactions, including the chemical rearrangement of the biochar, which makes the biochar less reactive. In the third step, the biochar decomposes at a very slow rate, and a carbon-rich residual solid forms. By varying the pyrolytic conditions, such as HTT, RT, and HR, biochar can be obtained with desired properties. Based upon the employed HR and RT, the pyrolysis process can be categorized as slow, fast, and flash, with biochar yields of 25–50%, 15–25%, and 5–15% respectively (Figure 10.3).

10.3.1.2 Hydrothermal Carbonization

In comparison to other thermochemical methods used to make biochar, hydrothermal carbonization (HTC), also referred to as wet torrefaction or wet pyrolysis, uses HTC biomass treated with water for a longer period in an inert atmosphere.

HTC has high conversion efficiency, does not require pre-drying of the feedstock, and operates at comparatively moderate temperatures of 180–260°C at 15–60 bar for 5 to 6 hours. This process produces biochar of different sizes, shapes, and surface functionalities. Based on the reaction mechanism and conditions, HTC can be classified as low-temperature HTC and high-temperature HTC. The high-temperature HTC technique is favorable for the production of carbon nanotubes, graphite, and activated carbon materials at high temperatures (300–800°C) and high pressures (100–200 MPa). The low-temperature HTC technique is widely used for the production of highly reactive microporous/mesoporous carbon-based materials with surface functional groups (C-H-O) (Hu et al., 2010).

10.3.1.3 Gasification

Gasification is the thermochemical conversion of organic material into a valuable gaseous product, called syngas, as the primary product, and this process produces a solid product, called char, as the by-product. This is an efficient process for the production of power and heat, and for the production of hydrogen and second-generation biofuels (Molino et al., 2016). Gasification is a systematic reaction that involves the partial oxidation and water-gas reaction of carbon, producing carbon monoxide (CO) and hydrogen (H_2). Further, the CO shift reaction and steam methane reforming reaction take place, producing CO_2 and H_2 (Higman 2008). Gasification takes place at 700–1500°C with moderate to very fast heat transfer in a reaction time of a few seconds to minutes at elevated atmospheres. The gasification of food waste (FW) was found to be a sustainable process compared to other processed biomass such as poultry litter or sewage sludge.

10.3.1.4 Torrefaction

Torrefaction (T_{RF}) is considered to be an important pre-treatment method to enhance fuel quality for efficient valorization of FW and is also known as mild pyrolysis (Huang et al., 2020). It takes place at relatively low temperatures of 200–300°C (or above) in an inert atmosphere. The biochar produced through T_{RF} is primarily used for direct energy generation. Alternatively, syngas generated from T_{RF} could be recycled into the reactor for purging (Yek et al., 2021). Torrefaction lowers the moisture affinity of biomass and renders it hydrophobic. The energy and mass yields decrease with an increase in the T_{RF} temperature (Poudel et al., 2015). The torrefied biomass pellets generally entail three basic operations: conditioning (pre-treatment like shredding, drying, ash separation, grinding, and sieving), torrefaction (mild pyrolysis), and pelletization (densification using high pressure) (Sippula et al., 2017). Biomass can be torrefied either in raw or pelletized form (also known as upstream or downstream mode, respectively). T_{RF} significantly increases the hydrophobicity of biomass and, with a further upsurge in temperature, thermal fragmentation of lignocellulosic polymers takes place similar to pyrolysis, resulting in carbon-rich biochar (coke/unsaturated biopolymer), bio-oil (tar) (condensable volatiles/heavy oxygenates), and non-condensable syngas. The advancement of T_{RF} is wet T_{RF} (subcritical hydrothermal treatment through water or diluted acids), steam T_{RF}, and dry T_{RF}, which can be subdivided into non-oxidative and oxidative T_{RF} (air, O_2, or CO_2 as purge gas) (Chen et al., 2021; Yek et al., 2021).

10.3.2 Advanced Methods

10.3.2.1 Solar Pyrolysis

Solar pyrolysis is a greener approach to pyrolysis that uses concentrated solar energy as the heating source. It is often performed over a wide range of temperatures (150–2000°C), heat flux intensities of 0.01–12 MW/m^2, and heating rates of 5–450°C/s. A solar pyrolysis system mainly constitutes three parts: solar concentrator, solar collector, and supporting structure. The solar concentrator focuses all the solar radiation on it into a solar collector. A solar collector/pyrolysis reactor, in turn, absorbs the radiation from the concentrator and transfers the heat energy to the feedstock (Parthasarathy et al., 2022).

10.3.2.2 Vacuum Pyrolysis

Vacuum pyrolysis is a process that involves biomass degradation at low pressure in vacuum conditions. The temperature range is 450–600°C, and the range of pressure is 0.05 to 0.20 MPa at a low heating rate, similar to conventional pyrolysis. Despite being similar to conventional pyrolysis, the properties of biochar differ due to the effective removal of vapor during vacuum pyrolysis. Similar to conventional pyrolysis, the HR in vacuum pyrolysis is low. Though the HTT and HR are similar to slow pyrolysis, end products may differ significantly due to the use of vacuum or low pressure. Vacuum pyrolysis is often used to produce high-quality biochar, which shows high porosity and is thus highly useful in the adsorption of minerals and nitrogen when applied as a soil amendment (Gabhane et al., 2020).

10.3.2.3 Microwave-Assisted Pyrolysis

This is a type of pyrolysis in which microwaves are used for biochar production. It is a more controllable and highly efficient pyrolysis method. Microwaves have wavelengths between 1 mm and 1 m, and their corresponding frequencies range from 300 GHz to 300 MHz. It has a selective heating mechanism in which the biomass is targeted based on its dielectric properties. It represents a non-contact heat transfer method. Chen et al., (2023) produced highly efficient banana peel biochar through microwave pyrolysis against malachite green dye with an adsorption capacity of 2297.83 mg/g within 120 min. MAP overcomes the limitation of temperature differences between biomass core and surface in traditional pyrolysis. In addition, microwave reactors are found to be more efficient than conventional reactors (Leng et al., 2021).

10.4 PHYSIOCHEMICAL PROPERTIES OF FOOD WASTE BIOCHAR

The process parameters and type of feedstock are the key players that dramatically influence the physicochemical properties of the biochar. Being a complex matrix, FW requires a higher temperature to decompose.

10.4.1 Elemental Composition

Biochar is not entirely a carbon product but is a blend of carbon with sufficient amounts of hydrogen, oxygen, nitrogen, and inorganic (ash) content. The amount of ash depends on the type of feedstock and HTT. An increase in HTT increases the ash content of the biochar irrespective of feedstock. However, processed biomasses such as poultry litter, sewage sludge, and rice husk itself contain a high amount of ash that subsequently increases with HTT. Regardless of the type of feedstock, the increase of PyrT increases the C content of the biochars and decreases the H, O, and N content. Higher process temperature causes morphological and structural changes and the reduction of hydrophilic surface functional groups, leading to low O/C and H/C molar ratios and a steady decrease in the amount of hydrogen and oxygen.

An elemental analyzer, SEM with energy-dispersive X-ray (SEM-EDX) spectroscopy, is often used for the elemental analysis of FW biochars. The presence of carbon, oxygen, calcium, magnesium, potassium, zirconium, and other elements was reported with SEM-EDX analysis compared to SEM-EDS. The elemental analyzer specifically quantifies the elemental C, H, N, and S content of the biochar, and usually, oxygen content is calculated by the difference. The differences in the elements present in the biochar arise from the composition of the feedstock (raw materials), indicating that the selection of appropriate raw materials is crucial to create the intended biochar for specific applications. Increased pyrolysis temperature decreased the oxygen and hydrogen content of the biochar while carbon content increased. At a pyrolysis temperature ≥700°C, significant structural transition of biomass takes place, which produces 80–90% of carbon and arranges the C molecules as well-structured graphene layers (Keiluweit et al., 2010). As per EBC, biochar must contain ≥50% of total carbon, while IBI made this requirement a minimum of 10% organic carbon and categorized biochar into three categories according to the carbon content of the biochar: Class 1: ≥60% carbon content; Class 2: ≥30% to <60% carbon content; and Class 3: ≥10% to <30% carbon content.

10.4.2 pH

Inorganic alkali concentration and extent of carbonate formation contribute to the pH value of biochar. pH ranging from 6.5 to 10.8 has been reported for the above-mentioned criteria of elemental composition. The higher the content of ash and oxygen, the higher the pH (Figure 10.4). At a temperature of 300°C and above, the alkali component starts to detach from the organic network, which ultimately increases the pH of the biochar. Inorganic constituents in food char were found to have a catalytic effect, and char reactivity increased with the degree of conversion (Ahmed and Gupta 2010). The pH of biochar is more governed by PyrT than by the type of feedstock. Zhao et al., (2013) produced biochar from 12 different organic biomasses, including various FW, and found that pH varied less among the different feedstocks (8.8–10.8) than among the PyrT (5.43–10.8). Interestingly, the authors reported a contradictory case for cation exchange capacity. Biochar produced at lower temperatures has higher densities of acidic functional groups (phenolic and carboxylic groups), which decreases the biochar pH.

10.4.3 Surface Area

The key factor that influences the surface area of biochar is temperature. The large surface area is useful for heavy metal adsorption, helps in the removal of toxic pollutants, and impacts other properties such as cation exchange as well as water holding capacity. With an increase in PyrT, more volatiles are released under pressure, creating voids in the structure and resulting in a porous surface of the biochar. With low to moderate PyrT (300–500°C), many studies showed a non-significant increase in SA, while at high PyrT (>600°C), a dramatic increase in SA and pore volume was reported (Shakya and Agarwal 2023). Opatokun et al., (2016) pyrolyzed FW at 300, 400, 500, and 700°C, and the SA of the 300–500°C biochars were 1.91, 0.79, 1.90 m^2/g respectively, while SA of the 700°C biochar was 63.83 m^2/g, much higher than rest of the biochars (Table 10.1). Increasing PyrT consecutively increases the SA of biochars because of the successive decomposition of biomass, leading to the removal of alcoholic (-OH), aliphatic, and ester (-CO-) groups from the surface of the biochars (Li et al., 2016).

10.4.4 Porosity

One of the most important features of biochar is its porous structure. During pyrolysis, the primary fragmentation of biomass takes place (secondary fragmentation occurs only in case of oxidation or gasification) due to the rapid release of volatiles, which might be non-monotonous. The strong internal pressure builds up due to the volatilization of biomass components, causing particle swelling and creating bubbles. The release of volatiles in a non-uniform manner ruptures the bubbles,

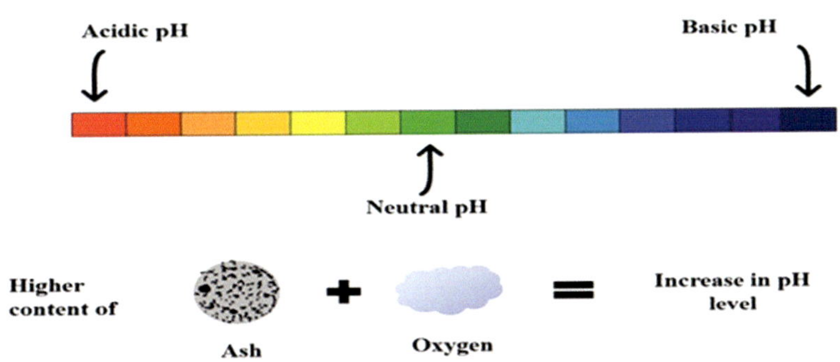

FIGURE 10.4 pH of biochar depends upon elemental composition.

creating pores in the char, and high pressure might lead to particle fragmentation (Wafiq et al., 2016). A successive increase of PyrT resulted in the removal of stronger aromatic –CO, –CH, and phenolic –OH cross-linked aromatic structures of the biomass, leading to the generation of a highly porous structure (Li et al., 2016).

According to the standard of the International Union of Pure and Applied Chemistry (IUPAC), there are three types of pores: micropores (<2 nm); mesopores (2–50 nm); and macropores (>50 nm). The structure of biochar pores is predominated by micropores (>80% of total pore volume), ranging from 0.05–0.0001 µm, and the percentage of the micropore volume in the total pore range is reported between 12.1–58.0%. On the other hand, mesopore size ranges between 18.9 and 31.7% (Kumar and Bhattacharya 2021; Leng et al., 2021). The porosity and SA of the biochar are interdependent on each other; the more micropores, the larger the porosity. A fall in porosity (which initially increases at moderate pyrolysis temperatures then decreases) at very high PyrT (>600 or 800–1000°C) is also reported, which might be due to the destruction of cell structures present in the feedstock. Porosity highly influences the water-holding capacity of the biochar.

The measurements of biochar SA and porosity (mainly micro-mesopores and part of macropores) are done by gas adsorption. However, macropore information may sometimes may be limited and inaccurate by gas adsorption; thus, additional methods such as mercury porosimetry are employed (Leng et al., 2021). The Brunauer–Emmett–Teller (BET) method is the most accepted and widely used method to analyze nitrogen adsorption isotherms for determining the surface area of biochar. The BET method is a certified and accepted method by IBI as well as EBC. The shape of the pore also influences the measurement of the SA.

10.4.5 Stability

Biochar is made up of aromatic carbon rings that are randomized and then condensed into polyaromatic sheets (amorphous and crystalline phases, respectively). Biochar that has a high degree of aromatic condensation, and aromaticity can withstand thermochemical and biological degradation, therefore, has high stability. The physicochemical characteristics of biochar, such as structure of pores, pH, SA, surface functionalities, ash content, and particle size distribution, exhibit effects on the biochar stability. During the process of thermochemical conversion, the most labile fraction of the biomass is released in the form of volatiles, and the degree of temperature increases this significantly. Studies carried out by Shakya et al., (2022) on different types of FW biochars clearly indicate the effect of PyrT on the thermostability of the biochars. The thermodegradable fraction (thermogravimetric analysis) of biochars significantly decreased with the increase of the PyrT irrespective of the types of feedstock. The higher fixed carbon content of the biochars is another indicator of their stability, and fixed carbon is directly correlated to PyrT. The molar ratios are another indicator of biochar stability, and Spokas (2010) linked them with the half-life of biochars (Box 10.1). The Van Krevelen diagram shows the effect of PyrT on biochar stability, and with high PyrT the biochar structure resembles graphene (Figure 10.5).

BOX 10.1. THE MOLAR RATIO AND BIOCHAR STABILITY

- Maximum molar H/C_{org} ratio: 0.7 (biochar must be less than or equal to it) (As per IBI and EBC)
- Molar O/C ratio: 0.4 (biochar must be less than or equal to it) (As per EBC)
- Half-life of biochars (Spokas, 2010)
  - O/C: <0.2: Half-life of biochar >1000 years
  - O/C: 0.2 to 0.6: Half-life of biochar >100 years to <1000 years
  - O/C: >0.6: Half life of biochar <100 years

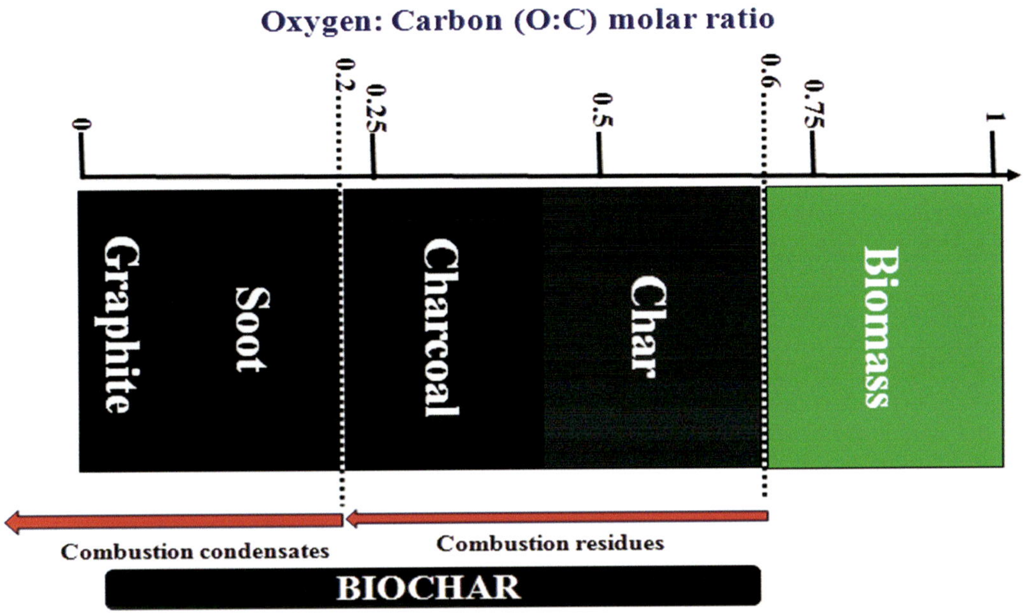

FIGURE 10.5 Spectrum of biomass conversion products and their O/C ratio. (Source: Spokas, 2010).

10.5 EFFECT OF PRODUCTION PARAMETERS ON THE PROPERTIES OF FOOD WASTE BIOCHARS

It is a well-known fact that the biochar production method and process parameters significantly influence biochar properties. Process conditions can be altered to produce tailor-made biochar with specific properties according to the intended end use.

10.5.1 Effect of High Treatment Temperature

Biochar is produced from thermochemical processes primarily via pyrolysis at a certain range of temperatures. This high end temperature is termed high treatment temperature (HTT). The difference in HTT leads to different carbonization degrees and generates highly organized and recalcitrant aromatic structures (Bolan et al., 2022). Despite the type of feedstock, either processed or unprocessed, the yield of biochar decreases, and this decline in biochar yield becomes more significant with higher PyrT. Besides biochar yield, volatile matter, nutrient content, cation exchange capacity, accessible P, and accessible cations in biochar decrease with a rise in PyrT, while their alkaline status and aromatization levels elevate (Zhang et al., 2019)

The complexity of FW increases the processing time under the same reaction conditions. Ahmed and Gupta (2010) used the gasification process for paper waste and FW under the same conditions and reported that FW needed more time to complete the gasification process. The HTT not only has a significant impact on the physicochemical properties of the biochar but also affects the energy storage properties of the char. The energy and mass yield of the produced char was found to decrease when the reaction temperature increased from 150 to 600°C during torrefaction. This increase in HTT had a positive impact on the higher heating value of the produced char (Poudel et al., 2015). FW biochar prepared at 300 to 900°C showed similar results; an increase in HTT led to an increase in C content, aromaticity, biochar pH, and hydrophobic nature of the biochar (Lee et al., 2019).

Interestingly, it was reported that an increase in HTT sometimes also has detrimental effects on the morphological attributes of the biochars. Shakya and Agarwal (2019) reported structural distortion (collapsed pore walls) when pineapple peel biochar was prepared at 650°C, while in lower PyrT biochars, the structural integrity remained the same. The additional heat creates a chemical rearrangement and pressure on the primary char during the last step of the pyrolysis process, which sometimes might lead to the collapse of the porous structure of the biochar. The authors also reported reduced Cr(VI) adsorption with increasing PyrT. At high PyrT (>300°C), the cleavage of the C-H and -OH groups is enhanced, resulting in the release of more volatiles as the gas and liquid fractions of the pyrolytic product (Demirbas 2004). Hence, the selection of optimum PyrT for biochar production should always be taken into account to get the desired product.

10.5.2 Effect of Residence Time

An increase in the residence time (RT) leads to a decrease in volatile substances and an increase in pH. At a low temperature, longer RT can decrease the yield of biochar. However, increasing the RT under high-temperature conditions primarily modifies the biochar surface and interior structure while having less of an impact on biochar yield (Sun et al., 2017). The selection of optimal RT will help to develop the surface morphology of the biochar. The selection of optimal RT can result in a biochar with high SA and porosity. The type of feedstock is also an important factor to consider when taking RT into account. Poudel et al., (2015) torrefied the FW at a temperature range of 150 to 600°C at heating rates of 10, 20, and 30°C per minute and reported that the torrefaction temperature had a more significant effect on the torrefaction of FW than the torrefaction RT. FW is generally a mixed, heterogeneous, and complex matrix that requires a longer time to achieve the desired morphological features in biochar. High-temperature treatment of the biomass for prolonged periods not only causes high mass loss as volatiles but can also damage the anatomical structure.

10.5.3 Effect of Mixing Ratio and Blending

FW itself is a complex matrix, and its constituents can alter the properties of the produced biochar. The nature of FW, whether salty, high sugar, high fat, moisture content, etc., can not only affect the mass yield but can also change the physicochemical characteristics of the produced biochar. The composition of the treated food waste and process conditions have significant impacts on the resulting products from the pyrolysis process (i.e., solid product, liquid, and gas) and their associated characteristics. Elkhalifa et al., (2022) studied various food wastes like carrots, cucumbers, and tomatoes as a single-component system (carrot), binary component system (carrots + cucumbers, 1:1 ratio), and ternary component system (carrots + cucumbers + tomatoes, 1:1:1 ratio) and reported that blending of biomass changed the product properties.

In addition, other natural or synthetic polymers and substances can be mixed with FW to produce biochar. Tang et al., (2018) co-pyrolyzed organic food waste (soybean protein) and polyvinyl chloride (PVC), a typical chlorine-enriched plastic waste, in protein and PVC ratios of 4:1, 1:1, and 1:4, and reported the highest char yield for the 4:1 mixing ratio. Park et al., (2021) mixed FW and wood bark with a weight ratios of 3:1, 1:1, and 1:3, and found that the purge gas (N_2) flow rate did not significantly affect the characteristics of the pyrolytic products, while the PyrT and mixing ratio of FW and wood bark greatly affected the yields of the pyrolytic products. Co-pyrolysis of rice husk with FW produced biochar with high nutrient content (Bian et al., 2022). Kane and Ryan (2022) blended biodegradable plastic polylactic acid (PLA) and recycled high-density polyethylene (HDPE) with FW to produce biochar and reported a reduction in the thermal stability of the polymers. The authors proposed the catalytic activity of inorganic compounds of biochar in the thermal degradation of PLA, resulting in shorter PLA chains and reducing the composite's mechanical properties. A reduction in SA was also reported when FW and wood bark were blended in a 3:1 ratio rather than a 1:3 ratio (Park et al., 2021).

10.6 POTENTIAL APPLICATIONS OF FOOD WASTE BIOCHAR FOR ENVIRONMENTAL MANAGEMENT

For decades, animal feed, waste landfills, composting, and fermentation have been the methods to dispose of FW. Despite being long in use, these methods have their limitations. To overcome the associated challenges with the above-mentioned methods, thermal conversion of FW to biochar is considered a win-win solution. Rajendran et al., (2022) conducted a techno-economic analysis for a biochar system and found that the annual plant revenue was $24,140,000 for a project lifetime of 20 years, the payback time was 3.16 years, the internal rate of return after tax was 39.92%, and the net value at 10.0% interest was $55,017,000. A minimum biodiesel selling price of $0.46/kg was used, and the environmental sustainability of the produced biodiesel was assessed by a life cycle assessment. The impact of global warming potential on the FW-based biodiesel scenario was –3.967 kg CO_2 equivalent. This section will discuss the role of biochar role in various environmental management applications.

10.6.1 Adsorbent

With inherent excellent textural properties like high surface area, better pore volume, and good pore structure, biochar has the potential to act as an adsorbent for various organic, inorganic, and volatile contaminants from soil and water. To improve the adsorption potential, many modification methods (chemical, physical, biological) have been developed and tested, such as those with metal, nanoparticle, and mineral impregnation; surface/microwave oxidation; ball milling; acid/alkali treatment; and zeolite and clay composition. To act as an adsorbent, certain factors besides biochar's own physicochemical properties affect the biochar adsorption capacity. Biochar dose, pH of the solution, ionic strength of the solution, time of the adsorption reaction, and nature and type of the adsorbate are the factors that affect the adsorption potential of the biochar. Texture properties of biochar make it a powerful adsorbent, akin to activated carbon. Biochar is different from traditional charcoal because biochar tends to have the ability to store carbon. Biochar can be used as a CO_2 adsorbent due to its exceptional surface area. The pore structure and surface area play a major role in the physisorption of CO_2 by biochar. FW-based biochar was prepared and successfully applied as an adsorbent for phenol removal from water with 14.61 mg/g of adsorption capacity (Lee et al., 2019). The adsorption of contaminants with biochar is a synergistic process that involves physical adsorption through the pores of biochar as well as chemical interaction between biochar surface functional groups and metal or organic contaminant ions. When biochar is produced at a low PyrT, a sufficient number of functional groups remain present due to incomplete carbonization. These groups give biochar a negative charge, leading to an increase in its cation exchange capacity that helps to bind with the contaminants. Additionally, biochar can absorb excess exchangeable cationic pollutants due to its high cation exchange capacity. In this manner, biochar application in soil will render the soil less soil polluted, resulting in clean, nutritious crops (Kumar et al., 2023). Interestingly, it was noticed that biochars prepared at a moderate HTT range (500–700°C) have a higher affinity toward organic contaminants than the biochars prepared at low HTT (300°C or less) (Vithanage et al., 2016). Possibly the degree of carbonization and aromaticity is not prominent enough at low HTT to make strong π-π interactions between the organic contaminant and biochar. Studies have shown that, therefore, it is important to find efficient methods or integrated techniques to improve the adsorption potential of the biochars.

10.6.2 Soil Amendment

The use of various FW as soil nutrients is an age-old practice. Being organic, FW contains C, N, P, K, Ca, and Mg, various macronutrients and essential elements (Mn, Zn, and Cu, etc.), and a few trace elements in low concentrations. However, the plant availability of nutrients varies depending on the type of feedstock, pH, the C/N ratio, microbiological activity, and modification in food waste. Also, direct use of FW as a soil amendment can cause contamination by heavy metals and foreign objects such as microplastics and pathogenic organisms. Biochar addition to soil is a promising approach to achieve waste management, nutrient delivery, contaminant immobilization, and climate change mitigation simultaneously. Soil amendment in the context of biochar arose from soils present in the Amazon region due to the high carbon content. Adjustments of these parameters can be made by mixing other products, such as sawdust, with digestate to increase the C/N ratio. Additionally, the high CEC and anion exchange capacity of biochar apparently play a role to make biochar a good sorbent, preventing nutrient loss (Li et al., 2023; Wang et al., 2024). The organic matrix of biochar acts as a habitat for plant growth, promoting bacteria and stimulating microbial activities, leading to improved soil health and an elevated plant biomass yield. The AC of biochar dissolves in soil water and delivers necessary inorganic nutrients to the plant. Ash content of biochar contains SiO_2, Al_2O_3, K_2O, P_2O_5, CaO, and MgO, the oxides of alkali and alkaline earth metals. The alkaline nature of biochar significantly helps to neutralize acidic soils up to the pH range of 10–12 at 500°C PyrT. Soil acidity neutralization will aid in the increase of agriculturally available land (goal 15 of the SDGs), which subsequently will help in increasing crop production, enabling the achievement of goal 2 of the SDGs (zero hunger) (Kumar and Bhattacharya 2021). A reduction of soil bulk density by 3 to 31%, increases in porosity by 14 to 64%, increases in available water by 4 to 130%, increase in wet aggregate stability by 3 to 226%, and limited or no effects on penetration resistance were reported after the addition of biochar in soil. Saturated hydraulic conductivity decreases in coarse-textured soils, and increases in fine-textured soils following biochar application (Blanco-Canqui 2017). With tomato waste, mixed FW, oily FW, and kitchen waste, biochars were prepared and explored as soil amendments with effective outcomes. Additionally, FW digestate produced from anaerobic digestion can also be used as feedstock for biochar production.

10.6.3 Catalyst

More than 35% of the world's GDP is attributed to catalytic technology, and 95% of industrial products are thought to be created using catalytic processes. The global market for catalysts is estimated to be worth $15 billion (Kumar et al., 2020). Carbon materials are the prime choice for catalysts due to their mechanical strength, high surface area and wider pore size distribution, chemical inertness, and thermal stability. Biochar comprises all these properties along with tunable functionalities. Biochar can be modified, designed, and impregnated with various metal ions, nanomaterials, and organic polymers that tune the properties of biochar and make it an excellent material as a catalyst. Rajendran et al., (2022) produced a biochar catalyst derived from citrus peel containing a rich carbon source and obtained a 96.3% biodiesel yield under the optimized reaction conditions (response surface methodology with a central composite design) of a 1:10 oil to methanol molar ratio, 4 w% catalyst loading, and 55°C reaction temperature for 52 min reaction time. Xiong et al., (2021) first used starch-rich rice waste to produce value-added chemicals by catalyzing through $AlCl_3$ and then converted the solid by-products (humans) to biochar-supported Lewis acid catalysts. During biochar production for the intended use as a catalyst, in addition to carbon, it is important to keep sulfur and nitrogen content in balance by converting them into pyridine (nitrogen) and thiophene sulfur.

The application of biochar as a catalyst is a relatively newer area than biochar research on soil. Researchers are using newer approaches, such as experimenting with complex matrixes like FW to synthesize tuneable and engineered biochars as efficient catalysts that must be reusable, unlike other catalysts.

10.6.4 Energy Storage

Apart from its use as an adsorbent, soil amendment, and in anaerobic digestion, biochar can be a precursor for activated carbon. Surface modification through various chemical treatments is used for the synthesis of biochar-based activated carbon. The biochar itself or the activated form of biochar could also be used. The porous structure and functional tuning of biochar have grabbed the attention of researchers to use it as an energy storage material, such as fuel cells, and as an additive to electrodes for energy in lithium-ion batteries. To improve the electrochemical performance of $Li–O_2$ batteries, various electrocatalysts have been used, and biochar is among them. The cost of electrodes (anode and cathode) for a typical microbial fuel cell varies from 20 to 50% of the overall fabrication costs, which can be significantly reduced by the use of FW-derived biochars with graphene-like structure and stability (Chakraborty et al., 2020).

10.6.5 Alternative Fuel

Escalating energy demand, over-exploitation of fossil fuels, and associated environmental impacts such as GHG emissions and global warming require alternative and renewable energy solutions. Carbon-based porous materials have received a lot of attention because of their high surface area, large pore volume, superior chemical stability, and easily customizable porosity. Torrefied mixed FW biochar showed high EY and calorific value (>50% enhancement) compared to the FW itself. It is also reported that torrefaction at high temperatures (>250°C) produces biochar with suitable characteristics to be used as an alternative fuel (Rago et al., 2018). Pahla et al., (2018) produced biochar through T_{RF} at 275°C and reported an upgrade in calorific value from 19.76 MJ/kg for dried raw food waste to 26.15 MJ/kg for torrefied FW; the HHV is comparable to that of bituminous coal.

10.7 FOOD WASTE BIOCHAR AND CIRCULAR ECONOMY

The gap in demand and supply, changing lifestyle patterns, irresponsible consumption behavior of consumers, and increased per capita income have imposed several stresses and burdens on Earth; FW is one of them. Biochar's multifaceted applications in various sectors show its great potential to address nutrient recycling, environmental and climate change issues, and to establish a circular economy model (Hu et al., 2021). Biochar production from FW itself contributes to carbon sequestration, and its capacity to absorb various VOCs and gases makes it a material of high value. High fixed carbon and low C_{org}/O ratio (highly dependent on PyrT) make it a highly recalcitrant material in the environment, leading to it being a reliable tool for direct carbon capture. The significant economic effectiveness of biochar production systems varies with respect to the feedstocks used, the conversion technology employed, and the inclusion of C sequestration subsidies/carbon or green credits reflecting the social value of greenhouse gas mitigation (Cely et al., 2015). Biochar can also be used as an environmentally sustainable substitute for activated carbon for the remediation of organic and inorganic pollutants. The addition of biochar in soil significantly improved ammonia volatilization and reduced N_2O emissions, and improved available P through enhancement of solubilization and reduction in P fixation on soil minerals, and high extractable K in biochar contributes to the K cycle in soil, leading to improved beneficial interactions with N, P, and K soil cycle processes (Jindo et al., 2020). Co-feeding animals with biochar can increase body weight and growth rate, improve microbial fermentation, reduce pathogens, and decrease enteric methane emissions. Additionally, this process indirectly impacts positively by improving the quality of fertilization and reducing GHG emissions, leading to mitigating climate change (Osman et al., 2022).

Focused and categorical research on biochar production and its characterization through FW can not only enhance its value but also strengthen the circular economy. Integration of biochar production with other processes such as green energy generation, biomass biorefinery, or biomass pelletization will be an additional advantage that can reduce the back cost and minimum selling price of biochar, making it more profitable. At present, a huge share of agriculture and allied sectors are the first choice of feedstock for biochar production, and its direct application in the soil is the first commercially acceptable sector worldwide. Application of 80% C-rich biochar and carbon dioxide offset worth is estimated to be \$2.93–\$90.83 per metric ton of biochar carbon sequestration (Khan et al., 2021b). The process of biochar production requires a separate plant, the input of the process energy, and the production of gaseous and volatile streams, entailing capital costs, operating costs, and emission control efforts (Poudel et al., 2015). Hence, it is of utmost importance to identify, manage, and operate the process sustainably in terms of less emissions and more output.

The concept of circular bioeconomy is a widely accepted and adopted concept by many policymakers, and their decisions have led to sustainable initiatives within the social, commercial, and industrial sectors of society. Yet, these decisions are subject to scrutiny, because valorization is still in its early stages; broad, detailed assessments relative to the life cycle of the product, life-cycle cost, and social-life cycle need to be made (Mak et al., 2020)

10.8 RESEARCH GAPS AND FUTURE SCOPE

Biochar production through FW is a promising approach to combat the stress of unused or overlooked food wastes. Unlike other biochemical and enzymatic waste valorization approaches, thermochemical treatments such as pyrolysis and gasification are the most economical and environmentally friendly methods with the intention of creating various value-added products with a lot of potential. Biochar produced from pyrolyzed FW can be used to improve the soil's ability to retain water and nutrients; hence, these biochars have specific and individual qualities. Additionally, biochar can be further processed to create additional higher value-added products, like activated carbons (Azargohar and Dalai 2006; Tan et al., 2017), nanochars, nanoparticles (Hosny et al., 2022; Khan et al., 2021a), biochar-based hydro/nanogels (Karakoyun et al., 2011; Zhao et al., 2023b), and many more. Biochar is always an intended product, and the composition of feedstock (food waste) and process conditions have a significant impact on its properties. Despite the bright perspectives, there is still much disagreement over the study of biochar's advantages. This has been related to the heterogeneity of biochar itself, whose inherent physical, chemical, and biological characteristics are greatly altered by production factors like feedstock types and treating conditions. Hence, it is a matter of great caution to select the feedstock and process parameters to obtain biochar of desired qualities. Nevertheless, in spite of the huge number of research publications every year worldwide, there are only a small number of pilot-scale plants available for biochar production from FW. At present, agricultural residue-based biochar plants are the most successful due to the continuous supply of biomass throughout the year. Also, pre-treatment of the feedstock, high moisture, and mixed components are challenges that require extra-economic or energy inputs, making the process less viable. Other than this, a few future key research directions in the following areas need attention: biomass density and efficient reactor design for maximum energy utilization and product recovery to make the pyrolysis process efficient, and identification of the optimum feedstock (alone/mixed/blended food waste) to get the biochar of desired qualities. Similar to feedstock optimization, process optimization, and its comprehensive influence is equally important, as co-pyrolysis of food waste with various blends, biomass, inorganic or other organic materials, or residue waste not only affects the desired product yield and by-product formation but can also generate undesired products such as PAHs, leachable metals, and VOCs. For a long time, most of the previous research focused only on the importance of single actions of the biochars, and for a long time, its correlation with various other applications and mechanisms has been ignored.

Biochar is a globally known tool for climate change mitigation and minimizing the global warming effect. However, optimization and alteration in the processing conditions for biochar production (e.g., high HTT and longer RT) to reduce labile/organic carbon is challenging due to the variability of FW. There is scope for further research in FW and biochar, made available through new thermal techniques such as microwave-assisted pyrolysis, solar pyrolysis, and vacuum pyrolysis. Nevertheless, process parameters such as HTT, RT, HR, pressure optimization and accordingly reactor design is a major research gap that requires attention for FW to biochar conversion. The implementation of food waste pyrolysis (and other thermal treatments) for biochar and other value-added products will be facilitated and strengthened by focused research on the technical difficulties involved. This will result in the production of various materials that can be used directly or further processed into other products. Economic assessment of the biochar system in terms of its feasibility, performance, and environmental implications is also necessary. Several studies are focused and dedicated to this. However, the diversification of biomass and its impact on processing techno-economic performance require more attention. The use of modeling, machine learning, and artificial intelligence for process and product characteristic optimization is another future avenue for automation and efficiency of the process. Despite continuous research and efforts, many environmental-related advantages of biochar are not well explored, leading to limited knowledge of its significant economic assistance, its potential market, and commercial value. Most importantly, lack of awareness, lack of firm policies, and lack of awareness about the beneficial applications of biochar among farmers are other areas of concern. A meaningful collaboration among researchers, academicians, biochar stakeholders, producers, consumers, and governments is required to advance the research and promote the biochar-based circular economy technology on a global scale.

10.9 CONCLUSIONS AND PERSPECTIVES

Food waste is a complex matrix that needs more attention to reduce the challenges associated with it, including social and environmental impacts. The traditional methods of disposing of FW, such as landfills or incinerators, pollute air, water, and land. Hence, FW valorization methods should be sustainable and must provide enterprises with economic opportunities. Traditional methods have their limitations; to overcome these limitations, thermal treatment of FW is a viable solution. Biochar production from FW is a win-win solution that simultaneously provides solutions for waste management and climate change mitigation, closing the loop of carbon emissions and based on the circular economy. Biochar produced from FW can be utilized as a soil amendment, adsorbent, energy storage material, catalyst, and in many more sectors. However, research is needed to explore more energy-efficient methods to produce biochar while understanding the complexity of FW.

REFERENCES

Ahmed II, Gupta AK (2010) Pyrolysis and gasification of food waste: Syngas characteristics and char gasification kinetics. *Applied Energy* 87:101–108. https://doi.org/10.1016/j.apenergy.2009.08.032

Azargohar R, Dalai AK (2006) Biochar as a Precursor of Activated Carbon. In: McMillan JD, Adney WS, Mielenz JR, Klasson KT (eds) *Twenty-Seventh Symposium on Biotechnology for Fuels and Chemicals*. Humana Press, Totowa, NJ, pp 762–773. https://doi.org/10.1007/978-1-59745-268-7_62

Ben Hassen-Trabelsi A, Kraiem T, Naoui S, Belayouni H (2014) Pyrolysis of waste animal fats in a fixed-bed reactor: Production and characterization of bio-oil and bio-char. *Waste Management* 34:210–218. https://doi.org/10.1016/j.wasman.2013.09.019

Bian R et al., (2022) Copyrolysis of food waste and rice husk to biochar to create a sustainable resource for soil amendment: A pilot-scale case study in Jinhua. *China Journal of Cleaner Production* 347:131269. https://doi.org/10.1016/j.jclepro.2022.131269

Blanco-Canqui H (2017) Biochar and soil physical properties. *Soil Science Society of America Journal* 81:687–711. https://doi.org/10.2136/sssaj2017.01.0017

Bolan N et al., (2022) Multifunctional applications of biochar beyond carbon storage. *International Materials Reviews* 67:150–200. https://doi.org/10.1080/09506608.2021.1922047

Cayuela ML, Sánchez-Monedero MA, Roig A, Hanley K, Enders A, Lehmann J (2013) Biochar and denitrification in soils: When, how much and why does biochar reduce N2O emissions? *Scientific Reports* 3:1732. https://doi.org/10.1038/srep01732

Cely P, Gascó G, Paz-Ferreiro J, Méndez A (2015) Agronomic properties of biochars from different manure wastes. *Journal of Analytical and Applied Pyrolysis* 111:173–182.

Chakraborty I, Sathe SM, Dubey BK, Ghangrekar MM (2020) Waste-derived biochar: Applications and future perspective in microbial fuel cells. *Bioresource Technology* 312:123587. https://doi.org/10.1016/j.biortech.2020.123587

Chen L, Mi B, He J, Li Y, Zhou Z, Wu F (2023) Functionalized biochars with highly-efficient malachite green adsorption property produced from banana peels via microwave-assisted pyrolysis. *Bioresource Technology* 376:128840. https://doi.org/10.1016/j.biortech.2023.128840

Chen W, Meng J, Han X, Lan Y, Zhang W (2019) Past, present, and future of biochar. *Biochar*. https://doi.org/10.1007/s42773-019-00008-3

Chen W-H et al., (2021) Progress in biomass torrefaction: Principles, applications and challenges. *Progress in Energy and Combustion Science* 82:100887. https://doi.org/10.1016/j.pecs.2020.100887

Demirbas A (2004) Effects of temperature and particle size on bio-char yield from pyrolysis of agricultural residues. *Journal of Analytical and Applied Pyrolysis* 72:243–248. http://dx.doi.org/10.1016/j.jaap.2004.07.003

Dhanda A, Raj R, Sathe SM, Dubey BK, Ghangrekar MM (2023) Graphene and biochar-based cathode catalysts for microbial fuel cell: Performance evaluation, economic comparison, environmental and future perspectives. *Environmental Research* 231:116143. https://doi.org/10.1016/j.envres.2023.116143

Elkhalifa S, Al-Ansari T, Mackey HR, McKay G (2019) Food waste to biochars through pyrolysis: A review. *Resources, Conservation and Recycling* 144:310–320. https://doi.org/10.1016/j.resconrec.2019.01.024

Elkhalifa S, Parthasarathy P, Mackey HR, Al-Ansari T, Elhassan O, Mansour S, McKay G (2022) Biochar development from thermal TGA studies of individual food waste vegetables and their blended systems. *Biomass Conversion and Biorefinery*. https://doi.org/10.1007/s13399-022-02441-0

Forbes, H. (2021) Food waste index report 2021. https://wedocs.unep.org/20.500.11822/35280

Gabhane JW, Bhange VP, Patil PD, Bankar ST, Kumar S (2020) Recent trends in biochar production methods and its application as a soil health conditioner: A review. *SN Applied Sciences* 2:1307. https://doi.org/10.1007/s42452-020-3121-5

Guo F et al., (2023) Biochar boosts nitrate removal in constructed wetlands for secondary effluent treatment: Linking nitrate removal to the metabolic pathway of denitrification and biochar properties. *Bioresource Technology* 379:129000. https://doi.org/10.1016/j.biortech.2023.129000

Gupta S, Kua HW, Koh HJ (2018a) Application of biochar from food and wood waste as green admixture for cement mortar. *Science of The Total Environment* 619–620:419–435. https://doi.org/10.1016/j.scitotenv.2017.11.044

Gupta S, Kua HW, Low CY (2018b) Use of biochar as carbon sequestering additive in cement mortar. *Cement and Concrete Composites* 87:110–129. https://doi.org/10.1016/j.cemconcomp.2017.12.009

Higman C (2008) Chapter 11 - Gasification. In: Miller BG, Tillman DA (eds) *Combustion Engineering Issues for Solid Fuel Systems*. Academic Press, Burlington, pp 423–468. https://doi.org/10.1016/B978-0-12-373611-6.00011-2

Hosny M, Fawzy M, Eltaweil AS (2022) Phytofabrication of bimetallic silver-copper/biochar nanocomposite for environmental and medical applications. *Journal of Environmental Management* 316:115238. https://doi.org/10.1016/j.jenvman.2022.115238

Hu B, Wang K, Wu L, Yu S-H, Antonietti M, Titirici M-M (2010) Engineering carbon materials from the hydrothermal carbonization process of biomass. *Advanced Materials* 22:813–828. https://doi.org/10.1002/adma.200902812

Hu Q et al., (2021) Biochar industry to circular economy. *Science of The Total Environment* 757:143820. https://doi.org/10.1016/j.scitotenv.2020.143820

Huang J, Qiao Y, Wang Z, Liu H, Wang B, Yu Y (2020) Valorization of food waste via torrefaction: Effect of food waste type on the characteristics of torrefaction products. *Energy & Fuels* 34:6041–6051. https://doi.org/10.1021/acs.energyfuels.0c00790

Huang J, Qiao Y, Wei X, Zhou J, Yu Y, Xu M (2019) Effect of torrefaction on steam gasification of starchy food waste. *Fuel* 253:1556–1564. https://doi.org/10.1016/j.fuel.2019.05.142

IBI. (2014). IBI_Biochar Standards 2. Standardized Product Definition and Product Testing Guidelines for Biochar 7 That is used in soil. Ref. Code: IBI-STD-2.1, International Biochar initiatives. https://biochar-international.org/wp-content/uploads/2023/01/IBI_Biochar_Standards_V2.1_Final.pdf (Accessed on 13 September 2024).

Jindo K et al., (2020) Role of biochar in promoting circular economy in the agriculture sector. Part 1: A review of the biochar roles in soil N, P and K cycles. *Chemical and Biological Technologies in Agriculture* 7:15. https://doi.org/10.1186/s40538-020-00182-8

Johari K, Saman N, Song ST, Cheu SC, Kong H, Mat H (2016) Development of coconut pith chars towards high elemental mercury adsorption performance – Effect of pyrolysis temperatures. *Chemosphere* 156:56–68. https://doi.org/10.1016/j.chemosphere.2016.04.114

Kane S, Ryan C (2022) Biochar from food waste as a sustainable replacement for carbon black in upcycled or compostable composites. *Composites Part C: Open Access* 8:100274. https://doi.org/10.1016/j.jcomc.2022.100274

Karakoyun N, Kubilay S, Aktas N, Turhan O, Kasimoglu M, Yilmaz S, Sahiner N (2011) Hydrogel–Biochar composites for effective organic contaminant removal from aqueous media. *Desalination* 280:319–325. https://doi.org/10.1016/j.desal.2011.07.014

Keiluweit M, Nico PS, Johnson MG, Kleber M (2010) Dynamic molecular structure of plant biomass-derived black carbon (biochar). *Environmental Science & Technology* 44:1247–1253.

Khan HA, Naqvi SR, Mehran MT, Khoja AH, Khan Niazi MB, Juchelková D, Atabani A (2021a) A performance evaluation study of nano-biochar as a potential slow-release nano-fertilizer from wheat straw residue for sustainable agriculture. *Chemosphere* 285:131382. https://doi.org/10.1016/j.chemosphere.2021.131382

Khan N, Chowdhary P, Gnansounou E, Chaturvedi P (2021b) Biochar and environmental sustainability: Emerging trends and techno-economic perspectives. *Bioresource Technology* 332:125102. https://doi.org/10.1016/j.biortech.2021.125102

Kumar A, Bhattacharya T (2021) Biochar: A sustainable solution. *Environment, Development and Sustainability* 23:6642–6680. https://doi.org/10.1007/s10668-020-00970-0

Kumar A, Bhattacharya T, Shaikh WA, Roy A, Chakraborty S, Vithanage M, Biswas JK (2023) Multifaceted applications of biochar in environmental management: A bibliometric profile. *Biochar* 5:11. https://doi.org/10.1007/s42773-023-00207-z

Kumar A, Saini K, Bhaskar T (2020) Advances in design strategies for preparation of biochar based catalytic system for production of high value chemicals. *Bioresource Technology* 299:122564. https://doi.org/10.1016/j.biortech.2019.122564

Lee C-G, Hong S-H, Hong S-G, Choi J-W, Park S-J (2019) Production of biochar from food waste and its application for phenol removal from aqueous solution. *Water, Air, & Soil Pollution* 230:70. https://doi.org/10.1007/s11270-019-4125-x

Leng L et al., (2021) An overview on engineering the surface area and porosity of biochar. *Science of The Total Environment* 763:144204. https://doi.org/10.1016/j.scitotenv.2020.144204

Li J et al., (2016) Biochar from microwave pyrolysis of biomass: A review. *Biomass and Bioenergy* 94:228–244. https://doi.org/10.1016/j.biombioe.2016.09.010

Li Y, Kumar Awasthi M, Sindhu R, Binod P, Zhang Z, Taherzadeh MJ (2023) Biochar preparation and evaluation of its effect in composting mechanism: A review. *Bioresource Technology* 384:129329. https://doi.org/10.1016/j.biortech.2023.129329

Liang S, Han Y, Wei L, McDonald AG (2015) Production and characterization of bio-oil and bio-char from pyrolysis of potato peel wastes. *Biomass Conversion and Biorefinery* 5:237–246. https://doi.org/10.1007/s13399-014-0130-x

Lin C-W, Jhan Y-C, Zhu T-J, Liu S-H (2023) Enhancement of chromium (VI) removal and power generation by adding biochar to a single-medium sediment-based microbial fuel cell. *Journal of Water Process Engineering* 53:103612. https://doi.org/10.1016/j.jwpe.2023.103612

Mak TMW, Xiong X, Tsang DCW, Yu IKM, Poon CS (2020) Sustainable food waste management towards circular bioeconomy: Policy review, limitations and opportunities. *Bioresource Technology* 297:122497. https://doi.org/10.1016/j.biortech.2019.122497

Molino A, Chianese S, Musmarra D (2016) Biomass gasification technology: The state of the art overview. *Journal of Energy Chemistry* 25:10–25. https://doi.org/10.1016/j.jechem.2015.11.005

Morone P, Koutinas A, Gathergood N, Arshadi M, Matharu A (2019) Food waste: Challenges and opportunities for enhancing the emerging bio-economy. *Journal of Cleaner Production* 221:10–16. https://doi.org/10.1016/j.jclepro.2019.02.258

O'Connor J et al., (2021) A review on the valorisation of food waste as a nutrient source and soil amendment. *Environmental Pollution* 272:115985. https://doi.org/10.1016/j.envpol.2020.115985

Opatokun SA, Kan T, Al Shoaibi A, Srinivasakannan C, Strezov V (2016) Characterization of food waste and its digestate as feedstock for thermochemical processing. *Energy & Fuels* 30:1589–1597. https://doi.org/10.1021/acs.energyfuels.5b02183

Osman AI et al., (2022) Biochar for agronomy, animal farming, anaerobic digestion, composting, water treatment, soil remediation, construction, energy storage, and carbon sequestration: A review. *Environmental Chemistry Letters* 20:2385–2485. https://doi.org/10.1007/s10311-022-01424-x

Pahla G, Ntuli F, Muzenda E (2018) Torrefaction of landfill food waste for possible application in biomass co-firing. *Waste Management* 71:512–520. https://doi.org/10.1016/j.wasman.2017.10.035

Parfitt J, Barthel M, Macnaughton S (2010) Food waste within food supply chains: Quantification and potential for change to 2050. *Philosophical Transactions of the Royal Society B: Biological Sciences* 365:3065–3081.

Park C, Lee N, Kim J, Lee J (2021) Co-pyrolysis of food waste and wood bark to produce hydrogen with minimizing pollutant emissions. *Environmental Pollution* 270:116045. https://doi.org/10.1016/j.envpol.2020.116045

Parthasarathy P, Al-Ansari T, Mackey HR, Sheeba Narayanan K, McKay G (2022) A review on prominent animal and municipal wastes as potential feedstocks for solar pyrolysis for biochar production. *Fuel* 316:123378. https://doi.org/10.1016/j.fuel.2022.123378

Poudel J, Ohm T-I, Oh SC (2015) A study on torrefaction of food waste. *Fuel* 140:275–281. https://doi.org/10.1016/j.fuel.2014.09.120

Qambrani NA, Rahman MM, Won S, Shim S, Ra C (2017) Biochar properties and eco-friendly applications for climate change mitigation, waste management, and wastewater treatment: A review. *Renewable and Sustainable Energy Reviews* 79:255–273. https://doi.org/10.1016/j.rser.2017.05.057

Qian K, Kumar A, Zhang H, Bellmer D, Huhnke R (2015) Recent advances in utilization of biochar. *Renewable and Sustainable Energy Reviews* 42:1055–1064.

Rago YP, Surroop D, Mohee R (2018) Assessing the potential of biofuel (biochar) production from food wastes through thermal treatment. *Bioresource Technology* 248:258–264. https://doi.org/10.1016/j.biortech.2017.06.108

Rajendran N, Kang D, Han J, Gurunathan B (2022) Process optimization, economic and environmental analysis of biodiesel production from food waste using a citrus fruit peel biochar catalyst. *Journal of Cleaner Production* 365:132712. https://doi.org/10.1016/j.jclepro.2022.132712

Safarian S (2023) To what extent could biochar replace coal and coke in steel industries? *Fuel* 339:127401. https://doi.org/10.1016/j.fuel.2023.127401

Senadheera SS, Gupta S, Kua HW, Hou D, Kim S, Tsang DCW, Ok YS (2023) Application of biochar in concrete – A review. *Cement and Concrete Composites* 143:105204. https://doi.org/10.1016/j.cemconcomp.2023.105204

Shakya A, Agarwal T (2019) Removal of Cr(VI) from water using pineapple peel derived biochars: Adsorption potential and re-usability assessment. *Journal of Molecular Liquids* 111497. https://doi.org/10.1016/j.molliq.2019.111497

Shakya A, Agarwal T (2020) Potential of Biochar for the Remediation of Heavy Metal Contaminated Soil. In: Singh JS, Singh C (eds) *Biochar Applications in Agriculture and Environment Management*. Springer International Publishing, Cham, pp 77–98. https://doi.org/10.1007/978-3-030-40997-5_4

Shakya A, Agarwal T (2023) Effect of pyrolysis temperature on physicochemical properties and hexavalent chromium adsorption of the corncob biochars. *Biomass Conversion and Biorefinery*. https://doi.org/10.1007/s13399-023-05018-7

Shakya A, Ahmad F (2020) Biochar: A Growing Sanguinity as a Combinatorial Tool for Remediation of Heavy Metals from Wastewaters and Solid Waste Management. In: Gothandam KM, Ranjan S, Dasgupta N, Lichtfouse E (eds) *Environmental Biotechnology* Vol. 1. Springer International Publishing, Cham, pp 87–111. https://doi.org/10.1007/978-3-030-38192-9_4

Shakya A, Vithanage M, Agarwal T (2022) Influence of pyrolysis temperature on biochar properties and Cr(VI) adsorption from water with groundnut shell biochars: Mechanistic approach. *Environmental Research* 114243. https://doi.org/10.1016/j.envres.2022.114243

Sippula O, Lamberg H, Leskinen J, Tissari J, Jokiniemi J (2017) Emissions and ash behavior in a 500kW pellet boiler operated with various blends of woody biomass and peat. *Fuel* 202:144–153. https://doi.org/10.1016/j.fuel.2017.04.009

Sirohi R et al., (2023) Emerging trends in role and significance of biochar in gaseous biofuels production. *Environmental Technology & Innovation* 30:103100. https://doi.org/10.1016/j.eti.2023.103100

Spokas KA (2010) Review of the stability of biochar in soils: Predictability of O: C molar ratios. *Carbon Management* 1:289–303.

Sun J, He F, Pan Y, Zhang Z (2017) Effects of pyrolysis temperature and residence time on physicochemical properties of different biochar types. *Acta Agriculturae Scandinavica, Section B — Soil & Plant Science* 67:12–22. https://doi.org/10.1080/09064710.2016.1214745

Tan S et al., (2023) Utilization of current pyrolysis technology to convert biomass and manure waste into biochar for soil remediation: A review. *Science of The Total Environment* 864:160990. https://doi.org/10.1016/j.scitotenv.2022.160990

Tan X-F et al., (2017) Biochar as potential sustainable precursors for activated carbon production: Multiple applications in environmental protection and energy storage. *Bioresource Technology* 227:359–372.

Tang Y, Huang Q, Sun K, Chi Y, Yan J (2018) Co-pyrolysis characteristics and kinetic analysis of organic food waste and plastic. *Bioresource Technology* 249:16–23. https://doi.org/10.1016/j.biortech.2017.09.210

United Nations Environment Programme, Food Waste Index Report 2024. Think Eat Save: Tracking Progress to Halve Global Food Waste. [online]. Available: https://wedocs.unep.org/20.500.11822/45230. (Accessed on 10 September 2024].

Vithanage M, Mayakaduwa SS, Herath I, Ok YS, Mohan D (2016) Kinetics, thermodynamics and mechanistic studies of carbofuran removal using biochars from tea waste and rice husks. *Chemosphere* 150:781–789. https://doi.org/10.1016/j.chemosphere.2015.11.002

Wafiq A, Reichel D, Hanafy M (2016) Pressure influence on pyrolysis product properties of raw and torrefied Miscanthus: Role of particle structure. *Fuel* 179:156–167. https://doi.org/10.1016/j.fuel.2016.03.092

Wang Y, Wang K, Wang X, Zhao Q, Jiang J, Jiang M (2024) Effect of different production methods on physicochemical properties and adsorption capacities of biochar from sewage sludge and kitchen waste: Mechanism and correlation analysis. *Journal of Hazardous Materials* 461:132690. https://doi.org/10.1016/j.jhazmat.2023.132690

Wen C et al., (2023a) Biochar as the effective adsorbent to combustion gaseous pollutants: Preparation, activation, functionalization and the adsorption mechanisms. *Progress in Energy and Combustion Science* 99:101098. https://doi.org/10.1016/j.pecs.2023.101098

Wen J, Wang B, Dai Z, Shi X, Jin Z, Wang H, Jiang X (2023b) New insights into the green cement composites with low carbon footprint: The role of biochar as cement additive/alternative. *Resources, Conservation and Recycling* 197:107081. https://doi.org/10.1016/j.resconrec.2023.107081

Xiong X, Yu IKM, Dutta S, Mašek O, Tsang DCW (2021) Valorization of humins from food waste biorefinery for synthesis of biochar-supported Lewis acid catalysts. *Science of The Total Environment* 775:145851. https://doi.org/10.1016/j.scitotenv.2021.145851

Yek PNY et al., (2021) Progress in the torrefaction technology for upgrading oil palm wastes to energy-dense biochar: A review. *Renewable and Sustainable Energy Reviews* 151:111645. https://doi.org/10.1016/j.rser.2021.111645

Yuan X et al., (2022) Sustainable food waste management: Synthesizing engineered biochar for CO2 capture. *ACS Sustainable Chemistry & Engineering* 10:13026–13036. https://doi.org/10.1021/acssuschemeng.2c03029

Yuan X et al., (2023) Recent advancements and challenges in emerging applications of biochar-based catalysts. *Biotechnology Advances* 67:108181. https://doi.org/10.1016/j.biotechadv.2023.108181

Zhang C, Sun S, He S, Wu C (2022) Direct air capture of CO2 by KOH-activated bamboo biochar. *Journal of the Energy Institute* 105:399–405. https://doi.org/10.1016/j.joei.2022.10.017

Zhang Z, Zhu Z, Shen B, Liu L (2019) Insights into biochar and hydrochar production and applications: A review. *Energy*. https://doi.org/10.1016/j.energy.2019.01.035

Zhao B, Xu X, Xu S, Chen X, Li H, Zeng F (2017) Surface characteristics and potential ecological risk evaluation of heavy metals in the bio-char produced by co-pyrolysis from municipal sewage sludge and hazelnut shell with zinc chloride. *Bioresource Technology* 243:375–383. https://doi.org/10.1016/j.biortech.2017.06.032

Zhao C, Xu Q, Gu Y, Nie X, Shan R (2023a) Review of advances in the utilization of biochar-derived catalysts for biodiesel production. *ACS Omega* 8:8190–8200. https://doi.org/10.1021/acsomega.2c07909

Zhao L, Cao X, Mašek O, Zimmerman A (2013) Heterogeneity of biochar properties as a function of feedstock sources and production temperatures. *Journal of Hazardous Materials* 256:1–9.

Zhao Y et al., (2023b) Enhanced reactive brilliant blue removal using chitosan–biochar hydrogel beads. *Molecules* 28:6137.

11 Bioactive Compounds from Food Waste

Taru Negi, Anupama Anand, Shubham Samkaria, and Ranjan Kaushik

11.1 INTRODUCTION

Food loss is the reduction in quantity or standard of food primarily due to choices and initiatives made by food suppliers (and not including retailers, consumers, and food service providers) in the food supply network, whereas food waste pertains to a decrease in food quantity or standard due to choices and initiatives made by retailers, consumers, and food service providers (FAO, 2019). Some authors define food disposal throughout the food supply network, i.e., from production to consumption, as food waste, whereas others differentiate it based on the stage of food discarded in the food supply network (Diekmann & Germelmann, 2021). However, the FAO (2013) has used the term "food wastage" to apply to both food loss and food waste produced across the food supply chain (Figure 11.1). Approximately 14% of the world's food is lost after its harvest and before it reaches retail outlets (FAO, 2019). A further 17% of global food production is wasted by retailers and customers (primarily in households) according to UNEP (2021). Both food loss and waste together account for 8–10% of worldwide global emissions of greenhouse gases, which has a significant role in climate change and global warming. According to FAO data, the food lost and wasted annually could potentially feed approximately 1.25 billion people (FAO, 2022, 2023). Food wastage also results in the waste of resources (land, labor, water, and energy) used throughout the entire food supply network. Seeing the negative effects of food waste on economy, environment and society, the UN General Assembly declared September 29 of every year to be the International Day of Awareness of Food Loss and Waste. It was first marked in 2000, and the theme of 2023 was "Reducing food loss and waste: Taking Action to Transform Food Systems" (FAO, 2023). Reducing food waste has also been included in the Sustainable Development Goals, whose motto is to "halve per capita global food waste at the retail and consumer levels and reduce food losses along production and supply chains by 2030" (Principato et al., 2021).

Different stages from where food loss and waste are generated are primary production, processing, manufacturing, marketing, and consumption (household and hospitality). Primary production covers the waste caused in the initial stages (i.e., before harvest and after harvest) of the food supply network. The before-harvest stage includes food waste caused by spoilage, adverse weather conditions, insect and pest infestation, and high production, whereas the after-harvest stage comprises food waste that occurs due to lack of storage facilities, transportation, and unsuitable appearance (shape, weight, etc.) of the food commodity (FAO, 2019; Jeswani et al., 2021).

Processing and manufacturing waste include food that is processed through a wide range of procedures throughout manufacturing, such as peeling, slicing, cutting, trimming, grinding, heating, boiling, and packing, which results in the generation of food waste. Food manufacturers also discard items that do not meet the standards as per the retailer specifications and products that consumers don't find visually appealing (WRAP, 2015). Distribution waste includes waste that occurs due to poor cold chain management, high-speed vehicles (damaged soft foodstuffs), broken packaging, overstocking due to erroneous forecasting, poor shelf rotation of products, and the expiration of "best before" dates. Lastly, consumption waste includes both household and hospitality waste. Household

DOI: 10.1201/9781032706030-11

FIGURE 11.1 Different stages of food wastage in food supply network.

food waste occurs due to excessive food servings, personal preferences, accidental spoilage (defective working of the freezer), confusion in reading food expiration labels, purchasing excessive quantities of food products (because of lucrative offers), lack of culinary skills, and improper storage. Waste from the hospitality sector occurs during food preparation, from spoilage and customer leftovers, excess stock ordering, and buffet-style canteens (WRAP, 2013). Different sources of food waste in Europe include manufacturing (by-products – bones from the meat industry, etc.), food services (like restrictions in taking leftovers home, not offering various portion sizes, etc.), wholesale and retail (defects in packaging, overstocking, etc.), and households (discarding skins of fruits and vegetables, buying too much, bad storage, etc.) (European Environment Agency, 2020).

Food commodities such as fruits and vegetables (39%), cereals (24%), root and tubers (19%), milk and eggs (9%), meat (5%), oilseeds and pulses (3%), and fish and seafood (1%) contribute to food waste (Roy et al., 2023). Fruits and vegetable waste such as skin, pulp, seeds, and stems are abundant in bioactive compounds like dietary fibers, vitamins, oils, polyphenols, essential oils, carotenoids, tannins, natural pigments, volatiles, fatty acids, and enzymes. Animal product processing includes milk and meat processing industries. Types of waste generated from dairy products are whey and colostrum, which consist of bioactive peptides, oligosaccharides, galactooligosaccharides, and lactoferrin, while waste generated from the meat industry includes blood, trimmings, cuttings, bones, horns, and skin, which contain bioactive peptides and collagen hydrolysate. Marine industry waste includes tails, skin, salmon nasal cartilage, and shrimp shells, which are known to contain proteoglycans and bioactive peptides as bioactive compounds (Ben-Othman et al., 2020). Generated food loss and waste are cheap and renewable sources of bioactive compounds that have a promising future in the preparation of functional food and nutraceuticals, as the market demand for these products has been increasing globally (Vilas-Boas et al., 2021).

11.2 EXTRACTION TECHNIQUES

Abundant bioactive compounds found in numerous food wastes and by-products have led to the establishment of a standard and dependable method for screening health-promoting substances. Identifying the optimal extraction process is critical to unlocking numerous opportunities for the separation, identification, and characterization of bioactive compounds that can provide top-notch products to consumers. The food industry is increasingly seeking ways to extract functional as well as bioactive compounds from food waste in large quantities, having outstanding bioavailability to enhance food delivery in various food applications. However, the sustainability issues of extraction methods are essential from the perspectives of health, the environment, and the economy (Santos-Buelga et al., 2019). Many of these extraction techniques and their objectives are common, such as effective separation of desired bioactive compounds, low cost, time efficiency, and easy sample preparation. The extraction of bioactive compounds is controlled by various factors, including the extraction technology, samples, and extraction solvents. Strategies for the isolation of bioactive compounds are divided into conventional and green/eco-friendly (Gil-Martín et al., 2022). Currently, the demand for solid-liquid extraction (SLE) techniques that employ both organic and petroleum solvents, along with leaching, is increasing. This method is well-known among households for producing tea and coffee at home. In the commercial sector, Soxhlet extraction is more commonly used to isolate numerous bioactive compounds and essential oils using non-polar solvents (hexane, etc.) (Yan et al., 2021). Additionally, traditional percolation and maceration methods have been employed with solvents such as methanol, ethanol, and acetone. However, the conventional extraction techniques that are commonly used are often time-consuming, energy-intensive, and require a substantial amount of solvents during processing, which can be a significant disadvantage. Emerging technologies are gaining popularity in the pharmaceutical, food, and medicinal industries due to their numerous advantages. These technologies offer faster processing times, increased extraction efficiency, and reduced solvent requirements compared to conventional methods (Alara et al., 2018). This section of the chapter deals with the conventional and novel extraction techniques used in different food waste matrices for the effective extraction of bioactive compounds.

11.2.1 Conventional Techniques

11.2.1.1 Solvent Extraction (SE)

In SE, the raw materials of the appropriate size are subjected to various organic solvents, which selectively extract desirable soluble components of interest, including substances like anthocyanins, which are known for their properties that combat cancer and reduce inflammation. Typically, the samples undergo centrifugation and filtration to eliminate solid biomass, and the resulting residue or extract can be converted into a food supplement or incorporated to make functional foods (Reynoso-Camacho et al., 2021). Solvent extraction is quite advantageous, as it is economical and the separation is easy; however, this method utilizes some toxic solvents required for concentration and typically requires abundant amounts of solvent and a prolonged time to complete. Additionally, the potential for the destruction of heat-sensitive bioactive components cannot be overlooked during prolonged extraction times. Although the efficacy of solvent extraction has been enhanced by combining with other methods, such as Soxhlet, ultrasonication, microwave extraction, and supercritical fluid extraction to achieve higher yields, it still has certain limitations that need to be addressed. Gan and Latif (2011) explored *Parkia speciosa* pod powder's bioactive compounds by the solvent extraction method. They discovered that a 50% acetone solution had the maximum polyphenol content compared to methanol, hexane, ethanol, and ethyl acetate. Moreover, Bandar et al., (2013) investigated the efficiency of different organic solvents in extracting bioactive compounds from plant materials. They found that ethanol yielded the highest extraction rates, while hexane had the lowest yields. Additionally, it was also observed that an increased extraction time increases the recovery of extracted compounds.

11.2.1.2 Maceration

Macerated household tonics have been prepared for a long time using a maceration process. This process has been found effective for temperature-sensitive bioactive compounds. In this process, a small amount of the finely ground sample is added to a closed container along with a solvent to increase the surface area. The resulting liquid rich in extract is strained. In some cases, pressing is used to extract the remaining solution. Gentle stirring is occasionally used to enhance the diffusivity and then effectively separate the concentrated bioactive compounds. However, to date, the processing time has been long, ranging from a few hours to a week, which limits the efficiency. Polyphenolics from kinnow (*Citrus reticulata* L.) peels were extracted and quantified. In a study conducted by Safdar et al., (2017), both ultrasound extraction with 80% methanol and maceration with 80% ethyl acetate were done. They found that the maximum amount of polyphenols was obtained with the ultrasound-assisted extraction method, yielding 32.48 mg GAE/g extract, while the maceration technique resulted in only 8.64 mg GAE/g extract.

11.2.1.3 Percolation and Decoction

Percolation is an efficient process where a solvent is continuously run through a sample, leading to the collection of the saturated solvent containing bioactive compounds. This method is considered more efficient than maceration due to its continuous approach. Firstly, the sample is dried and subjected to conditioning with the solvent for about 2–4 hours, allowing it to absorb the solvent. Subsequently, the sample is completely submerged in the solvent for 24 hours. After the completion of the cycle, the percolator is replenished with fresh solvent for successive recovery of compounds, ensuring maximum recovery. Conversely, the decoction technique is primarily used for seeds and stems (hard waste). In this method, powdered samples are boiled in a sealed container with a desirable amount of water for 30 minutes until the volume reduces by boiling, and the sample undergoes filtration through a sieve to remove any solid residue. This technique is generally employed for extracting bioactive compounds that are both resistant to heat and soluble in water (Belwal et al., 2020).

11.2.1.4 Hydro-Distillation

Hydro-distillation (HD) is widely used for obtaining essential and bioactive oils and typically involves the use of water, steam, or a combination of both to extract volatile oils from a sample. The process of HD involves diffusion and hydrolysis followed by decomposition by heat (Hikal et al., 2021). The sample is placed in a still vessel, and hot water or steam is applied to soften the tissue and extract bioactive compounds. The volatile oils are then collected through indirect cooling of the vapors using a condenser and separator. One of the main advantages of hydro-distillation is that it does not require organic solvents and is a relatively fast extraction process. However, the solvent used, the polarity of the bioactive compound, and the mass transfer rate are important factors to consider when choosing a solvent for this technique (Chemat et al., 2020; Wen et al., 2019).

11.2.2 Novel Extraction Techniques

The lack of eco-friendly, energy-efficient, and economically viable techniques for extracting valuable components from food waste and by-products is a common issue. Traditional extraction techniques have been limited by these factors, but recent innovations in non-thermal extraction have overcome these obstacles. This section will therefore discuss some of the advanced techniques that are being employed to recover bioactive compounds from food waste.

11.2.2.1 Ultrasound-Assisted Extraction (UAE)

Waves having frequencies greater than 20 kHz and wavelengths in the millimeter range are termed ultrasonic. The high-pressure oscillations create numerous nucleation points for vacuum nanobubbles, which accumulate energy through a continuous phenomenon of compression and rarefaction fluctuations until their resistance threshold is surpassed and the cell collapses vigorously. This simultaneous repetition of the phenomenon of numerous vacuum bubbles producing intense shearing forces (with speeds of approximately 280 m/s), namely microjets and high shock waves (with temperatures of 5000°C and pressures of 200 megapascals), is a phenomenon in fluid mechanics known as cavitation. The mechanical energy generated by sonication has intrinsic effects on the physical as well as chemical composition of the sample structure, including breaking the cell microstructures (cell walls and membranes) and reducing the particle size (More & Arya, 2021).

Additionally, the mechanical impact increases the surface area, enhances the extractability, and facilitates the diffusion and mass transfer of bioactive compounds.

UAE is quite a popular and relatively affordable extraction technique employing ultrasonic waves to improve solvent extraction efficiency. The technique is simple and requires the deposition of a sample with solvent contact in an ultrasonic bath system. The bath operates at a frequency of 40 kHz or 20 kHz, depending on the energy concentration (Talmaciu et al., 2015; Rodríguez de Luna et al., 2020). UAE promotes the effective diffusion of phenolics in the presence of GRAS solvents, making it a popular choice for polyphenol extraction from various plant materials. An ethanol-water UAE has been reported to mobilize polyphenolic compounds from defatted cocoa beans, restricting the use of non-GRAS non-polar phases and increasing the total amount extracted at 35 kHz for 30 min and recovered by 59.7 and 12.8%, respectively (Toro-Uribe et al., 2020). Likewise, in the UAE process, phenolics have been successfully extracted from cocoa shells, revealing a significant amount of phenolic content (51.1 mg GAE/g extract) (Grillo et al., 2019).

Various solvents demonstrate efficient extraction of plant phenolics, for instance, in red grape pomace where polyphenols and other pigments were recovered using 90% aqueous glycerol (66.70 mg GAE/g DW) while 90.07 mg GAE/g DW was extracted from onion solid wastes (Trasanidou et al., 2016). This efficacy is significant due to the dissolution of molecules of glycerol with low polarity, mainly phenolic compounds. Likewise, the ethylene glycol in a (48:32:20, v/v) ethanol:polyethylene glycol:water mixture enhanced the yield of trans-resveratrol by approximately 40% from red grape skin (Babazadeh et al., 2017). Also, UAE facilitated the extraction and characterization of polyphenols from habanero chili and its by-products (leaves, peduncles, and stems) across various soil types when a methanol:water (80:20) mixture was utilized (Chel-Guerrero et al., 2021). Notably, major differences in different concentrations were observed, revealing that the highest total polyphenolic compounds were found in the by-products, *viz.*, leaves and peduncles, allowing for comprehensive utilization of the plants. Further, choline chloride:malic acid-based NADES (water content 35.4%) effectively extracted anthocyanins from wine lees at an ultrasonic power of 341.5 W and a time of 30.6 min, as noted by Bosiljkov et al., (2017).

11.2.2.2 Microwave-Assisted Extraction (MAE)

In different food matrices, microwaves of 300 MHz to 300 GHz frequency are applied. These ionizing radiations create an electric field by ionic conduction and dipole rotation that forces the material to heat up due to intense alignment to the electric field (Zhi et al., 2017). The water molecules inside the matrices evaporate and create pressure due to evaporation, followed by rupturing cells and thereby releasing bioactive compounds. Various researchers reported that MAE recovered bioactive components from food wastes. Meini et al., (2021) successfully extracted bioactive compounds from grape pomace with a 75–80% increased yield percentage compared to the conventional solvent extraction method. In another report, the efficiency of the recovery of polyphenolics and antioxidant value of up to 85% was revealed by MAE (Alexandru et al., 2014) in vegetable waste. Similarly, Arrutia et al., (2020) isolated pectin-derived oligosaccharides from food waste, specifically potato waste. The results indicated a 40–45% increase in pectin yield in that solvent extraction method. This signified that a short heating time is sufficient to allow pectin hydrolysis, minimizing pectin degradation at high exposure temperatures in conventional extraction systems. Moreover, Alishahi et al., (2011) extracted chitosan from shrimp and prawn waste through MAE. The study recorded a yield efficiency of bioactive compounds increased by 15%, while the extraction time also reduced to 35 mins from 3 hours, along with higher antibacterial activity and high crystallinity. In addition to this, MAE has also been combined with other techniques for better extraction and cost-effectiveness. Kusuma et al., (2016) derived essential oils present in abundance in orange peels by MAE with steam distillation and reported that the extraction time decreased to 2.25 hours from 7 h with higher retention of bioactive compounds. Furthermore, MAE coupled with hydro-distillation, as reported by Megawati (2016) and Wu et al., (2019) for isolating essential oils from fennel seeds, resulted in a 1/10th reduction of time for extraction compared to conventional methods. Additionally, in apple pomace, bioactive compounds were extracted using microwave hydro-diffusion and gravity (MHG). It was revealed that the MHG technique dried 1.2 kg of apple pomace in 1–2.6 hours at a 5.1–13.9 mL/min evaporation rate compared to conventional drying, which took 3.6 to 9.9 hours. Also, the bioactive compounds were distinguished from apple pomace during the drying process through the microwave hydro-diffusion method, which is not possible in hot air drying. The yield of bioactive compounds was reported to be around 41.9% more than the conventional extraction (Fernandes et al., 2021).

11.2.2.3 Enzyme-Assisted Extraction (EAE)

EAE involves the application of enzymes, including hemicellulase, xylase, protease, cellulase, etc., from plant tissues. The enzyme plays a crucial part as it binds to the active site of the cell wall, which comprises a polysaccharide-lignin network, by breaking the glycosidic bond within and the proteolytic bond in the middle lamella which leads to the hydrolysis of polysaccharides and lipid structures (Grassino et al., 2020). Commonly utilized enzymes for this procedure include polygalacturonase, xylanase, cellulase, hemicellulase, amylase, etc. (Liu et al., 2016; Barbosa et al., 2020). Studies also report that such enzymes can effectively hydrolyze carbohydrate-rich products (such as date seeds) and convert them to sugars with yields of >30 g/L at 45°C with 120 FPU/g enzyme activity (Hamid and Ismail, 2020). Likewise, another investigation noted the effective recovery of pectin from passion fruit peel when proto-pectinase enzyme was used for hydrolysis, generating higher yield than the traditional extraction method (Vasco-Correa and Zapata, 2017). The research team also reported that the pectin obtained was extracted about 40% higher due to the application of the enzyme at 30 U/ml at pH 3 at 37°C. Furthermore, pomegranate seeds, a discarded product of the pomegranate juice industry, contain an abundance of fatty acids (22.9%), and it was revealed that the extraction using protease enzyme (at 50 U/g for 14 h, temperature 45°C, and pH 7.2), had a 4% increased recovery of oil when compared to Soxhlet extraction (Talekar et al., 2018). Another study on pomegranate peel by Roda et al., (2016) noted that enzymes combined with supercritical fluid extraction obtained higher amounts of polyphenolics, such as vanillic acid, ferulic acid, and syringic acid. The suitable conditions for this procedure involved enzymes in combination, *viz.*, pectinase, protease, and cellulase (25:25:50) at 3.8% at 49°C, for 85 minutes having a pH of 6.7.

In addition to these enzymes, neutrase, polygalacturonase, pectin esterase, alcalase, and endo polygalacturonase are commonly used (Nadar et al., 2018). EAE offers various advantages, including an increased yield, lower temperature and energy requirements, and improved product quality when compared to traditional methods. Moreover, it also eliminates equipment corrosion that occurs during high-temperature extraction processes, as observed by Roselló-Soto et al., (2016). Furthermore, the extraction of pectin from food waste such as kiwifruit pomace is also reported by Yuliarti et al., (2015), and bioactives from passion fruit peel were recorded by Liew et al., (2015). Teles et al., (2021) used a combination of EAE along with high-pressure processing (HPP) to extract phenolics from grape pomace and observed that this method is effective and environmentally friendly. However, antioxidant activity was increased by EAE with the application of pectinase, tannase, and cellulase enzymes, as reported by Fernandez et al., (2015), to effectively isolate proanthocyanidins from grape skin and seeds. Pectinase was most effective in releasing total phenols on grape skins, while all three enzymes were effective in enhancing phenolics in grape seeds. Also, Ferri et al., (2016) enhanced the recovery of polyphenols from the pomace of red grapes (*Vitis vinifera* L.) by utilizing commercial enzymes *viz.*, α-amylase, xylanase, pectinase, and cellulose-based enzymes. Further, Saad et al., (2019) observed an increase in yield (48%) and antioxidant activity (25%) from raspberry pomace when alkaline protease was added in an aqueous medium to extract.

11.2.2.4 Hot Pressurized Liquid Extraction (HPLE)

HPLE is not fully explored compared to other mentioned techniques for the extraction of phenolics and other compounds from food waste. Only a few researchers have explored this technique in food waste sources, such as orange peels (Barrales et al., 2018), grape marc (Pereira et al., 2019), and residual masses of blackberry and blueberry (Machado et al., 2017), demonstrating higher extractability of polyphenolics and anthocyanins. These investigations have employed diverse solvents, such as water, ethanol, ethanol with water, and composite solvents like ethyl acetate in methanol. In a study conducted by Barrales et al., (2018) maximum recovery of phenolics from orange peels was achieved using 75% ethanol. Similarly, Pereira et al., (2019) recorded the most monomeric anthocyanins and phenolic compounds from grape marc using a 50% (w/w) ethanol-water solvent ratio. Furthermore, Machado et al., (2015) noted that anthocyanin and phenolic compounds from blackberry, grumixama residues, and blueberry attained the highest yield by using water as a solvent at pH 2.0 and 70% ethanol combined with HPLE.

11.2.2.5 Pulsed Electric Field (PEF) and High Voltage Electric Discharges (HVED)

Electro-chemistry, which does not involve heat, uses electric fields to create tension in plant cell walls and membranes, causing them to become more permeable to solvents. This results in increased access to intracellular phytochemicals, which can improve the efficiency of mass transfer and reduce the energy, time, and solvent requirements. Recently, there has been an increase in the use of electrotechnical methods in green chemistry to utilize plant wastes and by-products. PEF applies short pulses of electricity ranging from 140 to 220 V for durations ranging from microseconds to various hundred seconds in closed, sealed equipment containing the sample. The pressure exerted by these electric potentials on the cell envelope can result in the formation of pores, which can be both reversible and irreversible based on the energy level, duration, and intensity of the pulses. Rodríguez de Luna et al., (2020) explained that as the cell envelope weakens, PEF allows the plant bioactives to evacuate by increasing the electrical conductivity, permeability, and diffusivity of the solute. The optimization of PEF extraction is mainly determined by the electric field/mass ratio, which influences the size of the pores and disintegration of the wall/membrane with increasing intensity of electric pulses (Peiro et al., 2019; Rodríguez de Luna et al., 2020). Notably, electric fields can trigger secondary metabolism resulting in increased levels of phenolics and other secondary metabolites in specific matrices.

Simultaneously, the photodissociation of water during electric discharge generates hydroxyl radicals, ozone, and hydrogen which can degrade polyphenolic compounds. Also, combining PEF with conventional extraction techniques provided promising results for pre-treatments of phenolics by assessing certain conditions such as temperature and pH, resulting in improved solvent consumption, extraction time, and yield. Research has demonstrated the potential of utilizing plant by-products such as potato peels, grape by-products (Boussetta et al., 2013), and vine shoots (Rahman et al., 2014). Additionally, studies have also demonstrated the effectiveness of this approach in plum and flaxseed hulls as per the reports of Boussetta et al., (2014). Other instances of waste exploration by PEF extraction for the extraction of phenolics involve orange and lemon peel waste (Peiro et al., 2019). . PEF is employed as a favored strategy to selectively permeabilize cell walls and enhance polyphenol concentrations during winemaking.

Another non-thermal technology, HVED, is effectively acting towards the recovery of bioactive compounds from food waste (Li et al., 2019). It shares similar procedures, rationale, and equipment with PEF, with the primary difference being that the electric fields (20–40 kV) are generated from the two electrodes to the sample inside the discharge chamber, resulting in electroporation of the outer cell wall and membrane beyond the threshold (Mahnič-Kalamiza et al., 2014; Saulis et al., 2015). Research conducted on the extraction of phenolic components from papaya peel revealed certain risks to chemicals and products from electrolysis after the application of HVED that destroy the chemical structure of the extracts. To mitigate oxidative damage to extracted phenolics, pH can be screened, as acidic pH stabilizes ozone and inhibits the breakdown into hydroxyl radicals caused due to the presence of water molecules (Parniakov et al., 2014). To obtain high-quality extracts, some features of HVED treatment time and solvent selection should also be addressed. Studies have shown that addressing these criteria can significantly affect the extractability of compounds

(Brianceau et al., 2016). HVED regimes have become increasingly popular because of their ability to extract phytochemicals from natural sources.

Several methods were explored for both laboratory and pilot-scale scenarios for extracting polyphenols from agricultural and food waste, such as grape by-products and orange peels (El Kantar et al., 2018). Boussetta et al., (2009a) and Boussetta et al., (2009b) isolated polyphenolic compounds from grape pomace and revealed higher recovery efficiency in the HVED method with shorter extraction time when applied with 40 kV of pulses for 160 s as compared to the conventional method which took 240 mins under similar conditions without HVED. Also, higher retention of catechins was obtained for similar conditions, respectively. Additionally, some different solvents like DES-6 (lactic acid:glucose) were found efficient in extracting naringin, the primary flavonoid from grapefruit peels, by El Kanter et al., (2019).

Additionally, some different solvents like DES-6 (lactic acid:glucose) were found efficient in extracting naringin, the primary flavonoid from grapefruit peels, by El Kanter et al., (2019).

11.2.2.6 Solid-Liquid Extraction

The use of deep eutectic solvents (DES) to extract phenolic compounds from food residuals, particularly agricultural waste, is widely recognized as one of the most effective approaches. However, this method has drawbacks, such as extended processing times, high capital costs, low extraction yields, and the use of organic solvents. Organic solvents are useful for solubilizing and extracting phenolic compounds; however, they pose challenges such as toxicity, temperature sensitivity, and non-biodegradability. DES preparation was first described by Abbott et al., (2003), in which a cutting-edge and eco-friendly solvent extraction method was developed to address these challenges and recover phenolic compounds. This cost-effective and easily accessible solvent utilizes a mixture of hydrogen bond donors (HBD) and acceptors (HBA) at a specific temperature. Several studies have demonstrated that DES offers several advantages over traditional organic solvents, such as low cost, ease of manufacturing, and availability. Further, the use of non-toxic and inexpensive salt choline chloride (ChCl) is the most common method for creating deep eutectic solvents (DES), with urea, glycerol, and ethylene glycol being the most frequently utilized hydrogen-bonding donors. Furthermore, alcohols, sugars, carboxylic acids, and amino acids have also been employed in the creation of DES by hyphenating primary metabolites with renewable starting ingredients. Naviglio Extractor R, another option for extracting bioactive compounds, is a relatively new solid-liquid extractor. This technique creates a negative pressure gradient in the outer as well as inner parts of a solid structure that is suspended in an appropriate solvent to reach equilibrium and has several advantages, such as the extraction ability at room temperature to reduce thermal stress in heat-destructing compounds. Also, employing high pressure decreases the extraction time and improves the extraction efficiency.

Recent research has reported advancements in eutectic systems for extracting active phenolic compounds from agricultural by-products. For instance, a choline-chloride-based non-aqueous solvent with a water content of 35.4%, along with malic acid as an H^+ donor and UAE, demonstrated better results (6.55 mg/g DW) in recovering anthocyanins from wine lees (Bosiljkov et al., 2017). This trend can be seen in the extraction of phenolic compounds from by-products during oil processing using different techniques mentioned in the reports of Bonacci et al., (2020) or to extract desired bioactive compounds (Paradiso et al., 2019). Furthermore, the recovery of flavonoids (222.97 mg GAE/g DW) from onion peel (Pal & Jadeja, 2019) and orange peels (3.61 mg GAE/g) has been reported, as well as from numerous other sources (Redha, 2021).

11.2.2.7 Supercritical Fluid Extraction

The destruction of natural heat-labile compounds by traditional extraction techniques has led various industries to shift to greener options such as SFE. This technology employs the use of different solvents under pressure ranging from 200 to 400 bar, while the temperatures vary from 40°C to 60°C, as reported by Gullón et al., (2020). The solvents commonly used in SFE technology include toluene, ethylene, ammonia, and carbon dioxide, which are preferred for their availability, environmental safety, non-explosiveness, non-toxicity, affordability, and ease of removal. The fluid used in SFE technology is subjected to the material where it behaves like both a liquid and a gas (supercritical conditions), in which mainly CO_2 reaches its supercritical stage, i.e., beyond its critical point, by applying a pressure of 7.38 MPa and a temperature of 31°C, as detailed by Ding et al., (2020). In this condition, the polarity decreases, while an increase in solubility allows the fluid to penetrate within the cell matrix quickly.

The bioactive compounds are dissolved and extracted alongside other components in SFE to enhance the extractability of bioactives with high polarity. Modifiers are sometimes added to the fluid ((Torres-Ossandón et al., 2018; Gallego et al., 2019). Cruz et al., (2017) extracted jelly palm seed bioactive compounds by comparing Soxhlet extraction, UAE, and SFE with CO_2 as a solvent, and ethanol as the co-solvent. The results indicated that SFE using CO_2, ethanol, and water as solvents improved the extraction yields and was found better than Soxhlet extraction while extracting oil composed of fatty acids and phenolic compounds at 52°C and 35.7 MPa, with a constant flow rate of 30 g CO_2/min for 150 min. On the other hand, linoleic and oleic acids were identified as the major fatty acid components of guava seed oil during extraction. Moreover, tocopherols, vanillin, cinnamaldehyde, vanillic acid, phytosterols, β-sitosterol, cinnamic acid, and γ-tocopherol were the major phenolic compounds found in the guava seed oil as observed by Narváez-Cuenca et al., (2020). Further, the reports of Cuco et al., (2019) revealed higher yields of phenolic compounds in pumpkin seed oil extracted using SFE technology at the optimal conditions of 22 MPa and 333 K, and this method was also found to be better when hyphenated with UAE (25 kHz) with n-hexane.

Supercritical fluid extraction (SFE) demonstrated greater oil content than Soxhlet extraction. A study on the solubility of oil extraction from pomegranate seeds with high phenolic content, attained due to punicic acid, was conducted by Natolino and Da Porto (2019) at 320 bars and a temperature of 60°C. The study reported enhanced solubility, higher oxygen stability, and enhanced solubility by SFE. Similarly, various studies have found that SFE using CO_2 was effective for the extraction of phenolic compounds from Brazilian cherry at high pressure and low temperature at a 2 g/min flow rate, namely terpenes, germacrene, flavonols, and ellagitannins etc. The extraction conditions were carried out at 25 MPa pressure, 40°C temperature, with a raw material mass of 300 g. The CO_2 gas flow rate of 6.0–7.0 kg resulted in enhanced recovery of phenolic components from the seeds of strawberries (Grzelak-Błaszczyk et al., 2017).

SFE has been instrumental in enhancing oil extraction from rosehip seeds under optimal conditions, specifically at 30 MPa at 40 °C at 0.75 mL/min, with particle size 355 μm for 150 min (Salgın et al., 2016). Furthermore, the SFE process at its optimized conditions of 300 bar, 50ºC, concentration of 0.3 kg CO_2/h, for 4 h was found to be the most effective for extracting oil from raspberry seeds, which are rich in omega-3 fatty acids, tocopherols, and antioxidants, as reported by Marić et al., (2020). Additionally, the successful recovery of oil rich in vitamin E from bilberry seeds with CO_2 as the solvent using SFE under optimal conditions of 20 MPa and 60 ºC for 80 minutes was reported by Gustinelli et al., (2018).

11.2.2.8 Subcritical Water Extraction (SWE)

SWE has risen in recent times as a cost-effective solvent extraction method. In this process, the solvent is used in its sub-critical condition, which lies between its boiling point and critical point (Zhang et al., 2019; Essien et al., 2020). Solvents of different polarity, including methanol, ethanol, and acetone, can be used for this purpose; however, water is the most commonly utilized solvent because of some properties such as high polarity, low basicity, high diffusion rate, and green technology/environmentally friendly (Essien et al., 2020). The polar as well as viscous regions of water decreases when the temperature rises to 374°C from 100°C, having a pressure of 1 to 22 MPa, which eventually enhances the solubility, facilitating higher extractability through higher diffusion, partitioning, equilibrium, or convection (Essien et al., 2020). Although SWE is an inexpensive and clean technique, some studies have reported cases of oxidation and hydrolysis of pectin chains (Gallego et al., 2019; Zhang et al., 2020).

Research conducted by Rodrigues et al., (2019) using SWE at a temperature of 150°C could effectively separate papaya seed extracts, resulting in the identification of some primary phenolic components such as vanillic acid, mandelic acid, and ferulic acid. This method outperformed the Soxhlet extraction, which was performed at 40°C for 6 h.

In addition to this, the reports of Pereira et al., (2017) provided a higher yield of unsaturated fatty acids and tocopherols (86.36%) with SWE at optimized conditions at 40°C for 6 h using subcritical propane extraction at 6 MPa pressure in sweet passion fruit seed oil, attributing towards a product of higher antioxidant potential. A study conducted by Duba et al., (2015) reported yields of 44.3–77 mg/g and 44–124 mg/g of polyphenols from grape seeds and peels, respectively, in semi-continuous mode when water was subjected to a temperature of 80–120°C and a pressure of 100 MPa. Liew et al., (2018) utilized SWE technology at 30 MPa pressure at 120°C in dynamic mode to extract pectin from pomelo fruit peel, achieving a yield of 18.8%. Meanwhile, Tanaka et al., (2012), utilized subcritical water for the effective extraction of 78% pectin and 80% of hemicelluloses and celluloses in optimized conditions of 200°C, 20 MPa, and water flow at 2.1, 3.5, and 7.0 mL/min from *Citrus juno* peels.

11.2.2.9 Microbial Fermentation in Dairy By-Products to Isolate Bioactive Components

The primary dairy by-products, mainly whey, can provide healthy bioactive peptides through microbial fermentation, which offers nutraceutical and functional food benefits (Kaur et al., 2020). As per Duan et al., (2021), galacto-oligosaccharides (GOSs) are produced by converting lactose using β-galactosidase, which is present in dairy by-products such as cheese whey and yogurt whey. GOSs are indigestible oligosaccharides made with a combination of 2 to 10 saccharide units with prebiotic properties beneficial for colon health, as they promote bifidobacterial sustainability, as per a study by Mano et al., (2019). Meanwhile, the limitation of this method can be the extreme commercial loss of enzymes, which can be addressed using immobilized enzymes. Additionally, there are many challenges faced in the microbial production of GOS, such as low yield, low producibility, and overall quality of the obtained product, as reported by Rico-Rodriguez et al., (2021). Nevertheless, microalgae are also potential organisms to extract GOS from dairy by-product whey effectively. For instance, sweet whey-containing media cultivated *Tetradesmus obliquus* within seven days, resulting in an abundant amount of β-galactosidase production, as per a study by Suwal et al., (2019). This species can hydrolyze lactose under both mixotrophic and heterotrophic conditions.

Pleurotus spp., including *Pleurotus eryngii*, *Pleurotus sajor-caju*, and *Pleurotus pulmonarius*, have been found to produce bioactive compounds that may offer potential health benefits. For example, these compounds may improve heart health, increase immunity, limit cancer-causing risks, and regulate blood glucose levels. Research on *Pleurotus djamor* PLO13, a specific species of Pleurotus, has demonstrated its potential as a nutritional supplement. Velez et al., (2019), in their latest study, investigated the mycelial growth of *Pleurotus djamor* PLO13 in whey and selenium-containing media and reported an increase in mycelial growth, which also signified high antioxidant activity as well as higher β-glucans and ergosterol concentrations. The study also points out that *Pleurotus djamor* can be effectively utilized as a nutritional supplement.

Some examples of bacteriocins produced in media prepared with whey are nisin and pediocin. Another prominent antibiotic for the treatment of anti-drug effects is bacteriocins, which possess high antimicrobial activity against both gram-positive and gram-negative bacteria. Research conducted by Ryan and Walsh (2016) and Panesar and Kennedy (2012) revealed that whey, if diluted, can serve as an appropriate substrate for the production of bacteriocins mainly by using *Lactococcus lactis* subsp. Lactis CECT 539 and *Pediococcus acidilactici* NRRL B-5627, respectively.

Taking into account the various extraction methods discussed above for the isolation of different bioactives from food waste, it is worth noting some other extraction scenarios, as outlined in Table 11.1.

TABLE 11.1
Different Extraction Methods to Isolate Bioactive Compounds from Food Waste

S.No.	Food Waste/ By-Product	Extraction Technique/ Conditions	Key Findings	References
1.	Red beet roots	UAE/solid:solvent ratio (g/mL) (1:20), time 16 min, power 90 W	Higher yield (4%) of betacyanins and betaxanthins than conventional extraction	Nutter et al., (2021)
2.	Grape skins	Enzymatic treatment / sample:solvent ratio (g/mL) (1:20), maceration conditions: pH 3.6, temperature: 20–30°C, enzyme loading: 15 mg/L, enzymes: cellulase, glucosidase, and pectinase	Successful extraction of anthocyanins and flavonols with higher yield than conventional techniques	Nogales-Bueno et al., (2020)
3.	Mango peel	Enzyme assisted ultrasound extraction (EAUE) and UAE / temperature: 63°C, time: 110 min, pH 5.5, power: 90 watt, alcalase enzyme	TPC (mg GAE/g) EAUE: 33.56 ± 1.04, UAE: 25.64 ± 0.98	Sharif et al., (2021)
4.	Olive pomace	Combine microwave and enzyme assisted extraction/ solid:liquid ratio (1:15 g/mL), time: 17 min, temperature: 60°C, ramp time: 5 min, tannase enzyme (2%)	Phenolics: 341 mg GAE/g	Macedo et al., (2021)
5.	Passion fruit pomace	Ultrasonication + cellulase enzyme/ solvent:solid ratio (50:1 mL/g, temperature: 30°C, enzyme loading: 6%, time: 47 min, pH 5.0, frequency: 50 kHz, power: 300 W	1.5 to 2 times higher phenolics and antioxidant activity	Wang et al., (2021)
6.	Spent coffee ground (SCG)	Ezyme treatment/(1 g SCG + 0.1 mL (0.2 M pectinase) + 200 mM sodium acetate buffer), 37°C, time 60 min; pH 5.5, enzyme: pectinase	TPC: 267.17 mg TAE/ mL) TFC: 58.30 mg QE/g)	Khairil Anuar et al., (2020)
7.	Blueberry waste	Supercritical fluid extraction/CO_2 (90%), ethanol (5%), water (5%)	TPC 134 mg GAE/g	Paes et al., (2014)
8.	*Moringa oleifera* leaves	Stirring/ solid:liquid (2.5:50 g/ mL), glycerol:sodium acetate (molar ratio 6:1), temperature: 50°C, time: 180 min	TPC 53.8 mg GAE/g TFC 16.5 mg RE/g	Karageorgou et al., (2017)
9.	Mustard seed meal	UAE/solvent: solid (50:1 g/L), 70% ethanol, 40°C temperature, 30 min time, 60 W power	TPC 13.8 mg /g sinapic acid	Dubie et al., (2013)
10.	Cocoa bean shell	DES couples with microwave extraction/solid:liquid (50:1 mg/ mL), temperature 50°C, time 60 min.	Threobromine: 2.50–5.00 mg/g, caffeine: 0.77–1.59 mg/g	Pavlović et al., (2020).

(*Continued*)

TABLE 11.1 (CONTINUED)
Different Extraction Methods to Isolate Bioactive Compounds from Food Waste

S.No.	Food Waste/ By-Product	Extraction Technique/ Conditions	Key Findings	References
11.	Meat industry waste	Enzyme assisted using microbial proteases/phosphate buffer, pH 6.0, fungal protease or bacteria derived protease	Bacterial derived protease hydrolyzes meat proteins and generates non-toxic, stable, gastrointestinal tract-friendly bioactive peptides	Ryder et al., (2016)
12.	Green soybean pods to soybean milk	UAE employed having power of 500 W at 20 kHz frequency for 15 min	Procyanidin 0.72 mg/100 g .	Leksawasdi et al., (2022)
13.	Cheese whey	Microbial fermentation by *Pseudomonas taetrolens* bacteria at 30ºC for 48 h	Production of lactobionic acid	Goderska et al., (2014)
14.	Meat by-product (duck skin)	Enzymatic treatment/ sample:buffer solution (1:100 kg/ mL), enzyme: 20 g, time: 8 h, enzymes used: proteases, papain, pepsin, trypsin, and α-chymotrypsin	Antioxidative peptide produced which protects DNA from oxidative damage caused by H_2O_2	Lee et al., (2010)
15.	Pomegranate peel	UAE/ 140W power, signal-to-noise ratio 1:70, frequency: 30–40 kHz, time: 30 minutes.	67% antioxidant activity, TFC 7 mg/g, TPC 69.87 mg/g	Motikar et al., (2020)

11.3 BIOACTIVE COMPOUNDS FROM VARIOUS FOOD WASTE STREAMS

Agricultural products, including fruits, vegetables, cereals, tubers, and roots are perishable commodities that produce substantial waste. Figure 11.2 shows the different kind of food waste generated from various industries like fruits & vegetables, coffee & tea, dairy, brewer, nuts, animal & marine. The discarded items contain various bioactive compounds such as vitamins (including A and E), polyphenols, dietary fiber, flavonoids, essential minerals, fatty acids, tannins, etc. Each type of fruit and vegetable waste contains unique bioactive compounds. Peels are often a significant by-product of the processing of fruits and vegetables. Banana peels contain carotenoids such as palmitate, caprate, xanthophylls, and laurate, whereas citrus peels contain eriocitrin, hesperidin, and naringin. Research on the Badami variety of mangoes has revealed the presence of noteworthy phenolic compounds, like protocatechuic acid, gentisic acid, and gallic acid, in its peel. Moreover, mangiferin, quercetin, and acids like gentisyl-protocatechuic, syringic, ellagic, and gallic were detected in both the raw and ripe peels. These compounds are known for their antioxidant properties, suggesting that incorporating mango peel or extracts rich in these compounds into diets or products may offer associated health benefits (Ajila et al., 2010). Oven drying pomegranate peel at 60°C led to a high concentration of punicalin (888.04 ± 141.03 mg CE/kg). Freeze-dried peel exhibited more catechin content (i.e., 674.51 mg/kg) in comparison to peel dried in the oven. Higher total phenolic, flavonoid, and tannin values were also noted in freeze-dried peels across the studied temperature range. Additionally, oven-dried pomegranate peel at 40°C shows a higher content of vitamin C (i.e., 31.19 μg AAE/g) (Mphahlele et al., 2016). Apple pomace from four different cultivars underwent microwave-assisted extraction with different solvents to recover bioactive compounds. Analyses

FIGURE 11.2 Various kinds of food waste from different industries.

showed the presence of numerous terpenic and phenolic compounds. Among the terpenes, ursolic acid is the most abundant, along with triterpenic acid derivatives (Grigoras et al., 2013).

The chemical makeup of olive pomace comprises tocopherol (α and β), γ-tocopherol, and α tocotrienol, with the highest concentration of α-tocopherol (i.e. 2.63 mg/100 g). Other vitamers were in concentrations of less than 0.1 mg/100 g. An abundant amount of lipid fractions, i.e. oleic acid (75%), palmitic acid (10%), linoleic acid (9%), and stearic acid (3%), were present in the lipid fraction of olive pomace. Approximately 79% (of total phenolics in olive pomace originated from hydroxytyrosol and comsegoloside, with hydroxytyrosol content measured at 83.6 mg/100 g and tyrosol present in smaller quantities (3.4 mg/100 g) (Antónia Nunes et al., 2018). Olive pomace can be successfully used as a natural additives or as an ingredient in food and cosmetic formulations due to the antioxidant characteristics of the above-mentioned compounds. Grape pomace, particularly Cabernet Sauvignon pomace, is an excellent source of fatty acids (polyunsaturated) and phenolic compounds. Among grape varieties, the highest levels of total phenolic compounds, flavonoids, and monomeric anthocyanins were reported in Cabernet Sauvignon pomace. Additionally, it exhibited the lowest EC50 value, indicating superior antioxidant activity compared with the other samples analyzed (Iora et al., 2015).

Coffee silverskin (CS) as well as spent coffee grounds (SCG) are the primary by-products produced during roasting and brewing processes of coffee, respectively. These by-products contain bioactive compounds, such as caffeine, chlorogenic acid, phenolic compounds, lignocellulose, and dietary fiber (Arya et al., 2022). Other coffee waste products, such as pulp, husk, flour, and parchments, are also known to contain anthocyanins, dietary fiber, and polyphenols (Arya et al., 2022). Spent tea leaves are the solid matter that remains after the making of black tea, which consists of protein, fiber, amino acids, and functional components, such as saponins, phenols, steroids, and tannins (Negi et al., 2022). The brewing industry produces three types of waste: spent barley grain, hot trub, and spent brewer's yeast. Spent barley grain is the insoluble material left after the removal of wort and is rich in proteins, fiber, and lipids. Hot trub is the sediment formed during the boiling of wort and contains proteins, polyphenols, pectins, starch, glucans, minerals, fatty acids, and non-isomerized bitter substances. Spent brewer's yeast (SBY) is yet another waste produced during the brewing process, comprising 15% of all the waste. At the end of second fermentation and following maturation, SBY is recovered by sedimentation.

Puligundla et al., (2020) reported that tree nuts, such as almonds, walnuts, and cashews, contain significant amounts of proteins, minerals, vitamins, β-glucans, and saccharides (mono and oligo). According to Sari et al., (2022), these tree nuts also contain bioactive compounds such as bioactive peptides, phenolic compounds, enzymes, amino acids, vitamins, flavonoids, natural pigments, and dietary fibers. However, castor seed cake, a waste of castor oil extraction, cannot be used as a source of protein because of the presence of toxic substances such as ricin, as stated by Jain and Naik (2018). Nevertheless, this by-product can be utilized for the production of enzymes through solid-state fermentation with the help of *Aspergillus oryzae* and *Aspergillus japonicus*.

Animal product processing primarily focuses on the meat and milk processing industries. Within the meat industry, the largest amount of waste is generated during slaughter, including skin, gastrointestinal contents, bones, blood, internal organs, and tendons. Waste materials are a significant source of proteins, bioactive peptides, fatty acids, vitamins, and minerals. Beef and lamb livers, for instance, are particularly rich in manganese and copper (Rahman et al., 2014). Additionally, beef tongue, an underutilized by-product of the meat industry, can be used to produce protein-rich and unsaturated fatty acid-rich beef tongue powder, thereby enhancing the nutrient content of food products (Long & Mohan, 2021). The fish-processing industry generates a substantial amount of waste, accounting for over 60% of the by-products. This waste includes frames, fins, heads, skin, viscera, trimmings, and roes. Fish protein hydrolysates, which break down fish proteins into smaller peptides, can be produced from various fish by-products, including head, skin, muscle, liver, bone, frame, viscera, and egg. These hydrolysates can serve as easily accessible sources of amino acids for human nutrition (Chalamaiah et al., 2012). Shrimp waste is also a rich source of fat-soluble astaxanthin, a xanthophyll carotenoid that can be utilized as a food colorant in products such as cheese, butter, and ice cream (Prameela et al., 2017). The crab processing industry generates food waste, including shells, viscera, undersized claws, and other materials, which are a significant source of high-quality proteins, lipids, pigments, antioxidants, enzymes, vitamins, and minerals. These waste materials have great potential to be utilized in the food industry (Nanda et al., 2021).

In milk processing industries, whey and buttermilk are the main wastes produced from these industries. Whey (produced during casein and cheese manufacturing) is a source of protein, lactose, minerals, fats, soluble vitamins, and minerals. Whey can be utilized for the preparation of ricotta or cottage cheeses; however, secondary cheese whey is produced during its manufacturing process. From whey, whey powder, whey protein, and lactose can be obtained via ultrafiltration (Sar et al., 2022). Whey protein is comprised of thermosensitive fractions such as β-lactoglobulin (β-Lg), blood serum albumin, α-lactalbumin (α-La), and immunoglobulins. Buttermilk, which is derived from butter production, contains the membrane of milk fat globule, minerals, proteins, lecithin, and lactose. The membrane of the milk fat globule is present in buttermilk and formed when the milk fat globule is disrupted during churning. It is high in phospholipids, resulting in buttermilk having seven times more content of phospholipids than whole milk. The major MFGM proteins include xanthine oxidase/dehydrogenase, mucin-like glycoproteins, adipophilin, butyrophilin, and breast cancer type 1 and type 2 susceptibility proteins. Polar lipids and sphingolipids are the most abundant polar lipids isolated from buttermilk in MFGM, and both are known for their various health benefits (Barukčić et al., 2019). Another by-product of the dairy industry is sediment waste of clarified butter, which is derived from the production of clarified ghee or butter, and comprises milk fats and milk proteins (De Jesus et al., 2015).

11.4 BIOACTIVITY AND HEALTH BENEFITS

Bioactive compounds have various health-improving properties and have proven to lessen the chances of development of various diseases like cardiovascular diseases, cataracts, Alzheimer's disease, Parkinson's disease, and cancers (Figure 11.3). These compounds hold great interest due to their antioxidant, anti-diabetic, anti-microbial, anti-cancer, and anti-inflammatory properties. The antioxidant and radical scavenging properties either prevent or postpone the oxidative damage to DNA, proteins, and lipids and therefore have been primarily linked to health-promoting benefits. There is enough epidemiological data that establishes a positive relationship between good human health and consumption of foods rich in bioactive compounds such as flavonoids, phenolic acids, carotenoids, etc., which lowers the risk of development of various diseases as well as age-dependent disorders (Santos et al., 2019). The by-products and the recovered biomolecules from food wastes can be used for the purpose of development of functional foods, nutraceuticals, and pharmaceuticals preparations (Baiano, 2014). For instance, bioactive compounds derived from by-products like sterols, carotenes, polyphenols, and terpenes contain high antioxidant activities and therefore have the potential to be utilized as natural sources of antioxidants for the development of functional foods and also serve as food additives to extend the shelf life of products (Kalogeropoulos et al., 2012).

11.4.1 Antioxidant Activity

The oxidative breakdown of reactive oxygen species (ROS) and free radicals has been associated with chronic diseases; as a result, antioxidant compounds have garnered substantial interest from researchers in combating these diseases. Numerous studies have shown that phenolic compounds have significant antioxidant properties (Bei et al., 2017). Cereals contain diverse range of phenolic

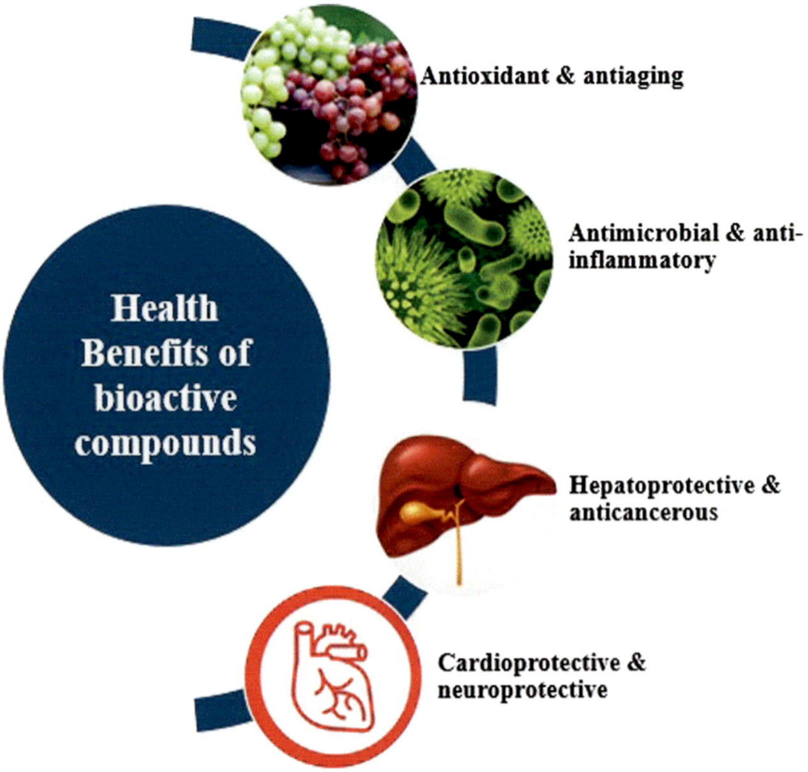

FIGURE 11.3 Health benefits of bioactive compounds derived from food waste.

compounds, including saponins, phytoestrogens, phenolic acids, and flavonoids, among various others. Different extracts from different parts of wheat, barley, rice, and sorghum have shown high radical scavenging and antioxidant activities (Ramakrishna et al., 2019). Many pigmented root vegetables such as sweet purple potatoes, chicory, and carrots etc. demonstrate high antioxidant activities owing to the presence of anthocyanins (Frond et al., 2019; Zhang et al., 2016). Cyanidin and petunidin are the commonly found anthocyanins in root vegetables. These compounds enhance cellular redox conditions of the gut region, prevent oxidative damage, and hence protect from acute GI diseases related to oxidative stress (Zhang et al., 2016). Fruits also represent an abundant source of bioactive compounds like phenolic acids, carotenoids, and flavonoids that act as antioxidants and serve as functional foods (Canan et al., 2016; Farhadi et al., 2016). The flavonoid concentration in apple peel is about three to six times more than the inner part and contains unique compounds such as quercetin glycosides, which is not present in the inner part. The major polyphenols identified in apple pomace include phloridzin, quercetin, and chlorogenic acid. The antioxidant activity values of these compounds are comparable to ascorbic acid (Khalid et al., 2021; Hobbi et al., 2023). In a study, apple peel extracts inhibited lipid and protein oxidation of minced rainbow trout (*Oncorhynchus mykiss*) in refrigerated storage conditions (4°C). The oxidation was inhibited due to the lowering of the peroxides and TBARS (thiobarbituric acid) (Bitalebi et al., 2019). Similarly, olive oil waste extracts have been found to be rich in polyphenols and were used as natural antioxidants to delay the oxidation of lipids as well as proteins in lamb meat patties. The use of the natural antioxidants extended the product shelf life by 3 days and also maintained acceptable meat color for a prolonged time (Muíño et al., 2017). Blueberries, chokeberries, raspberries, and blackberries find their use in commonly processed food products like juices, jams, and jellies. Substantial quantities of waste are produced during the manufacturing process, which contains valuable health-promoting compounds (such as anthocyanins, flavonoids, and phenolic acids) and can be recovered for further utilization across various industries. Seed pomace, a by-product of blackberry (*Rubus fruticosus*) and raspberry (*Rubus idaeus*) is produced in large quantities and can serve as an excellent raw material for oil extraction. These oils are not only rich in linoleic (omega-6) and α–linolenic (omega-3) acids but also contain bioactive compounds like tocopherols, carotenoids, phenols, and sterols, which possess antioxidant properties. Consequently, the oil composition derived from blackberry and raspberry seed pomace remains stable even after prolonged freezing, attributed to the presence of bioactive compounds which act as antioxidants (Radočaj et al., 2014).

11.4.2 Anti-Cancer Activity

Phenolic compounds have been proven to reduce the chances of developing cancer and also inhibit the various stages of the cancer formation process (Tabish, 2017). The various polyphenolic compounds present in tea, coffee, and cocoa are helpful in the prevention of cancer. For instance, the effect of ferulic acid on tongue carcinomas as well as preneoplastic lesions was studied in rats by Tanaka et al., (1994). The results showed that rats fed with ferulic acid and exposed to 4-nitroquinoline-1-oxide in drinking water for a period of 5 weeks significantly reduced the impact of the carcinogen. Mangiferin is another natural bioactive compound found in many plant species, such as mango fruit tree (*Mangiferin indica*), which has been studied by different research groups for cancer treatment.

Mangiferin administration has potential advantages against leukemia, brain, lung, breast, cervical, and prostate cancers, either alone or in combination with anti-cancer medications (Núñez-Selles et al., 2016). Wang et al., (2019) conducted an experiment in which HepG2 cells in mice were treated with grape seed proanthocyanidin extract (GSPE). It was observed that doses as high as 200 mg/kg of the extract significantly hindered the growth of HepG2 cells in mice without causing any toxicity or autophagy. Additionally, GSPE led to the phosphorylation of proteins associated with the mitogen-activated protein kinase (MAPK) pathway. These findings suggest that the extract possesses potential anti-liver cancer properties. In another study, the seed extracts showed anti-inflammatory and antiproliferative activities against the HCT-116 (colorectal carcinoma) and HepG-2 (liver cancer) cell lines. Meanwhile, avocado peel extracts were reported to induce apoptosis in MDA-MB-231 cells as a result of increased activation of caspase 3 target protein PARP (Velderrain-Rodriguez et al., 2021). In a study, the phenolics extracted from apple waste were reported to beneficially influence the key stages of colon carcinogenesis (McCann et al., 2007). The Caco-2 colon cancer cells exhibited inhibition of cell proliferation when exposed to apple extracts up to a concentration of 50 mg/ml, with 40% inhibition being reported for apple extract with skin and 29% without skin (Tarozzi et al., 2004).

11.4.3 Antimicrobial Activity

Food wastes have antimicrobial activities attributed to the phenolic compounds and/or essential oils present in the extracts. Phenolic compounds like flavonols, found in fruits and green tea, possess antibacterial activity (Nishino et al., 2000). There are various natural antimicrobial compounds that can be added to foods, which mainly include phenolic compounds, terpenes, peptides, and polysaccharides, among others. The agro-food by-products are a major source of bioactive compounds having potential antimicrobial activity. For instance, olive leaf extracts have been used to reduce bacterial spoilage in shrimp and organic leafy greens. The bioactive compounds as natural antimicrobials have also been used in meat products to improve meat quality as well as prolong shelf life (Hayes et al., 2010). Pomegranate peel extracts have been explored by different groups of researchers as natural antimicrobials in chicken and meat products. The peel extracts showed significant antimicrobial activity against *Staphylococcus aureus* and *Bacillus cereus* when used in chicken products (Kanatt et al., 2010). Pomegranate peel extract was also reported to prolong the shelf life of chicken and meat products by 2–3 weeks during chilling temperatures (Zavala et al., 2011). Bioactive compounds have also been reported to inhibit microbial spoilage of fish against *Escherichia coli*, *Listeria monocytogenes*, and *S. aureus*, extending the shelf life of the product (Martínez et al., 2019). Bioactive compounds found in essential oils have antimicrobial activity against common pathogenic microbes like *S. aureus* and *E. coli*, and *Pseudomonas aeruginosa* (Yamani et al., 2016).

11.4.4 Anti-Inflammatory Activity

Inflammation is the natural reaction of the body to tissue damage, occurring as the immune system works to address various pathogens, toxins, and damaged cells. These stimuli can induce inflammatory reactions which may be acute or chronic, leading to tissue damage or immune-mediated diseases (Chen et al., 2018). Wineries produce a lot of grape skin waste, which is rich in resveratrol (3,5,40-trihydroxystilbene). Resveratrol is a powerful antioxidant also having good anti-inflammatory properties. This compound improves the anti-inflammatory response of NF-κB cells, hence inhibiting the inflammatory reaction. The compound also enhances the cytochrome P-450 enzyme activity and helps in the detoxification of the liver (Ortega & Campos, 2019). Raffaele et al., (2021) investigated the effect of bioactive compounds contained in pomegranate waste in an *in vitro* study of intestinal inflammation induced by exposing Caco-2 cells to lipopolysaccharides (LPS). The pomegranate waste extracts were efficient in preventing inflammatory bowel disease. The anti-inflammatory response was attributed to the activation of HO-1 (cytoprotective enzyme), buffering the harmful effect of oxidative stress. The study demonstrated the potential of the extract to lower intestinal inflammation, maintain colon length, and lessen IL-6 levels.

11.4.5 Hepatoprotective Activity

The liver holds a central role as the primary metabolic organ of the human body, which regulates all metabolic processes crucial for bodily functions. Hepatocytes, comprising over 75% of the liver's parenchyma, play a pivotal role in maintaining the liver's multifaceted functions and supporting overall physiological processes in the body. Numerous hepatotoxic substances, including alcohol, carbon tetrachloride (CCl_4), paracetamol, certain antibiotics, organophosphates, mycotoxins, and heavy metals, have been identified as capable of triggering the production of free radicals within the liver. This oxidative stress can result in liver damage and impairment. Bioactive compounds play an important role in ROS and RNS metabolism and in the inhibition of uncontrolled oxidation of essential biomolecules (Ganesan, Jayachandran, and Xu, 2018). Extracts obtained from the peels, pulps, and seeds of red oranges contain high amounts of bioactive compounds like C3G, which demonstrate good antioxidant activities and are efficient in reducing lipid accumulation in FFA-exposed HepG2 cells; hence, they can be useful to prevent or counteract hepatic steatosis (Burò et al., 2022). In one study, it was observed that bilberry and blackberry pomace extracts contained high amounts of polyphenols and were effective in scavenging OH^- and O_2^- radicals. Hussain et al., (2017) studied the protective effects of cranberry extract against liver damage induced by carbon tetrachloride (CCl_4) in Wistar albino rats. The findings revealed that oral administration of cranberry extract demonstrated significant potential in protecting the liver from damage caused by chemicals such as CCl_4. Additionally, the extract exhibited strong antioxidant properties, effectively defending against oxidative stress induced by free radicals.

11.4.6 Anti-Aging Activity

Nowadays, agro-industrial waste is being exploited for active ingredients to obtain skin care products since it is an excellent source of bioactive compounds, polyphenols, vitamins, and antioxidants (Pinto et al., 2021). It is for the same reason that citrus fruit wastes find their extensive utilization in the cosmetics industry. The natural antioxidants present in citrus peel reduce oxidative damage and aid in delaying skin aging and skin-related issues like wrinkles, acne, dark spots etc. (Sebghatollahi et al., 2022). Similarly, orange peel waste has good potential application in the cosmetics industry owing to the anti-tyrosinase activity of the orange peel extract. The orange peel extracts were used in a study with the aim of development of whitening cream that aids in lowering melanin synthesis. The formulated cream reduced melanin pigment by 17.33% (Wuttisin et al., 2017). In another study, Kim et al., (2019) reported that when mice were orally administered grape peel extract at a dose of 2 mg/kg body weight, it significantly reduced skin thickening induced by UV-B radiation by 63%. Additionally, the extract stimulated the Nrf2/HO-1 signaling pathway, which played a role in reducing the damage caused by UVB radiation. Chestnut shells are another by-product of chestnut peeling processes that can be utilized for their bioactive compounds by the cosmetics industry. Extracts from chestnut shells contain phenolic compounds that regulate the expression of aquaporin-3 (AQP-3) genes. AQP-3 is a protein channel crucial for transporting glycerol and water across membranes, vital for skin hydration. This hydration is key in minimizing the signs of aging. UV-B radiation diminishes AQP-3 gene expression, but studies indicate that phenolic compounds can counteract this effect by modulating the synthesis of this protein (Squillaci et al., 2018).

11.4.7 Neuroprotective Activity

Neurodegeneration is a multifaceted process with various causative factors, such as genetic predisposition, environmental factors, aging, etc., primarily targeting neurons in the human brain, resulting in the gradual decline of neural tissues and neuronal death. It is intricately linked to aging and characterized by disruptions in protein balance, leading to the buildup of insoluble deposits and inclusion bodies across different regions of the brain (Newton et al., 2019; Kurtishi et al., 2019). Spent coffee grounds (SCGs) are rich in bioactive compounds such as dietary fiber, polysaccharides, minerals, phenolics, volatile compounds, amino acids, proteins, and alkaloids (Kovalcik et al., 2018). Methanol extract showed significant effectiveness in mitigating the oxidative stress induced by H_2O_2 in SH-SY5Y cells. This effect was attributed to the ability of the extract to enhance the levels of endogenous antioxidant enzymes such as thioredoxin reductase, heme oxygenase 1, and glutathione reductase (Angeloni et al., 2021). Olive leaves are rich source of two bioactive compounds:

oleuropein and hydroxytyrosol. Hydroxytyrosol is a strong human monoamine oxidase (hMAO) inhibitor. The MAO enzymes are target spots for the treatment of neurodegeneration and even depression. They are involved in the catalytic deamination of endogenous as well as dietary amines and the regulation of dopamine metabolism. Therefore, bioactive compounds like hydroxytyrosol that inhibit MAOs have the potential to be used for the treatment of various neurological diseases like Alzheimer's and Parkinson's (Martínez-Zamora et al., 2021; Bertelli et al., 2020). Similarly, another bioactive compound, kaempferol, present in abundance in grape vine (*Vitis vinifera*) is also a selective inhibitor of the MAO-A isoform, and therefore has potential use in neurological conditions (Gidaro et al., 2016). Alawadi et al., (2023) studied the effect of 10% calorie substitution of apple pomace in a Western diet fed to Sprague-Dawley rats and found successful results in the prevention of neurodegeneration and nurturing brain functioning.

11.4.8 Cardioprotective Activity

Bioactive compounds can enhance cardiovascular health through various functions such as inhibiting platelet aggregation, decreasing vascular inflammation, controlling apoptotic actions, restricting low-density lipoprotein (LDL) oxidation, and improving lipid profiles (Bhuyan et al., 2017). Quercetin compound has been extracted from apple, plum pomace, and from peel wastes of banana, kiwi, tomato, and guava fruits (Barreira et al., 2019; Sójka et al., 2015; Vu et al., 2018; Wang et al., 2008). Skin wastes of onion and garlic have also been reported to be excellent sources of quercetin (Lee et al., 2014; Marrelli et al., 2019). The antioxidative and antiplatelet properties of quercetin promote actions such as inhibition of smooth muscle cell proliferation and migration, which results in enhanced cardiac cell mitochondrial function (Ay et al., 2016). The cardioprotective function of quercetin was demonstrated in rats by Sengul et al., (2021). Quercetin dosage (50 mg/kg and 100 mg/kg) had significant protective effects on 5-FU-induced cardiotoxicity in rats after 14 days of administration. Rodriguez-Rodriguez et al., (2012) investigated the cardioprotective properties of enzymatic extract of grape pomace in an ex-vivo model utilizing male Wistar rats. They treated the rat arteries with varying concentrations of phenolic compound extract ranging from 0.0001 to 0.03 g/L. The results demonstrated a dose-dependent relaxation effect on the aortic rings, implicating the activation of endothelial nitric oxide synthase (eNOS) via a nitric oxide (NO)-dependent mechanism. Furthermore, when the arteries were exposed to endothelin-1 (ET-1), which typically induces contraction and superoxide anion (O_2^-) production, treatment with the phenolic compound extract inhibited the contraction of the aortic rings, likely due to the antioxidant activity conferred by the phenolic compounds within the extract.

11.5 APPLICATIONS OF FOOD WASTE-DERIVED BIOACTIVES

Foods made from materials or products that would otherwise be regarded as waste or by-products are often referred to as "upcycled" foods. Bioactive compounds, sustainable packaging, single-cell protein, animal feed, bioethanol, biodiesel, biohythane, and biohydrogen are examples of upcycling food waste (Bangar et al., 2024). Extracts (rich in bioactive compounds) obtained from food waste material can be used as natural antioxidants (to delay lipid and protein oxidation mainly in meat products), preservatives (to delay microbiological spoilage), colorants (berries for anthocyanins, citrus peel for carotenoids etc.), thickeners (citrus peel rich in pectin), and as anti-browning agents to inhibit polyphenol oxidase and peroxidase activity (Vilas-Boas et al., 2021). Negi et al., (2021) used apple pomace powder food waste for the preparation of a fiber-enriched muffin premix that can be stored for up to 6 months. Kultys and Moczkowska-Wyrwisz (2022) successfully prepared pasta by incorporating beetroot-apple pomace and carrot pomace food waste, which enhanced the dietary fiber content and improved the technological properties such as swelling index and water absorption of the prepared pasta. Mala et al., (2024) noted moderate to high satisfaction in sensory aspects by incorporating 10% of dried pineapple peel (PP) powder in crackers, which resulted in an increase in total phenol content, total flavonoid content, and antioxidant properties of crackers compared to the control (without adding PP powder in crackers). Khan et al., (2022) obtained the maximum yield of single-cell protein (SCP) at 30°C for 7 days from potato peels and the minimum yield from citrus and banana peels. Further, they successfully prepared protein-enriched bread by incorporating SCP (acceptable at 4% concentration) into wheat flour without affecting the organoleptic properties of the prepared bread.

Karakuş et al., (2023) developed packaging film from processing waste of orange and pomegranate peels that showed high thermal degradation between 160 and 300°C, antimicrobial activity, and antioxidant activity. Barzan et al., (2024) prepared active film packaging by incorporating extracts of food waste (grape marc, olive pomace, and moringa leaf) as coatings or in between layers in packaging. They noted the high radical scavenging property of the prepared food packaging film as it prevented lipid peroxidation (more than 15%) in ground beef over 16 days. Zhang et al., (2024) developed active food packaging with outstanding characteristics (such as UV light blocking, high radical scavenging, improved water and gas barriers, and antibacterial capabilities against *L. monocytogenes* and *E. coli.*) from carboxymethyl cellulose film amalgamated with extract (polyphenol-rich) and carbon dots from coffee husk waste. The developed active composite film slowed down the decay of fresh-cut apples and, therefore, extended the shelf-life to 7 days at 4°C.

Whey, a by-product of the cheese industry, has high biological value proteins that enhance the value and nutritional content of other food products when added (Ahmad et al., 2019). Bayomy et al., (2024) used resistant starch obtained from leftover rice, chickpea flour, and whey (a by-product obtained from mozzarella cheese) for the preparation of extruded snacks and observed an increase in the content of protein, fiber, resistant starch, essential amino acids, minerals, and antioxidant activity in all formulated samples. Elkot et al., (2024) developed a fermented sports beverage by utilizing ricotta cheese whey fortified with spirulina powder and fermented by probiotic (*Lactobacillus acidophilus* + *Bifidobacterium* spp. + *Streptococcus thermophilus*) culture. Zhao et al., (2021) added tofu whey powder (TWP) to wheat flour pan bread and observed an increase in antioxidant activities, phenolic content, flavonoid content, and specific volume of bread samples. Overall, TWP improved the physicochemical and functional properties of bread. Dong et al., (2019) observed an enhancement in functional properties (solubility, emulsification, foaming, oil holding capacity) of chicken bone (by-product of poultry processing) protein pretreated by ultrasonication and prepared by the enzymatic method. Gao et al., (2022) prepared low-fat beef patties (as meat patties contain high levels of fat) by replacing beef fat with ultrahigh pressure-assisted cowhide (waste of the beef industry) gelatin. Along with a decrease in fat content (saturated fatty acid decreased, and polyunsaturated fatty acid increased) and an increase in protein content, cowhide gelatin successfully inhibited lipid oxidation of patties for 28 days (4°C) in cold storage.

11.6 CONCLUSIONS AND PERSPECTIVES

Various bioactive compounds are present in food waste, and it is of utmost importance to upcycle the food waste into value-added items to decrease the load of environmental pollution and save resources. Scientists have discovered new and innovative extraction techniques that can increase the extractability of bioactive compounds from food waste. Industries such as fruit and vegetable, coffee and tea, brewing, dairy, meat, marine, and seed and nut oil generates lot of food waste per day. Scientifically, it has been proven that this waste is a storehouse of bioactive compounds that have health-beneficial properties, so it is important to utilize these compounds. However, there are many challenges that need to be addressed before using food waste bioactive compounds in functional food or nutraceuticals, mainly with regard to safety issues and toxicity. Apart from that, it is important to change the perspective of consumers towards food waste and create awareness regarding the relevance of food waste and its utilization.

The shift to a circular economy process, which is "restorative and regenerative by design," is essential for effective food waste management. Repurposing waste streams to create new, consumer-friendly products is the goal of the circular economy. Various challenges need to be addressed to achieve a circular economy. It is of utmost importance to ensure streamlined sourcing, quality, and microbial safety of food waste. The technological capabilities for food waste sorting, secure storage, and distribution for additional processing are not well-established. Substantial resource investment is required to upgrade these critical areas for efficient food waste processing. Studies related to consumer acceptance are important to ensure the economic success of upcycled foods. Visual appeal, cost savings, ecological advantages, and personal fulfillment are the four types of benefits that consumers seek from upcycled foods. The main hindrance to the acceptance of upcycled food by consumers is neophobia (fear of trying new food products) and disgust (emotional reaction, perception of uncleanliness, contamination). Carbon trust or carbon footprint labels can increase the acceptability of upcycled food products, especially among eco-conscious consumers. In order to allay consumer fears and promote consumer participation, it is essential to communicate effectively and positively about the advantages of upcycled items by creating awareness and educating consumers.

REFERENCES

Abbott, A. P., Capper, G., Davies, D. L., Rasheed, R. K., & Tambyrajah, V. (2003). Novel solvent properties of choline chloride/urea mixtures. *Chemical Communications*, 1, 70–71.

Agcam, E., Akyıldız, A., & Balasubramaniam, V. M. (2017). Optimization of anthocyanins extraction from black carrot pomace with thermosonication. *Food Chemistry*, 237, 461–470. https://doi.org/10.1016/j.foodchem.2017.05.098

Ahmad, T., Aadil, R. M., Ahmed, H., Rahman, U., Soares, B., Souza, S., Pimentel, T. C., Scudino, H., Guimarães, J. H., Esmerino, E. A., Freitas, M. Q., Almada, R. B., Vendramel, S., Silva, M. C., & Cruz, A. G. (2019). Treatment and utilization of dairy industrial waste: A review. *Trends in Food Science & Technology,* 88, 361–372. https://doi.org/10.1016/j.tifs.2019.04.003

Ahmad-Qasem, M. H., Ahmad-Qasem, B. H., Barrajon-Catalan, E., Micol, V., Carcel, J. A., & Garcia-Perez, J. V. (2016). Drying and storage of olive leaf extracts. Influence on polyphenols stability. *Industrial Crops and Products*, 79, 232–239.

Ajila, C. M., Aalami, M., Leelavathi, K., & Prasada Rao, U. J. S. (2010). Mango peel powder: A potential source of antioxidant and dietary fiber in macaroni preparations. *Innovative Food Science & Emerging Technologies*, 11, 219–224.

Alara, O. R., Abdurahman, N. H., Abdul Mudalip, S. K., & Olalere, O. A. (2018). Microwave-assisted extraction of Vernonia amygdalina leaf for optimal recovery of total phenolic content. *Journal of Applied Research on Medicinal and Aromatic Plants*, 10, 16–24. https://doi.org/ 10.1016/j.jarmap.2018.04.004

Alawadi, A. A., Benedito, V. A., Skinner, R. C., Warren, D. C., Showman, C., & Tou, J. C. (2023). RNA-sequencing revealed apple pomace ameliorates expression of genes in the hypothalamus associated with neurodegeneration in female rats fed a Western diet during adolescence to adulthood. *Nutritional Neuroscience*, 26(4), 332–344.

Alexandru, L., Binello, A., Mantegna, S., Boffa, L., Chemat, F., & Cravotto, G. (2014). Efficient green extraction of polyphenols from post-harvested agro-industry vegetal sources in Piedmont. *Comptes Rendus Chimie*, 17(3), 212–217.

Alishahi, A., Mirvaghefi, A., Tehrani, M. R., Farahmand, H., Shojaosadati, S. A., Dorkoosh, F. A., & Elsabee, M. Z. (2011). Enhancement and characterization of chitosan extraction from the wastes of shrimp packaging plants. *Journal of Polymers and the Environment*, 19, 776–783.

Angeloni, S., Freschi, M., Marrazzo, P., Hrelia, S., Beghelli, D., Juan-García, A., & Angeloni, C. (2021). Antioxidant and anti-inflammatory profiles of spent coffee ground extracts for the treatment of neurodegeneration. *Oxidative Medicine and Cellular Longevity*, 1–19.

Antónia Nunes, M., Costa, A. S. G., Bessada, S., Santos, J., Puga, H., Alves, R. C., Freitas, V., & Oliveira, M. B. P. P. (2018). Olive pomace as a valuable source of bioactive compounds: A study regarding its lipid- and water-soluble components. Science of the Total Environment, 644, 229–236. https://doi.org/10.1016/j.scitotenv.2018.06.350

Arrutia, F., Adam, M., Calvo-Carrascal, M. Á., Mao, Y., & Binner, E. (2020). Development of a continuous-flow system for microwave-assisted extraction of pectin-derived oligosaccharides from food waste. Chemical Engineering Journal, 395, 125056.

Arya, S. S., Venkatram, R., More, P. R., & Vijayan, P. (2022). The wastes of coffee bean processing for utilization in food: A review. *Journal of Food Science Technology*. 59(2), 429–444. https://doi.org/10.1007/s13197-021-05032-5

Ay, M., Charli, A., Jin, H., Anantharam, V., Kanthasamy, A., & Kanthasamy, A. G. (2016). Quercetin. In R. C. Gupta (Ed.), *Book Nutraceuticals*, 1st ed., Academic Press; Elsevier: New York, NY, USA; Volume 32, pp. 447–452.

Ayala-Zavala, J. F. N., Vega-Vega, V., Rosas-Domínguez, C., Palafox-Carlos, H., Villa-Rodriguez, J. A., Siddiqui, M. W., & González-Aguilar, G. A. (2011). Agro-industrial potential of exotic fruit byproducts as a source of food additives. *Food Research International*, 44(7), 1866–1874.

Babazadeh, A., Taghvimi, A., Hamishehkar, H., & Tabibiazar, M. (2017). Development of new ultrasonic–solvent assisted method for determination of trans-resveratrol from red grapes: Optimization, characterization, and antioxidant activity (ORAC assay). *Food Bioscience*, 20, 36–42. https://doi.org/10.1016/j.fbio.2017.08.003

Baiano, A. (2014). Recovery of biomolecules from food wastes—A review. *Molecules*, 19(9), 14821–14842.

Bandar, H., Hijazi, A., Rammal, H., Hachem, A., Saad, Z., & Badran, B. (2013). Techniques for the extraction of bioactive compounds from Lebanese Urtica Dioica. *American Journal of Phytomedicine and Clinical Therapeutics*, 6, 507–513.

Bangar, S. P., Chaudhary, V., Kajla, P., Balakrishnan, G., & Phimolsiripol, Y. (2024). Strategies for upcycling food waste in the food production and supply chain. *Trends in Food Science & Technology*, 143, 104314. https://doi.org/10.1016/j.tifs.2023.104314

Barbosa, J. R., dos Santos Freitas, M. M., da Silva Martins, L. H., & de Carvalho Junior, R. N. (2020). Polysaccharides of mushroom Pleurotus spp.: New extraction techniques, biological activities and development of new technologies. *Carbohydrate Polymers*, 229, 115550.

Barrales, F. M., Silveira, P., Barbosa, P. P. M., Ruviaro, A. R., Paulino, B. N., Pastore, G. M., Alves Macedo, G., & Martínez, J. (2018). Recovery of phenolic compounds from citrus by-products using pressurized liquids—an application to orange peel. *Food and Bioproducts Processing*, 112, 9–21. https://doi.org/10.1016/j.fbp.2018.08.006

Barreira, J. C. M., Arraibi, A. A., & Ferreira, I. C. F. R. (2019). Bioactive and functional compounds in apple pomace from juice and cider manufacturing: Potential use in dermal formulations. *Trends in Food Science and Technology*, 90, 76–87.

Barukčić, I., Lisak Jakopović, K., & Božanić, R. (2019). Valorisation of whey and buttermilk for production of functional beverages - An overview of current possibilities. *Food Science and Biotechnology*, 57(4), 448–460. https://doi.org/10.17113/ftb.57.04.19.6460

Barzan, G., Sacco, A., Giovannozzi, A. M., Portesi, C., Schiavone, C., Salafranca, J., Wrona, M., Nerín, C., & Rossi, A. M. (2024). Development of innovative antioxidant food packaging systems based on natural extracts from food industry waste and Moringa oleifera leaves. *Food Chemistry*, 432, 137088.

Bayomy, H. M., Alamri, E. S., Albalawi, A. N., Alharbi, R., Ozaybi, N. A., Rozan, M. A., & Shamsia, S. M. (2024). Production of extruded functional snacks based on resistant starch using waste rice and whey milk. *LWT*, 197, 115871. https://doi.org/10.1016/j.lwt.2024.115871

Bei, Q., Liu, Y., Wang, L., Chen, G., & Wu, Z. (2017). Improving free, conjugated, and bound phenolic fractions in fermented oats (Avena sativa L.) with Monascus anka and their antioxidant activity. *Journal of Functional Foods*, 32, 185–194.

Belwal, T., Chemat, F., Venskutonis, P. R., Cravotto, G., Jaiswal, D. K., Bhatt, I. D., et al., (2020). Recent advances in scaling-up of non-conventional extraction techniques: Learning from successes and failures. *TrAC, Trends in Analytical Chemistry*, 127, Article 115895. https://doi.org/10.1016/j.trac.2020.115895

Ben-Othman, S., Jõudu, I., & Bhat, R. (2020). Bioactives from agri-food wastes: Present insights and future challenges. *Molecules*, 25(3), 510. https://doi.org/10.3390/molecules25030510

Bertelli, M., Kiani, A. K., Paolacci, S., Manara, E., Kurti, D., Dhuli, K., & Michelini, S. (2020). Hydroxytyrosol: A natural compound with promising pharmacological activities. *Journal of Biotechnology*, 309, 29–33.

Bhuyan, D. J., & Basu, A. (2017). Phenolic compounds: potential health benefits and toxicity. In Q. V. Vuong (Ed.), Utilisation of Bioactive Compounds from Agricultural and Food Waste (pp. 27-59). CRC Press.

Bitalebi, S., Nikoo, M., Rahmanifarah, K., Noori, F., & Ahmadi Gavlighi, H. (2019). Effect of apple peel extract as natural antioxidant on lipid and protein oxidation of rainbow trout (Oncorhynchus mykiss) mince. *International Aquatic Research*, 11, 135–146.

Bonacci, S., Di Gioia, M. L., Costanzo, P., Maiuolo, L., Tallarico, S., & Nardi, M. (2020). Natural deep eutectic solvent as extraction media for the main phenolic compounds from olive oil processing wastes. *Antioxidants (Basel)*, 9(6), 513. https://doi.org/10.3390/antiox9060513

Bosiljkov, T., Dujmić, F., Bubalo, M. C., Hribar, J., Vidrih, R., Brnčić, M., … Jokić, S. (2017). Natural deep eutectic solvents and ultrasound-assisted extraction: Green approaches for extraction of wine lees anthocyanins. *Food and Bioproducts Processing*, 102, 195–203. https://doi.org/10.1016/j.fbp.2016.12.005

Boussetta, N., Lanoisellé, J. L., Bedel-Cloutour, C., & Vorobiev, E. (2009a). Extraction of soluble matter from grape pomace by high voltage electrical discharges for polyphenol recovery: Effect of sulphur dioxide and thermal treatments. *Journal of Food Engineering*, 95, 192–198. https://doi.org/10.1016/j.jfoodeng.2009.04.030

Boussetta, N., Lebovka, N., Vorobiev, E., Adenier, H., Bedel-Cloutour, C., & Lanoisellé, J. L. (2009b). Electrically assisted extraction of soluble matter from chardonnay grape skins for polyphenol recovery. *Journal of Agricultural and Food Chemistry*, 57(4), 1491–1497. https://doi.org/10.1021/jf802579x

Boussetta, N., Lesaint, O., & Vorobiev, E. (2013). A study of mechanisms involved during the extraction of polyphenols from grape seeds by pulsed electrical discharges. *Innovative Food Science and Emerging Technologies*, 19, 124–132. https://doi.org/10.1016/j.ifset.2013.03.007

Boussetta, N., Soichi, E., Lanoisellé, J. L., & Vorobiev, E. (2014). Valorization of oilseed residues: Extraction of polyphenols from flaxseed hulls by pulsed electric fields. *Industrial Crops and Products*, 52, 347–353. https://doi.org/10.1016/j.indcrop.2013.10.048

Brianceau, S., Turk, M., Vitrac, X., & Vorobiev, E. (2016). High voltage electric discharges assisted extraction of phenolic compounds from grape stems: Effect of processing parameters on flavan-3-ols, flavonols and stilbenes recovery. *Innovative Food Science and Emerging Technologies*, 35, 67–74. https://doi.org/10.1016/j.ifset.2016.04.006

Burò, I., Consoli, V., Castellano, A., Vanella, L., & Sorrenti, V. (2022). Beneficial effects of standardized extracts from wastes of red oranges and olive leaves. *Antioxidants*, 11(8), 1496.

Canan, İ., Gündoğdu, M., Seday, U., Oluk, C. A., Karaşahin, Z., Eroğlu, E. Ç., & Ünlü, M. (2016). Determination of antioxidant, total phenolic, total carotenoid, lycopene, ascorbic acid, and sugar contents of Citrus species and mandarin hybrids. *Turkish Journal of Agriculture and Forestry*, 40(6), 894–899.

Chalamaiah, M., Dinesh Kumar, B., Hemalatha, R., & Jyothirmayi, T. (2012). Fish protein hydrolysates: proximate composition, amino acid composition, antioxidant activities and applications: A review. *Food Chemistry*, 2012;135(4):3020–3038. https://doi.org/10.1016/j.foodchem.2012.06.100

Chel-Guerrero, L. D., Oney-Montalvo, J. E., & Rodríguez-Buenfil, I. M. (2021). Phytochemical characterization of by-products of habanero pepper grown in two different types of soils from Yucat´an. Mexico. *Plants (Basel)*, 10(4), 779. https://doi.org/10.3390/plants10040779

Chemat, F., Huma, Z., & Khan, M. K. (2010). Applications of ultrasound in food technology: Processing, preservation and extraction. *Ultrason Sonochem*, 18, 813–835. https://doi.org/10.1016/j.ultsonch.2010.11.023

Chen, C. Y., Kao, C. L., & Liu, C. M. (2018). The cancer prevention, anti-inflammatory and anti-oxidation of bioactive phytochemicals targeting the TLR4 signaling pathway. *International Journal of Molecular Sciences*, 19(9), 2729.

Corrales, M., Toepfl, S., Butz, P., Knorr, D., & Tauscher, B. (2008). Extraction of anthocyanins from grape by-products assisted by ultrasonics, high hydrostatic pressure or pulsed electric fields: A comparison. *Innovative Food Science and Emerging Technologies*, 9(1), 85–91. https://doi.org/10.1016/j.ifset.2007.06.002

Cruz, P. N., Pereira, T. C., Guindani, C., Oliveira, D. A., Rossi, M. J., & Ferreira, S. R. (2017). Antioxidant and antibacterial potential of butia (Butia catarinensis) seed extracts obtained by supercritical fuid extraction. *The Journal of Supercritical Fluids*, 119, 229–237.

Cuco, R. P., Cardozo-Filho, L., & da Silva, C. (2019). Simultaneous extraction of seed oil and active compounds from peel of pumpkin (Cucurbita maxima) using pressurized carbon dioxide as solvent. *The Journal of Supercritical Fluids*, 143, 8–15.

De Jesus, C. S. A., Elba Ruth, V. G., Daniel, S. F. R., & Sharma, A. (2015). Biotechnological alternatives for the utilization of dairy industry waste products. *Advances in Bioscience and Biotechnology*, 6(3), 223–235.

Diekmann, L., & Germelmann, C. C. (2021). Leftover consumption as a means of food waste reduction in public space? Qualitative insights from online discussions. *Sustainability,* 13, 13564. https://doi.org/10.3390/su132413564

Ding, L., Zhao, Q., Zhou, X., Tang, C., Chen, Y., & Cai, Z. (2020). Changes in protein structure and physicochemical properties of egg white by super critical carbon dioxide treatment. *Journal of Food Engineering*, 284, 110076.

Dong, Z. Y., Li, M. Y., Tian, G., Zhang, T. Hua., Ren, H., & Quek, S. Y. (2019). Effects of ultrasonic pretreatment on the structure and functionality of chicken bone protein prepared by enzymatic method. *Food Chemistry*, 299, 125103. https://doi.org/10.1016/j.foodchem.2019.125103

Duan, F., Zhao, R., Yang, J., Xiao, M., Lu, L., (2021). Integrated utilization of dairy whey in probiotic β-galactosidase production and enzymatic synthesis of galacto-oligosaccharides. *Catalysts,* 11(6). https://doi.org/10.3390/catal11060658

Duba, K. S., Casazza, A. A., Mohamed, H. B., Perego, P., & Fiori, L. (2015). Extraction of polyphenols from grape skins and defatted grape seeds using subcritical water: Experiments and modeling. *Food and Bioproducts Processing*, 94, 29–38.

Dubie, J., Stancik, A., Morra, M., & Nindo, C. (2013). Antioxidant extraction from mustard (Brassica juncea) seed meal using high-intensity ultrasound. *Journal of Food Science*, 78(4), E542–E548.

El Kantar, S., Boussetta, N., Lebovka, N., Foucart, F., Rajha, H. N., Maroun, R. G., Louka, N., & Vorobiev, E. (2018). Pulsed electric feld treatment of citrus fruits: Improvement of juice and polyphenols extraction. *Innovative Food Science and Emerging Technologies*, 46, 153–161.

Elkot, W. F., Elmahdy, A., El-Sawah, T. H., Alghamdia, O. A., Alhag, S. K., Al-Shahari, E. A., Al-Farga, A., & Ismail, H. A. (2024). Development and characterization of a novel flavored functional fermented whey-based sports beverage fortified with Spirulina platensis. *International Journal of Biological Macromolecules*, 258, 128999. https://doi.org/10.1016/j.ijbiomac.2023.128999

Essien, S. O., Young, B., & Baroutian, S. (2020). Recent advances in subcritical water and supercritical carbon dioxide extraction of bioactive compounds from plant materials. *Trends in Food Science and Technology*, 97, 156–169.

European Environment Agency. (2020). What are the sources of food waste in Europe? https://www.eea.europa.eu/media/infographics/wasting-food-1/view

FAO. (2013). *Food Wastage Footprint: Impacts on Natural Resources*. FAO: Rome, Italy. https://www.fao.org/3/i3347e/i3347e.pdf

FAO. (2019). *The State of Food and Agriculture 2019. Moving forward on food loss and waste reduction.* Rome. https://doi.org/10.4060/CA6030EN

FAO. (2022). Tackling food loss and waste: A triple win opportunity. https://www.fao.org/newsroom/detail/FAO-UNEP-agriculture-environment-food-loss-waste-day-2022/en

FAO. (2023). Technical Platform on the Measurement and Reduction of Food Loss and Waste. https://www.fao.org/platform-food-loss-waste/flw-events/international-day-food-loss-and-waste/en

Farhadi, K., Esmaeilzadeh, F., Hatami, M., Forough, M., & Molaie, R. (2016). Determination of phenolic compounds content and antioxidant activity in skin, pulp, seed, cane and leaf of five native grape cultivars in West Azerbaijan province, Iran. *Food Chemistry*, 199, 847–855.

Fernandes, P. A., Bastos, R., Calvão, J., Neto, F., Coelho, E., Wessel, D. F., & Passos, C. P. (2021). Microwave hydrodiffusion and gravity as a sustainable alternative approach for an efficient apple pomace drying. *Bioresource Technology*, 333, 125207.

Fernandez, K., Vega, M., & Aspé, E. (2015). An enzymatic extraction of proanthocyanidins from país grape seeds and skins. *Food Chemistry*, 168(2), 7–13.

Ferri, M., Bin, S., Vallini, V., Fava, F., Michelini, E., Roda, A., … Tassoni, A. (2016). Recovery of polyphenols from red grape pomace and assessment of their antioxidant and anti-cholesterol activities. *New Biotechnology*, 33(3), 338–344. https://doi.org/ 10.1016/j.nbt.2015.12.004

Frond, A. D., Iuhas, C. I., Stirbu, I., Leopold, L., Socaci, S., Andreea, S., & Carmen, S. (2019). Phytochemical characterization of five edible purple-reddish vegetables: Anthocyanins, flavonoids, and phenolic acid derivatives. *Molecules*, 24(8), 1536.

Gallego, R., Bueno, M., & Herrero, M. (2019). Sub-and supercritical fuid extraction of bioactive compounds from plants, food-by-products, seaweeds and microalgae—an update. *Trends in Analytical Chemistry*, 116, 198–213.

Gan, C.-Y., & Latif, A. A. (2011). Optimisation of the solvent extraction of bioactive compounds from Parkia speciosa pod using response surface methodology. *Food Chemistry*, 124, 1277–1283.

Ganesan, K., Jayachandran, M., & Xu, B. (2018). A critical review on hepatoprotective effects of bioactive food components. *Critical Reviews in Food Science and Nutrition*, 58(7), 1165–1229.

Gao, Y., Qiu, Y., Nan, H., Wang, L., Yang, D., Zhang, L., & Yu, Q. (2022). Ultra-high pressure-assisted preparation of cowhide gelatin as a promising fat substitute: Improve the nutrition ratio and antioxidant capacity of beef patties. *Food Research International,* 157, 111260. https://doi.org/10.1016/j.foodres.2022.111260

Gidaro, M. C., Astorino, C., Petzer, A., Carradori, S., Alcaro, F., Costa, G., & Alcaro, S. (2016). Kaempferol as selective human MAO-A inhibitor: Analytical detection in calabrian red wines, biological and molecular modeling studies. *Journal of Agricultural and Food Chemistry*, 64(6), 1394–1400.

Gil-Martín, E., Forbes-Hernández, T., Romero, A., Cianciosi, D., Giampieri, F., & Battino, M. (2022). Influence of the extraction method on the recovery of bioactive phenolic compounds from food industry by-products. *Food Chemistry*, 378, 131918.

Goderska, K., Szwengiel, A., & Czarnecki, Z. (2014). The utilization of Pseudomonas taetrolens to produce lactobionic acid. *Applied Biochemistry and Biotechnology*, 173, 2189–2197.

Grassino, A. N., Ostojić, J., Miletić, V., Djaković, S., Bosiljkov, T., Zorić, Z., Ježek, D., Brnčić, S. R., & Brnčić, M. (2020). Application of high hydrostatic pressure and ultrasound-assisted extractions as a novel approach for pectin and polyphenols recovery from tomato peel waste. *Innovative Food Science and Emerging Technologies,* 64, 102424.

Grigoras, C. G., Destandau, E., Fougère, L., & Elfakir, C. (2013). Evaluation of apple pomace extracts as a source of bioactive compounds. *Industrial Crops and Products*, 49, 794–804. https://doi.org/10.1016/j.indcrop.2013.06.026

Grillo, G., Boffa, L., Binello, A., Mantegna, S., Cravotto, G., Chemat, F., et al., (2019). Cocoa bean shell waste valorisation; Extraction from lab to pilot-scale cavitational reactors. *Food Research International*, 115, 200–208. https://doi.org/10.1016/j.foodres.2018.08.057

Grzelak-Błaszczyk, K., Karlińska, E., Grzęda, K., Rój, E., & Kołodziejczyk, K. (2017). Defatted strawberry seeds as a source of phenolics, dietary fber and minerals. *LWT,* 84, 18–22.

Gullón, B., Gagaoua, M., Barba, F. J., Gullón, P., Zhang, W., & Lorenzo, J. M. (2020). Seaweeds as promising resource of bioactive compounds: Overview of novel extraction strategies and design of tailored meat products. *Trends in Food Science and Technology*, 100, 1–18.

Gustinelli, G., Eliasson, L., Svelander, C., Alminger, M., & Ahrné, L. (2018). Supercritical CO2 extraction of bilberry (Vaccinium myrtillus L.) seed oil: Fatty acid composition and antioxidant activity. *The Journal of Supercritical Fluids*, 135, 91–97.

Hamid, H. S. H. B., & Ismail, K. S. K. (2020). Optimization of enzymatic hydrolysis for acid pretreated date seeds into fermentable sugars. *Biocatalysis and Agricultural Biotechnology*, 24, 101530.

Hayes, J. E., Stepanyan, V., Allen, P., O'grady, M. N., & Kerry, J. P. (2010). Effect of lutein, sesamol, ellagic acid and olive leaf extract on the quality and shelf-life stability of packaged raw minced beef patties. *Meat Science*, 84(4), 613–620.

Hikal, W. M., Said-Al Ahl, H. A. H., Tkachenko, K. G., Bratovcic, A., Szczepanek, M., Rodriguez, R. M. (2021). Sustainable and environmentally friendly essential oils extracted from pineapple waste. *Biointerface Research in Applied Chemistry*, 12(5), 6833–6844. https://doi.org/10.33263/BRIAC125.68336844

Hobbi, P., Okoro, O. V., Hajiabbas, M., Hamidi, M., Nie, L., Megalizzi, V., & Shavandi, A. (2023). Chemical composition, antioxidant activity and cytocompatibility of polyphenolic compounds extracted from food industry apple waste: potential in biomedical application. *Molecules*, 28(2), 675.

Hussain, F., Malik, A., Ayyaz, U., Shafique, H., Rana, Z., & Hussain, Z. (2017). Efficient hepatoprotective activity of cranberry extract against CCl4-induced hepatotoxicity in Wistar albino rat model: Down-regulation of liver enzymes and strong antioxidant activity. *Asian Pacific Journal of Tropical Medicine*, 10(11), 1054–1058.

Iora, S. R. F., Maciel, G. M., Zielinski, A. A. F., da Silva, M. V., Pontes, P. V. d. A., Haminiuk, C. W. I., & Granato, D. (2015). Evaluation of the bioactive compounds and the antioxidant capacity of grape pomace. *International Journal of Food Science and Technology*, 50(1), 62–69. https://doi.org/10.1111/ijfs.12583

Jain, R., and Naik, S. N. (2018). Adding value to the oil cake as a waste from oil processing industry: Production of lipase in solid state fermentation. *Biocatalysis and Agricultural Biotechnology,* 15, 181–184.

Jeswani, H. K., Figueroa-Torres, G., & Azapagic, A. (2021). The extent of food waste generation in the UK and its environmental impacts. *Sustainable Production and Consumption,* 26, 532–547. https://doi.org/10.1016/j.spc.2020.12.021

Kalogeropoulos, N., Chiou, A., Pyriochou, V., Peristeraki, A., & Karathanos, V. T. (2012). Bioactive phytochemicals in industrial tomatoes and their processing byproducts. *LWT-Food Science and Technology*, 49(2), 213–216.

Kanatt, S. R., Chander, R., & Sharma, A. (2010). Antioxidant and antimicrobial activity of pomegranate peel extract improves the shelf life of chicken products. *International Journal of Food Science & Technology*, 45(2), 216–222.

Karageorgou, I., Grigorakis, S., Lalas, S., & Makris, D. P. (2017). Enhanced extraction of antioxidant polyphenols from Moringa oleifera Lam. leaves using a biomolecule-based low-transition temperature mixture. *European Food Research and Technology*, 243, 1839–1848.

Karakuş, E., Ayhan, Z., & Haskaraca, G. (2023). Development and characterization of sustainable-active-edible-bio based films from orange and pomegranate peel waste for food packaging: Effects of particle size and acid/plasticizer concentrations. *Food Packaging and Shelf Life,* 37, 101092.

Kaur, J., Kumar, V., Sharma, K., Kaur, S., Gat, Y., Goyal, A., & Tanwar, B. (2020). Opioid peptides: An overview of functional significance. *International Journal of Peptide Research and Therapeutics,* 26(1), 33–41. https://doi.org/10.1007/s10989-019-09813-7

Khairil Anuar, M., Mohd Zin, Z., Juhari, N. H., Hasmadi, M., Smedley, K. L., & Zainol, M. K. (2020). Influence of pectinase–assisted extraction time on the antioxidant capacity of Spent Coffee Ground (SCG). *Food Research*, 4, 2054–2061.

Khalid, M. U., Shabbir, M. A., Mustafa, S., Hina, S., Quddoos, M. Y., Mahmood, S., & Rafique, A. (2021). Effect of apple peel as an antioxidant on the quality characteristics and oxidative stability of mayonnaise. *Applied Food Research*, 1(2), 100023.

Khan, Muhammad Kashif Iqbal, et al., (2022). Sustainable food industrial waste management through single cell protein production and characterization of protein enriched bread. *Food Bioscience,* 46, 101406.

Kim, J., Oh, J., Averilla, J. N., Kim, H. J., Kim, J. S., & Kim, J. S. (2019). Grape peel extract and resveratrol inhibit wrinkle formation in mice model through activation of Nrf2/HO-1 signaling pathway. *Journal of Food Science*, 84(6), 1600–1608.

Kovalcik, A., Obruca, S., & Marova, I. (2018). Valorization of spent coffee grounds: A review. *Food and Bioproducts Processing*, 110, 104–119.

Kultys, E., & Moczkowska-Wyrwisz, M. (2022). Effect of using carrot pomace and beetroot-apple pomace on physicochemical and sensory properties of pasta. *LWT,* 168, 113858. https://doi.org/10.1016/j.lwt.2022.113858

Kurtishi, A., Rosen, B., Patil, K. S., Alves, G. W., & Møller, S. G. (2019). Cellular proteostasis in neurodegeneration. *Molecular Neurobiology*, 56, 3676–3689.

Kusuma, H. S., Putra, A. F. P., & Mahfud, M. (2016). Comparison of two isolation methods for essential oils from orange peel (Citrus auranticum L.) as a growth promoter for fish: Microwave steam distillation and conventional steam distillation. *Journal of Aquaculture Research & Development*, 7, 409.

Lee, K. A., Kim, K. T., Kim, H. J., Chung, M. S., Chang, P. S., Park, H., & Pai, H. D. (2014). Antioxidant activities of onion (Allium cepa L.) peel extracts produced by ethanol, hot water and subcritical water extraction. *Food Science and Biotechnology*, 23, 615–621.

Lee, S. J., Kim, E. K., Hwang, J. W., Oh, H. J., Cheong, S. H., Moon, S. H., & Park, P. J. (2010). Purification and characterisation of an antioxidative peptide from enzymatic hydrolysates of duck processing by-products. *Food Chemistry*, 123(2), 216–220.

Leksawasdi, N., Taesuwan, S., Prommajak, T., Techapun, C., Khonchaisri, R., Sittilop, N., & Khemacheewakul, J. (2022). Ultrasonic extraction of bioactive compounds from green soybean pods and application in green soybean milk antioxidants fortification. *Foods*, 11(4), 588.

Li, Z., Fan, Y., & Xi, J. (2019). Recent advances in high voltage electric discharge extraction of bioactive ingredients from plant materials. *Food Chemistry*, 277, 246–260. https://doi.org/10.1016/j.foodchem.2018.10

Liew, S. Q., Chin, N. L., Yusof, Y. A., & Sowndhararajan, K. (2015). Comparison of acidic and enzymatic pectin extraction from passion fruit peels and its gel properties. *Journal of Food Process Engineering*, 39(5), 501–511.

Liew, S. Q., Teoh, W. H., Tan, C. K., Yusof, R., & Ngoh, G. C. (2018). Subcritical water extraction of low methoxyl pectin from pomelo (Citrus grandis (L.) Osbeck) peels. *International Journal of Biological Macromolecules*, 116, 128–135.

Long, J. M., & Mohan, A. (2021). Development of meat powder from beef byproduct as value-added food ingredient. *LWT-Food Science and Technology*, 146: 111460.

Macedo, G. A., Santana, Á. L., Crawford, L. M., Wang, S. C., Dias, F. F. G., & de Moura Bell, J. M. L. N. (2021). Integrated microwave- and enzymeAssisted extraction of phenolic compounds from olive pomace. *Lebensmittel-wissenschaft + i.e. Technologie,* 138, 110621. https://doi.org/10.1016/j.lwt.2020.110621

Machado, A. P. F., Pasquel-Reategui, J. L., Barbero, G. F., & Martínez, J. (2015). Pressurized liquid extraction of bioactive compounds from blackberry (Rubus fruticosus L.) residues: A comparison with conventional methods. *Food Research International*, 77, 675–683. https://doi.org/10.1016/j.foodres.2014.12.042

Mahnic-Kalamiza, S., Vorobiev, E., & Miklavcic, D. (2014). Electroporation in food processing and biorefinery. *The Journal of Membrane Biology*, 247, 12, 1279–304. https://doi.org/10.1007/s00232-014-9737-x
Mala, T., Piayura, S., & Itthivadhanapong, P. (2024). Characterization of dried pineapple (Ananas comosus L.) peel powder and its application as a novel functional food ingredient in cracker product. *Future Foods*, 9, 100322. https://doi.org/10.1016/j.fufo.2024.100322.
Mano, M. C. R., Paulino, B. N., & Pastore, G. M. (2019). Whey permeate as the raw material in galacto-oligosaccharide synthesis using commercial enzymes. *Food Research International*, 124, 78–85. https://doi.org/10.1016/j.foodres.2018.09.019
Marić, B., Pavlić, B., Čolović, D., Abramović, B., Zeković, Z., Bodroža-Solarov, M., Ilić, N., & Teslić, N. (2020). Recovery of high-content ω–3 fatty acid oil from raspberry (Rubus idaeus L.) seeds: Chemical composition and functional quality. *LWT*, 130, 109627.
Marrelli, M., Amodeo, V., Statti, G., & Conforti, F. (2019). Biological properties and bioactive components of L.: Focus on potential benefits in the treatment of obesity and related comorbidities. *Molecules*, 24, 119.
Martínez, L., Castillo, J., Ros, G., & Nieto, G. (2019). Antioxidant and antimicrobial activity of rosemary, pomegranate and olive extracts in fish patties. *Antioxidants*, 8(4), 86.
Martínez-Zamora, L., Peñalver, R., Ros, G., & Nieto, G. (2021). Olive tree derivatives and hydroxytyrosol: Their potential effects on human health and its use as functional ingredient in meat. *Foods*, 10(11), 2611.
McCann, M. J., Gill, C. I. R., O'brien, G., Rao, J. R., McRoberts, W. C., Hughes, P., & Rowland, I. R. (2007). Anti-cancer properties of phenolics from apple waste on colon carcinogenesis in vitro. *Food and Chemical Toxicology*, 45(7), 1224–1230.
Megawati, M. (2016). Ekstraksi Minyak Atsiri Kulit Jeruk Manis Dengan Metode Vacuum Microwave Asissted Hydrodistillation. *Jurnal Bahan Alam Terbarukan*, 4(2), 61–67.
Meini, M. R., Cabezudo, I., Galetto, C. S., & Romanini, D. (2021). Production of grape pomace extracts with enhanced antioxidant and prebiotic activities through solid-state fermentation by Aspergillus niger and Aspergillus oryzae. *Food Bioscience*, 42, 101168.
More, P. R., & Arya, S. S. (2021). Intensification of bio-actives extraction from pomegranate peel using pulsed ultrasound: Effect of factors, correlation, optimization and antioxidant bioactivities. *Ultrasonics Sonochemistry*, 72, Article 105423. https://doi.org/10.1016/j.ultsonch.2020.105423
Motikar, P. D., More, P. R., & Arya, S. S. (2020). A novel, green environment-friendly cloud point extraction of polyphenols from pomegranate peels: A comparative assessment with ultrasound and microwave-assisted extraction. *Separation Science and Technology*, 56(6), 1014–1025. https://doi.org/10.1080/01496395.2020.1746969
Mphahlele, R. R., Fawole, O. A., Makunga, N. P., & Opara, U. L. (2016). Effect of drying on the bioactive compounds, antioxidant, antibacterial and antityrosinase activities of pomegranate peel. *BMC Complementary and Alternative Medicine*, 16, 143. https://doi.org/10.1186/s12906-016-1132-y
Muíño, I., Díaz, M. T., Apeleo, E., Perez-Santaescolastica, C., Rivas-Cañedo, A., Pérez, C., Cañeque, V., Lauzurica, S., & de la Fuente, J. (2017). Valorisation of an extract from olive oil waste as a natural antioxidant for reducing meat waste resulting from oxidative processes. *Journal of Cleaner Production*, 2017, 140, 924–932.
Muíño, I., Díaz, M. T., Apeleo, E., Pérez-Santaescolástica, C., Rivas-Cañedo, A., Pérez, C., & de la Fuente, J. (2017). Valorisation of an extract from olive oil waste as a natural antioxidant for reducing meat waste resulting from oxidative processes. *Journal of Cleaner Production*, 140, 924–932.
Nadar, S. S., Rao, P., & Rathod, V. K. (2018). Enzyme assisted extraction of biomolecules as an approach to novel extraction technology: A review. *Food Research International*, 108, 309–330.
Nanda, P. K., Das, A. K., Dandapat, P., Dhar, P., Bandyopadhyay, S., Dib, A. L., Lorenzo, J., & Gagaoua, M. (2021). Nutritional aspects, flavour profile and health benefits of crab meat based novel food products and valorisation of processing waste to wealth: A review. Trends in Food Science and Technology. https://doi.org/10.1016/j.tifs.2021.03.059
Natolino, A., & Da Porto, C. (2019). Supercritical carbon dioxide extraction of pomegranate (Punica granatum L.) seed oil: Kinetic modelling and solubility evaluation. *Innovative Food Science and Emerging Technologies*, 151, 30–39.
Negi, T,. Kumar, Y., Sirohi, R., et al., (2022). Advances in bioconversion of spent tea leaves to value-added products. *Bioresource Technology*, 346, 126409. https://doi.org/10.1016/j.biortech.2021.126409
Negi, T., Vaidya, D., Tarafdar, A. et al., (2021). Physico-functional evaluation, process optimization and economic analysis for preparation of muffin premix using apple pomace as novel supplement. *Systems Microbiology and Biomanufacturing*, 1, 302–310. https://doi.org/10.1007/s43393-021-00026-y
Newton, T. M., Duce, J. A., & Bayle, E. D. (2019). The proteostasis network provides targets for neurodegeneration. *British Journal of Pharmacology*, 176(18), 3508–3514.

Nishino, H., Tokuda, H., Satomi, Y., Masuda, M., Onozuka, M., Yamaguchi, S., & Murakoshi, M. (2000). Cancer chemoprevention by phytochemicals and their related compounds. *Asian Pacific Journal of Cancer Prevention*, 1(1), 49–55.

Nogales-Bueno, J., Baca-Bocanegra, B., Heredia, F. J., & Hernández-Hierro, J. M. (2020). Phenolic compounds extraction in enzymatic macerations of grape skins identified as low-level extractable total anthocyanin content. *Journal of Food Science*, 85(2), 324–331. https://doi.org/10.1111/1750-3841.15006

Núñez Selles, A. J., Daglia, M., & Rastrelli, L. (2016). The potential role of mangiferin in cancer treatment through its immunomodulatory, anti-angiogenic, apoptopic, and gene regulatory effects. *Biofactors*, 42(5), 475–491.

Nutter, J., Fernandez, M. V., Jagus, R. J., & Agüero, M. V. (2021). Development of an aqueous ultrasound-assisted extraction process of bioactive compounds from beet leaves: A proposal for reducing losses and increasing biomass utilization. *Journal of the Science of Food and Agriculture*, 101(5), 1989–1997. https://doi.org/10.1002/jsfa.10815

Ortega, A. M. M., & Campos, M. R. S. (2019). Bioactive compounds as therapeutic alternatives. In *Bioactive compounds Health Benefits and Potential Applications, 1st ed.;* Campos, M.R.S. (Ed.). Amsterdam, The Netherlands: Elsevier, pp. 247–264.

Paes, J., Dotta, R., Barbero, G. F., & Martínez, J. (2014). Extraction of phenolic compounds and anthocyanins from blueberry (Vaccinium myrtillus L.) residues using supercritical CO2 and pressurized liquids. *The Journal of Supercritical Fluids*, 95, 8–16.

Pal, C. B. T., & Jadeja, G. C. (2019). Deep eutectic solvent-based extraction of polyphenolic antioxidants from onion (Allium cepa L.) peel. *Journal of the Science of Food and Agriculture*, 99, 4, 1969–1979. https://doi.org/10.1002/jsfa.9395

Panesar, P. S., & Kennedy, J. F. (2012). Biotechnological approaches for the value addition of whey. *Critical Reviews in Biotechnology,* 32(4), 327–348. https://doi.org/10.3109/07388551.2011.640624

Papoutsis, K., Pristijono, P., Golding, J. B., Stathopoulos, C. E., Bowyer, M. C., Scarlett, C. J., & Vuong, Q. V. (2018). Screening the effect of four ultrasound-assisted extraction parameters on hesperidin and phenolic acid content of aqueous citrus pomace extracts. *Food Bioscience*, 21, 20–26. https://doi.org/10.1016/j.fbio.2017.11.001

Paradiso, V. M., Squeo, G., Pasqualone, A., Caponio, F., & Summo, C. (2019). An easy and green tool for olive oils labelling according to the contents of hydroxytyrosol and tyrosol derivatives: Extraction with a natural deep eutectic solvent and direct spectrophotometric analysis. *Food Chemistry*, 291, 1–6. https://doi.org/10.1016/j.foodchem.2019.03.139

Parniakov, O., Barba, F. J., Grimi, N., Lebovka, N., & Vorobiev, E. (2014). Impact of pulsed electric fields and high voltage electrical discharges on extraction of high-added value compounds from papaya peels. *Food Research International*, 65, Part C, 337–343. https://doi.org/10.1016/j.foodres.2014.09.015

Parniakov, O., Barba, F. J., Grimi, N., Lebovka, N., & Vorobiev, E. (2016). Extraction assisted by pulsed electric energy as a potential tool for green and sustainable recovery of nutritionally valuable compounds from mango peels. *Food Chemistry*, 192, 842–848. https://doi.org/10.1016/j.foodchem.2015.07.096

Pavlović, N., Jokić, S., Jakovljević, M., Blažić, M., & Molnar, M. (2020). Green extraction methods for active compounds from food waste—Cocoa bean shell. *Foods*, 9(2), 140.

Peiro, S., Luengo, E., Segovia, F., Raso, J., & Almajano, M. P. (2019). Improving polyphenol extraction from lemon residues by pulsed electric eields. *Waste and Biomass Valorization*, 10, 889–897. https://doi.org/10.1007/s12649-017-0116-6

Pereira, D. T. V., Tarone, A. G., Cazarin, C. B. B., Barbero, G. F., & Martínez, J. (2019). Pressurized liquid extraction of bioactive compounds from grape marc. *Journal of Food Engineering*, 240, 105–113. https://doi.org/10.1016/j.jfoodeng.2018.07.019

Pereira, M. G., Hamerski, F., Andrade, E. F., Scheer, A. dP., & Corazza, M. L. (2017). Assessment of subcritical propane, ultrasound-assisted and Soxhlet extraction of oil from sweet passion fruit (Passifora alata Curtis) seeds. *The Journal of Supercritical Fluids*, 128, 338–348.

Pinto, D., de la Luz Cádiz-Gurrea, M., Vallverdú-Queralt, A., Delerue-Matos, C., & Rodrigues, F. (2021). Castanea sativa shells: A review on phytochemical composition, bioactivity and waste management approaches for industrial valorization. *Food Research International*, 144, 110364.

Prameela, K., Venkatesh, K., Immandi, S. B., Kasturi, A. P. K., Rama Krishna, C., & Murali Mohan, C. Next generation nutraceutical from shrimp waste: The convergence of applications with extraction methods. *Food Chemistry*, 237, 121–132. https://doi.org/10.1016/j.foodchem.2017.05.097

Principato, L., Mattia, G., Leo, A. D., & Pratesi, C. A. (2021). The household wasteful behaviour framework: A systematic review of consumer food waste. *Industrial Marketing Management,* 93, 641–649. https://doi.org/10.1016/j.indmarman.2020.07.010

Puligundla, P., Mok, C., & Park, S. (2020). Advances in the valorization of spent brewer's yeast. *Innovative Food Science & Emerging Technologies*, 62, 102350.

Radočaj, O., Vujasinović, V., Dimić, E., & Basić, Z. (2014). Blackberry (Rubus fruticosus L.) and raspberry (Rubus idaeus L.) seed oils extracted from dried press pomace after longterm frozen storage of berries can be used as functional food ingredients. *European Journal of Lipid Science and Technology*, 116(8), 1015–1024.

Raffaele, M., Greish, K., Vanella, L., Carota, G., Bahman, F., Bindayna, K. M., & Sorrenti, V. (2021). Potential health benefits of a pomegranate extract, rich in phenolic compounds, in intestinal inflammation. *Current Nutrition & Food Science*, 17(8), 833–843.

Rahman, U., Sahar, A., & Khan, M. A. (2014). Recovery and utilization of effluents from meat processing industries. *Food Research International,* 65, 322–328.

Ramakrishna, R., Sarkar, D., & Shetty, K. (2019). Metabolic stimulation of phenolic biosynthesis and antioxidant enzyme response in dark germinated barley (Hordeum vulgare L.) sprouts using bioprocessed elicitors. *Food Science and Biotechnology*, 28, 1093–1106.

Redha, A. A. (2021). Review on extraction of phenolic compounds from natural sources using green deep eutectic solvents. *Journal of Agricultural and Food Chemistry*, 69(3), 878–912. https://doi.org/10.1021/acs.jafc.0c06641

Reynoso-Camacho, R., Rodríguez-Villanueva, L. D., Sotelo-Gonzalez, A. M., RamosGomez, M., & P´erez-Ramírez, I. F. (2021). Citrus decoction by-product represents a rich source of carotenoid, phytosterol, extractable and non-extractable polyphenols. *Food Chemistry*, 350, Article 129239. https://doi.org/10.1016/j.foodchem.2021.129239

Rico-Rodriguez, F., Strani, L., Grassi, S., Lancheros, R., Serrato, J. C., & Casiraghi, E. (2021). Study of galactooligosaccharides production from dairy waste by FTIR and chemometrics as process analytical technology. *Food and Bioproducts Processing*, 126, 113–120. https://doi.org/10.1016/j.fbp.2020.12.009

Roda, A., De Faveri, D. M., Giacosa, S., Dordoni, R., & Lambri, M. (2016). Efect of pretreatments on the saccharifcation of pineapple waste as a potential source for vinegar production. *Journal of Cleaner Production*, 112, 4477–4484.

Rodrigues, L. G. G., Mazzutti, S., Vitali, L., Micke, G. A., & Ferreira, S. R. S. (2019). Recovery of bioactive phenolic compounds from papaya seeds agroindustrial residue using subcritical water extraction. *Biocatalysis and Agricultural Biotechnology*, 22, 101367.

Rodríguez De Luna, S. L., Ramírez-Garza, R. E., & Serna Saldívar, S. O. (2020). Environmentally friendly methods for flavonoid extraction from plant material: Impact of their operating conditions on yield and antioxidant properties. *The Scientific World Journal*, 2020. https://doi.org/10.1155/2020/6792069

Rodriguez-Rodriguez, R., Justo, M. L., Claro, C. M., Vila, E., Parrado, J., Herrera, M. D., & De Sotomayor, M. A. (2012). Endothelium-dependent vasodilator and antioxidant properties of a novel enzymatic extract of grape pomace from wine industrial waste. *Food Chemistry*, 135(3), 1044–1051.

Roy, P., Mohanty, A. K., Dick, P., & Misra, M. (2023). A review on the challenges and choices for food waste valorization: Environmental and economic impacts. *ACS Environmental Au Journal*, 3(2), 58–75. https://doi.org/10.1021/acsenvironau.2c00050

Ryan, M. P., & Walsh, G. (2016). The biotechnological potential of whey. *Reviews in Environmental Science and Bio/Technology*, 15(3), 479–498. https://doi.org/10.1007/s11157-016-9402-1

Ryder, K., Bekhit, A. E. D., McConnell, M., & Carne, A. (2016). Towards generation of bioactive peptides from meat industry waste proteins: Generation of peptides using commercial microbial proteases. *Food Chemistry*, 208, 42–50.

Saad, N., Louvet, F., Tarrade, S., Meudec, E., Grenier, K., Landolt, C., … Bressollier, P. (2019). Enzyme-assisted extraction of bioactive compounds from raspberry (Rubus idaeus L.) pomace. *Journal of Food Science*, 84, 6, 1371–1381. https://doi.org/10.1111/1750-3841.14625

Safdar, M. N., Kausar, T., Jabbar, S., Mumtaz, A., Ahad, K., & Saddozai, A. A. (2017). Extraction and quantification of polyphenols from kinnow (Citrus reticulate L.) peel using ultrasound and maceration techniques. *Journal of Food and Drug Analysis*, 25(3), 488–500.

Salgın, U., Salgın, S., Ekici, D. D., & Uluda, Ł. G. (2016). Oil recovery in rosehip seeds from food plant waste products using supercritical CO2 extraction. *The Journal of Supercritical Fluids*, 118, 194–202.

Santos, D. I., Saraiva, J. M. A., Vicente, A. A., & Moldão-Martins, M. (2019). Methods for determining bioavailability and bioaccessibility of bioactive compounds and nutrients. In *Innovative thermal and non-thermal processing, bioaccessibility and bioavailability of nutrients and bioactive compounds* (pp. 23–54). Woodhead Publishing.

Sar, T., Harirchi, S., Ramezani, M., et al., (2022). Potential utilization of dairy industries by-products and wastes through microbial processes: A critical review. *Science of the Total Environment*, 810, 152253. https://doi.org/10.1016/j.scitotenv.2021.152253

Sari, T. P., Sirohi, R., Krishania, M., et al., (2022). Critical overview of biorefinery approaches for valorization of protein rich tree nut oil industry by-product. *Bioresource Technology*, 362, 127775. https://doi.org/10.1016/j.biortech.2022.127775

Saulis, G., Rodaite-Riˇseviˇciene, R., Dainauskaite, V. S., & Saule, R. (2015). Electrochemical processes during high-voltage electric pulses and their Importance in food processing technology. In R. Rai (Ed.), *Advances in food biotechnology*. https://doi.org/10.1002/9781118864463.ch35

Sebghatollahi, Z., Ghanadian, M., Agarwal, P., Ghaheh, H. S., Mahato, N., Yogesh, R., & Hejazi, S. H. (2022). Citrus flavonoids: Biological activities, implementation in skin health, and topical applications: A review. *ACS Food Science & Technology*, 2(9), 1417–1432.

Sengul, E., Gelen, V., & Gedikli, S. (2021). Cardioprotective activities of quercetin and rutin in sprague dawley rats treated with 5-fluorouracil. *Journal of Animal and Plant Sciences*, 31, 423–431.

Sharif, T., Bhatti, H. N., Bull, I. D., & Bilal, M. (2021). Recovery of high-value bioactive phytochemicals from agro-waste of Mango (Mangifera Indica L.) using enzyme-assisted ultrasound pretreated extraction. *Biomass Conversion and Biorefinery*. https://doi.org/10.1007/s13399-021-01589-5

Sójka, M., Kołodziejczyk, K., Milala, J., Abadias, M., Viñas, I., Guyot, S., & Baron, A. (2015). Composition and properties of the polyphenolic extracts obtained from industrial plum pomaces. *Journal of Functional Foods*, 12, 168–178.

Squillaci, G., Apone, F., Sena, L. M., Carola, A., Tito, A., Bimonte, M., & Morana, A. (2018). Chestnut (Castanea sativa Mill.) industrial wastes as a valued bioresource for the production of active ingredients. *Process Biochemistry*, 64, 228–236.

Suwal, S., Bentahar, J., Marciniak, A., Beaulieu, L., Deschênes, J.-S., Doyen, A. (2019). Evidence of the production of galactooligosaccharide from whey permeate by the microalgae Tetradesmus obliquus. *Algal Research*, 39, 101470. https://doi.org/10.1016/j.algal.2019.101470

Tabish, S. A. (2017). Lifestyle diseases: Consequences, characteristics, causes and control. *Journal of Cardiology & Current Research*, 9(3), 00326.

Talekar, S., Patti, A. F., Singh, R., Vijayraghavan, R., & Arora, A. (2018). From waste to wealth: High recovery of nutraceuticals from pomegranate seed waste using a green extraction process. *Industrial Crops and Products*, 112, 790–802.

Talmaciu, A. I., Volf, I., & Popa, V. I. (2015). A comparative analysis of the "green" techniques applied for polyphenols extraction from bioresources. *Chemistry & Biodiversity*, 12, 1635–1651. https://doi.org/10.1002/cbdv.201400415

Tanaka, M., Takamizu, A., Hoshino, M., Sasaki, M., & Goto, M. (2012). Extraction of dietary fber from Citrus junos peel with subcritical water. *Food and Bioproducts Processing*, 90(2), 180–186.

Tanaka, T., Kawamori, T., Ohnishi, M., Okamoto, K., Mori, H., & Hara, A. (1994). Chemoprevention of 4-nitroquinoline 1-oxide-induced oral carcinogenesis by dietary protocatechuic acid during initiation and postinitiation phases. *Cancer Research*, 54(9), 2359–2365.

Tarozzi, Andrea, Marchesi, Alessandra, Cantelli-Forti, Giorgio, & Hrelia, Patrizia. (2004). Cold-storage affects antioxidant properties of apples in Caco-2 cells. *The Journal of Nutrition*, 134(5), 1105–1109.

Teles, A. S. C., Chavéz, D. W. H., Coelho, M. A. Z., Rosenthal, A., Gottschalk, L. M. F., & Tonon, R. V. (2021). Combination of enzyme-assisted extraction and high hydrostatic pressure for phenolic compounds recovery from grape pomace. *Journal of Food Engineering*, 288(1), 110128.

Toro-Uribe, S., Ibanez, ˜E., Decker, E. A., Villamizar-Jaimes, A. R., & Lopez-Giraldo, L. J. (2020). Food-safe process for high recovery of flavonoids from cocoa beans: Antioxidant and HPLC-DAD-ESI-MS/MS analysis. *Antioxidants (Basel)*, 9(5), 364. https://doi.org/10.3390/antiox9050364

Torres-Ossandón, M. J., Vega-Gálvez, A., López, J., Stucken, K., Romero, J., & Di Scala, K. (2018). Efects of high hydrostatic pressure processing and supercritical fuid extraction on bioactive compounds and antioxidant capacity of Cape gooseberry pulp (Physalis peruviana L.). *The Journal of Supercritical Fluids*, 138, 215–220.

Trasanidou, D., Apostolakis, A., & Makris, D. P. (2016). Development of a green process for the preparation of antioxidant and pigment-enriched extracts from winery solid wastes using response surface methodology and kinetics. *Chemical Engineering Communications*, 203, 1317–1325. https://doi.org/10.1080/00986445.2016.1189416

UNEP. (2021). Food waste index report. https://www.unep.org/resources/report/unep-food-waste-index-report-2021

Vasco-Correa, J., & Zapata, A. D. Z. (2017). Enzymatic extraction of pectin from pas - sion fruit peel (Passifora edulis f. favicarpa) at laboratory and bench scale. *LWT*, 80, 280–285.

Velderrain-Rodríguez, G. R., Quero, J., Osada, J., Martín-Belloso, O., & Rodríguez-Yoldi, M. J. (2021). Phenolic-rich extracts from avocado fruit residues as functional food ingredients with antioxidant and antiproliferative properties. *Biomolecules*, 11(7), 977.

Velez, M. E. V., da Luz, J. M. R., da Silva, M. D. C. S., Cardoso, W. S., Lopes, L. D. S., Vieira, N. A., & Kasuya, M. C. M. (2019). Production of bioactive compounds by the mycelial growth of Pleurotus djamor in whey powder enriched with selenium. *LWT,* 114, 108376. https://doi.org/10.1016/j.lwt.2019.108376

Vilas-Boas, A. A., Pintado, M., & Oliveira, A. L. S. (2021). Natural bioactive compounds from food waste: Toxicity and safety concerns. *Foods,* 10(7), 1564. Published 2021 July 6. https://doi.org/10.3390/foods10071564

Vu, H. T., Scarlett, C. J., & Vuong, Q. V. (2018). Phenolic compounds within banana peel and their potential uses: A review. *Journal of Functional Foods*, 40, 238–248.

Wang, L., Huang, W., & Zhan, J. (2019). Grape seed proanthocyanidins induce autophagy and modulate survivin in HepG2 cells and inhibit xenograft tumor growth in vivo. *Nutrients*, 11(12), 2983.

Wang, W., Gao, Y.-T., Wei, J.-W., Chen, Y.-F., Liu, Q.-L., & Liu, H.-M. (2021). Optimization of ultrasonic cellulase-assisted extraction and antioxidant activity of natural polyphenols from passion fruit. *Molecules,* 26(9), 2494. https://doi.org/10.3390/molecules26092494

Wang, Y. C., Chuang, Y. C., & Hsu, H. W. (2008). The flavonoid, carotenoid and pectin content in peels of citrus cultivated in Taiwan. *Food Chemistry*, 106, 277–284.

Wen, P., Hu, T. G., Linhardt, R. J., Liao, S. T., Wu, H., & Zou, Y. X. (2019). Mulberry: A review of bioactive compounds and advanced processing technology. *Trends in Food Science & Technology*, 83(November 2018), 138–158. https://doi.org/10.1016/j.tifs.2018.11.017

WRAP. (2013). Household food and drink waste in the United Kingdom 2012. *The Waste and Resources Action Programme.* https://wrap.org.uk/sites/default/files/2020-12/Household-Food-and-Drink-Waste-in-the-United-Kingdom-2012.pdf

WRAP. (2015). Strategies to achieve economic and environmental gains by reducing food waste. *The Waste and Resources Action Programme.* https://wrap.org.uk/resources/report/strategies-achieve-economic-and-environmental-gains-reducing-food-waste

Wu, Z., Xie, L., Li, Y., Wang, Y., Wang, X., Wan, N., & Yang, M. (2019). A novel application of the vacuum distillation technology in extracting Origanum vulgare L. essential oils. *Industrial Crops and Products*, 139, 111516.

Wuttisin, N., Boonmak, J., Thaipitak, V., Thitilertdecha, N., & Kittigowittana, K. (2017). Anti-tyrosinase activity of orange peel extract and cosmetic formulation. *International Journal of Pharmacology.* https://doi.org/10.3923/ijp.2023.834.841

Yamani, H. A., Pang, E. C., Mantri, N., & Deighton, M. A. (2016). Antimicrobial activity of Tulsi (Ocimum tenuiflorum) essential oil and their major constituents against three species of bacteria. *Frontiers in Microbiology*, 7, 195885.

Yan, B., Chen, Z. S., Hu, Y., & Yong, Q. (2021). Insight in the recent application of polyphenols from biomass. *Frontiers in Bioengineering and Biotechnology*, 9, Article 753898. https://doi.org/10.3389/fbioe.2021.753898

Yuliarti, O., Goh, K.K.T., Matia-Merino, L., Mawson, J., & Brennan, C. (2015). Extraction and characterisation of pomace pectin from gold kiwifruit (Actinidia chinensis). *Food Chemistry*, 187(11), 290–296.

Zhang, H., Liu, R., & Tsao, R. (2016). Anthocyanin-rich phenolic extracts of purple root vegetables inhibit pro-inflammatory cytokines induced by H2O2 and enhance antioxidant enzyme activities in Caco-2 cells. *Journal of Functional Foods*, 22, 363–375.

Zhang, J., Wen, C., Zhang, H., Duan, Y., & Ma, H. (2020). Recent advances in the extraction of bioactive compounds with subcritical water: A review. *Trends in Food Science and Technology*, 95, 183–195.

Zhang, R., Li, S., Zhu, Z., & He, J. (2019). Recent advances in valorization of Chaeno - meles fruit: A review of botanical profle, phytochemistry, advanced extraction technologies and bioactivities. *Trends in Food Science & Technology,* 91, 467–482.

Zhang, W., Khan, A., Ezati, P., Priyadarshi, R., Sani, M. A., Rathod, N. B., Goksen, G., & Rhim, J. (2024). Advances in sustainable food packaging applications of chitosan/polyvinyl alcohol blend films. *Food Chemistry*, 443, 138506.

Zhao, C., Lu, J. K., & Ameer, K. (2021). Effects of tofu whey powder on the quality attributes, isoflavones composition and antioxidant activity of wheat flour pan bread. *LWT,* 143, 11166. https://doi.org/10.1016/j.lwt.2021.111166

Zhi, W. J., Wang, L. F., & Hu, X. J. (2017). Recent advances in the effects of microwave radiation on brains. *Military Medical Research*, 4, 1–14.

12 Food Waste Management through Machine Learning, IoT, and Blockchain

Abdullah Bilal Öztürk, Ida Eyi Heathcote-Fumador, Isaac Arnold McSey, Elaine M'Nkubitu, Daniel Thomi, Steven Wainaina, and Mohammad J Taherzadeh

12.1 INTRODUCTION

The Food and Agriculture Organization (FAO) estimated that 32% of the food produced globally was wasted in 2009 (Girotto et al., 2015). That translates to approximately 1.3 billion tons per year being lost or wasted (Paul van der & Jason, 2017). Annually, and broken down by region, the United States contributes about 61 million tons, Australia produces about 4 million tons, China contributes 92.4 million tons, and Europe's contribution is estimated at 90 million tons (Girotto et al., 2015). The magnitude of food waste is similar in both developed and developing countries. However, there are significant differences in terms of the stages of the food supply chain (FSC) where food losses and waste occur. In developing nations, about 40% of the waste happens during post-harvest and processing stages, while in developed nations, about 40% of the waste occurs in retail and consumer stages (Girotto et al., 2015; Gustavsson et al., 2011). The factors contributing to food waste in developing countries include financial inadequacies, lack of proper infrastructure for harvesting, storage, and processing, and poorly developed marketing systems, while in developed countries, food losses and waste are mostly due to consumer behavior and a lack of coordination between actors (Girotto et al., 2015).

12.2 SOURCES OF FOOD WASTE

Food waste can be looked at from different angles, such as the sectors from which the waste comes, the processes that lead to food waste, or the type of product that was wasted or lost. According to the European Commission, food waste sources is classified into three groups: (a) food losses, which are food products lost during the production stage; (b) unavoidable food waste, which is food products lost during the consumption phase such as orange peels; and (c), avoidable food waste, which is food products that could have been consumed but were lost during the consumption phase (Thi et al., 2015). To gain a comprehensive overview of food waste sources, this section examines the entire food supply chain, starting from food production all the way to the consumer end. To do this, the food supply chain is broken down into two main parts: the pre-consumer stage and the post-consumer stage.

The pre-consumer stage involves all the processes and activities from food production to the point before it is distributed. This stage is further broken down into agricultural production, post-harvest handling and storage, processing, and packaging. The post-consumer part encompasses the activities after processing and packaging until the end of the life of the product. This involves distribution, retail, and consumption (Paul van der & Jason, 2017). Transportation is common in both pre- and post-consumer stages, for example, transporting harvested produce from the field to

DOI: 10.1201/9781032706030-12

storage facilities or directly to processing facilities, and transporting processed food from factories to retail shops or the consumer.

12.2.1 Pre-Consumer

12.2.1.1 Production

The food supply chain starts with the production of food from the agricultural sector. This consists of both farming and husbandry. At this stage of the FSC, waste comes from non-consumable sub-products such as cornstalks, and from low-quality or damaged products which are often left in the field. Sometimes good products might also be left in the field due to lack of or low commercial value. Poor harvesting techniques can also contribute to food loss as some of the product becomes damaged. Pests and diseases can also lead to food loss in the field as the products become unviable for human consumption (Julian et al., 2010). Lastly, weather conditions during harvest can lead to spoilage. For example, rainy weather during harvest can result in the rotting of legumes and grains (Shafiee-Jood & Cai, 2016).

12.2.1.2 Post-Harvest Handling and Storage

After crops are harvested, they need to be handled and stored properly to prevent deterioration and spoilage. A lack of proper transport and storage infrastructure leads to food loss at this stage. Many agricultural areas, especially in developing countries, are in rural regions, which lack good service roads. They also lack proper transportation and storage equipment and facilities, such as trucks fitted with cold units or storage units with humidity control. This leads to spoilage of perishable commodities such as milk, meat, and fresh vegetables. Poor storage facilities can also contribute to food loss, as stored products can become infested with pests and other pathogens (Julian et al., 2010; Shafiee-Jood & Cai, 2016).

12.2.1.3 Processing and Packaging

Food loss and waste also occur during processing and packaging for various reasons. Lack of proper or adequate manufacturing facilities causes delays or accumulation of food products at the facilities, causing spoilage. Processing errors can also lead to defective products that are not fit for consumption. Additionally, quality control issues can cause food waste even for perfectly edible food due to irregularities or aesthetic imperfections. In processing, poor packaging technology could also lead to food waste. Lastly, overestimating demand for different products can lead to a production surplus that is not consumed and is eventually wasted (Julian et al., 2010; Shafiee-Jood & Cai, 2016).

12.2.2 Post-Consumer

12.2.2.1 Distribution and Retail

This entails the stages between food processing and when it becomes available in retail shops, such as supermarkets and other food outlets. Several factors can lead to food loss and waste during distribution and retail. For example, during the transportation of food to storage facilities or processing facilities, food waste occurs due to inefficient transport systems. Transporting products in vehicles that lack optimum preservation conditions, especially those that need cold storage, leads to spoilage. Delays during transportation can also cause spoilage during transit. Moreover, poor handling during distribution can lead to visual defects, which make the food items unsellable.

During retail, inaccurate stock rotation can lead to food being spoiled on the shelf. Regular replenishment of stock also causes customers to pick the most recent products, and older stock ends up discarded. Inaccurate demand forecasting also causes retailers to overstock products that end up being wasted (Shafiee-Jood & Cai, 2016).

12.2.2.2 Consumer-Level

The reasons that consumers waste food are diverse and complex. It is estimated that two-thirds of discarded food happens at the household level (Buzby et al., 2014). At the consumer level, food waste happens within many interconnected practices of everyday life such as shopping routines, storing, cooking, and eating (Hebrok & Boks, 2017). During shopping, consumers might buy more food than they can consume within a given period. Poor storage of food, or lack thereof, can also lead to food spoilage. At the cooking stage, consumers might prepare food that becomes inedible and thereby becomes waste. When it comes to eating food, consumers may serve themselves portions that they cannot consume, resulting in leftovers that are eventually discarded.

Consumers also may misinterpret food date labels such as sell by, use by, or best before dates. This can lead to the disposal of food that has not yet expired. Other reasons for consumers wasting food might be based merely on preferences in terms of appearance and freshness of the food, and they could simply discard edible food.

Given the numerous factors causing food loss and waste at the various stages, each stage offers an intervention point to reduce food waste. The solutions at each stage might also vary based on attributes such as socio-economic status. For example, in developing countries, solutions might heavily lean towards providing proper infrastructure, while in developed countries, the main points of focus could be influencing consumer behavior and improving retail and distribution logistics. A summary of the stages and factors causing food loss and waste at each stage is provided in Table 12.1.

12.2.3 Conventional Food Waste Management Options

The most desirable way to reduce the impacts of food waste is to avoid its generation in the first place. High organic strength and water content generally typify food waste from either pre- or post-consumer stages. If left untreated or if mismanaged, food waste can cause undesirable impacts on humans as well as on the ecosystem as a result of, e.g., odor and proliferation of pathogens (Ojha et al., 2020). According to recent statistics, the annual global greenhouse gas (GHG) emissions from food waste reached 4.4 Gton CO_2 eq., representing about an 8% share of total anthropogenic GHG emissions (Mokrane et al., 2023). Several food waste management options have therefore been developed and applied with the main aims of eliminating the potential challenges that would

TABLE 12.1
A Summary of Food Loss and Waste through the Food Supply Chain (Adopted from Shafiee-Jood & Cai, 2016)

Pre-Consumer				Post-Consumer
Production and Harvesting	**Post-Harvest Handling and Storage**	**Processing and Packaging**	**Distribution @ Retail**	**Consumption**
Pests and diseases ; improper harvesting techniques ; surplus harvest ; low or no market value ; poor weather during harvest	Poor road networks ; lack of proper transport infrastructure ; lack of proper storage facilities ; pests and pathogens	Lack of processing equipment ; processing errors ; quality control ; lack of proper packaging	Poor logistical management ; stock forecast mistakes ; mishandling during distribution	Excess purchase ; poor storage at home ; inedible food preparation waste ; confusion over the labeling of dates ; food preferences ; simply discarding food

otherwise arise from untreated food waste and recovering valuable products that could be reused within the circular economy framework. Some of the established food waste management options are briefly highlighted in the sections that follow.

12.2.3.1 Re-Use by Humans or Feeding to Animals

The existence of poverty is an inevitable fact in many societies. Spare food that may eventually end up as waste could be donated to the less privileged as long as it is still acceptable for human consumption (Eriksson & Spångberg, 2017). Countries such as Canada and the United States offer tax incentives to organizations to promote such practices (Kinach et al., 2020). The utilization of food waste as a cost-effective animal feed has been a norm for many years. It involves little or no processing and directly feeding animals, such as swine, cattle, and chickens. However, due to potential microbial contamination, inconsistent nutrient availability, and harmful wrapping materials, there has been a decline in the direct application of food waste as animal feed, with some regions even enacting prohibition laws (San Martin et al., 2016; Truong et al., 2019).

12.2.3.2 Open Dumping and Landfilling

Open dumping sites are a common food waste final destination for some developing countries, partially due to rapid urbanization, insufficient financial resources, and irregular waste collection. In such cases, food waste is piled in heaps or is openly burned alongside other organic and inorganic wastes, both of which pose health and environmental risks (Dladla et al., 2016). The disposal of food waste in landfills has also been prevalent in different parts of the world. Due to the natural decay of this kind of waste that causes uncontrolled emission of GHGs, the release of leachates that could contaminate underground water sources, as well as the production of strong odors, landfilling is not considered the optimal management method for food waste (Krause et al., 2016; Lee et al., 2006). Landfilling of food waste has therefore been banned in countries such as Sweden and Korea (Eriksson et al., 2015; Lee et al., 2007).

12.2.3.3 Thermal Treatment

Food waste, especially post-consumer, ends up mixed with other organic and non-organic wastes in the municipal solid waste matrix unless well sorted at the source (Slorach et al., 2019). For such waste, thermal treatment is one of the prevalent management approaches that can also be a source of heat and power (Grycová et al., 2016). Thermal waste treatment processes can be classified as either incineration, pyrolysis, or gasification depending mainly on the supply of oxygen. It should, however, be considered that the thermal treatment of food waste can be an energy-intensive process due to the high moisture content (Mohammed et al., 2017).

12.2.3.4 Composting

Food waste can also be biologically treated via composting due to its loose structure and high organic-to-ash ratio (Al-Rumaihi et al., 2020). This technology can be defined as the biological breakdown of organic matter under aerobic conditions (Al-Rumaihi et al., 2020; Wei et al., 2017). The result of composting is a stabilized material that can improve agriculture due to its immense physical, biological, and chemical benefits on the soil, such as moisture retention, abundance of microbial flora, and presence of key elements like carbon (Al-Rumaihi et al., 2020; Liu et al., 2022). The application of food waste-derived compost would imply a reduction in the use of artificial fertilizers, which can be environmentally beneficial in the long term (Al-Rumaihi et al., 2020; Wei et al., 2017).

12.2.3.5 Anaerobic Digestion

Food waste can also be biologically processed using anaerobic digestion technology to produce biogas (an energy carrier) alongside digestate that can be applied as a biofertilizer. In addition to food waste, this technology is widely established for the recovery of resources from various organic wastes such as animal manure, lignocelluloses, and industrial residues in either low or high solid loading conditions (Wainaina et al., 2019). Anaerobic digestion also helps to stabilize food waste through the destruction of pathogens (at higher temperature conditions) and volume reduction (Holm-Nielsen et al., 2009).

12.3 FUNDAMENTALS OF MACHINE LEARNING, INTERNET OF THINGS, AND BLOCKCHAIN IN FOOD WASTE MANAGEMENT

Considering the constant rise in food waste generation, infrastructural systems such as modern technologies are important in coordinating, monitoring, simulating, and executing innovative solutions to the waste problem (Said et al., 2023). Presently, there is a major global transition towards automation through artificial intelligence (AI). The application of AI-based systems can facilitate the use of real-time data for the management of food waste (Wainaina & Taherzadeh, 2023). This section will therefore cover the basic principles of AI tools that are relevant for food waste management systems, namely machine learning, Internet of Things, and blockchain.

Machine learning (ML) is a subfield of AI that allows computer systems to automatically improve their performance on a specific task through experience or data (Choi et al., 2020). ML is a specific method of achieving the goal of automating intellectual tasks normally performed by humans (Choi et al., 2020). ML is a field that focuses on the learning aspect of AI by developing algorithms that best represent a set of data. The goal of machine learning is to enable computers to learn from data and make predictions or decisions without being explicitly programmed. The process involves training a machine learning model on a dataset and then using it to make predictions or decisions on new, unseen data.

There are several types of ML algorithms, including supervised learning, unsupervised learning, semi-supervised learning, and reinforcement learning, as summarized in Figure 12.1 (Choi et al., 2020; Jovel & Greiner, 2021; Wainaina & Taherzadeh, 2023). Supervised learning involves training a model on labeled data, where each data point has a known output, label, or target, i.e., the feature to be predicted. Unsupervised learning involves finding patterns in unlabeled data, where there is no known output or label. Semi-supervised learning is a combination of both supervised learning and unsupervised learning: the training dataset used consists of both labeled and unlabeled data. Reinforcement learning involves training a model to make decisions based on feedback received from the environment (Choi et al., 2020).

12.3.1 Supervised Machine Learning

Supervised learning is a type of machine learning algorithm where the input data is labeled with known outputs, or target variables, which are used to train a model to make predictions on new, unseen data (Choi et al., 2020; Jovel & Greiner, 2021). The goal is to learn a mapping function from the input variables to the output variables such that the model can accurately predict the output variables for new input data.

The basic steps of supervised machine learning are:

1. Acquire a dataset and split it into separate training, validation, and test datasets.
2. Use the training and validation datasets to inform a model of the relationship between features and target.
3. Evaluate the model using the test dataset to determine how well it predicts unseen instances.

In each iteration, the performance of the algorithm on the training data is compared with the performance on the validation dataset (Choi et al., 2020). The training process in supervised learning involves providing the model with a set of input-output pairs, called the training dataset, and

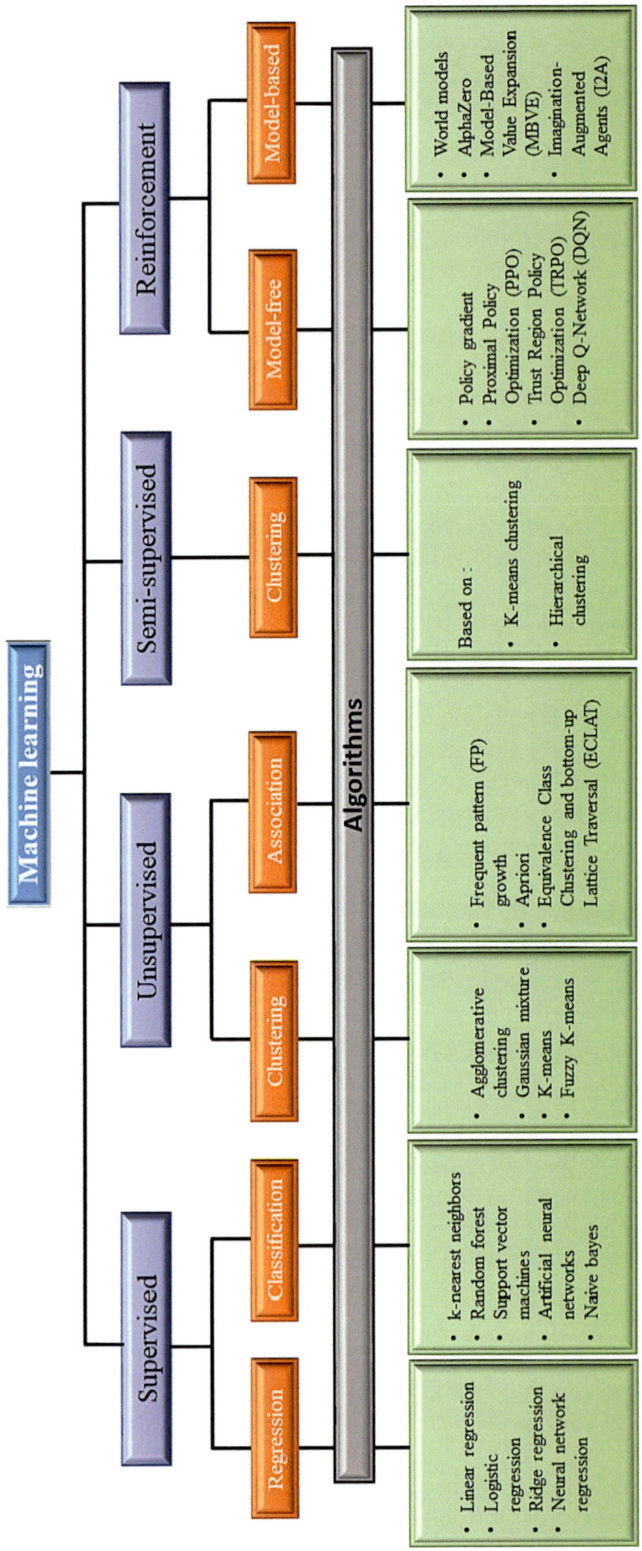

FIGURE 12.1 Types of machine learning algorithms. Adapted from Wainaina & Taherzadeh (2023).

adjusting the model parameters to minimize the difference between the predicted output and the actual output for each input (Choi et al., 2020; Jovel & Greiner, 2021). This is typically done using an optimization algorithm, such as gradient descent, which iteratively adjusts the model parameters to reduce the error between the predicted output and the actual output.

Various machine algorithms that are considered supervised include:

1. Linear models – the goal is to predict a continuous output variable (Choi et al., 2020; Jovel & Greiner, 2021; Zaki & Meira, 2014). For example, a linear regression model can be used to predict the price of a house based on its size, location, and other features.
2. Decision trees – used for classification and regression problems. They work by recursively partitioning the data into smaller subsets based on the features that are most informative for the task at hand (Jovel & Greiner, 2021; Zaki & Meira, 2014). For example, a decision tree can be used to classify emails as spam or not spam based on their content and other features.
3. Neural networks – modeled after the structure and function of the human brain. They are particularly effective for tasks that involve image or speech recognition, natural language processing, and other complex tasks. For example, a neural network can be used to identify objects in an image or transcribe spoken words into text.
4. Support vector machines – they work by finding the hyperplane that maximally separates the data into different classes (Zaki & Meira, 2014). For example, a support vector machine can be used to classify images of handwritten digits into their corresponding numbers.

Other algorithms include random forest, gradient boosting, k-nearest neighbor, and naïve Bayes classification (Jovel & Greiner, 2021; Zaki & Meira, 2014).

12.3.2 Unsupervised Learning

Unsupervised learning is a type of machine learning algorithm where the input data is not labeled with known outputs or target variables. Instead, the algorithm tries to identify patterns, relationships, and structure in the data without any guidance or supervision. The goal is to discover hidden patterns or groupings in the data that can provide insights or help in making decisions (Choi et al., 2020; Jovel & Greiner, 2021). There are several types of unsupervised learning algorithms, including clustering, anomaly detection, and dimensionality reduction (Bengio et al., 2012; Jovel & Greiner, 2021). Clustering algorithms aim to group similar data points based on some similarity measure, while anomaly detection algorithms aim to identify unusual or rare data points that do not fit into any of the identified groups. Dimensionality reduction algorithms aim to reduce the complexity of the data by representing it in a lower-dimensional space while preserving important information.

One of the challenges of unsupervised learning is that there is no objective metric to evaluate the performance of the algorithm, as there are no labeled outputs to compare the predictions with. This can make it difficult to determine whether the identified patterns or groups are meaningful or not. However, unsupervised learning is still a powerful tool for exploring and analyzing complex datasets and discovering hidden insights.

12.3.3 Reinforcement Learning

Reinforcement learning (RL) is a type of machine learning where an agent learns to take actions in an environment in order to maximize a reward signal. Unlike supervised learning, where the agent is given labeled data to learn from, and unsupervised learning, where the agent tries to find patterns in unlabeled data, in reinforcement learning, the agent learns through trial-and-error interactions with the environment (Choi et al., 2020; Jovel & Greiner, 2021; Kober et al., 2013).

The RL process typically involves the following elements:

1. An agent interacts that with the environment and takes actions based on its current state and policy.
2. An environment that responds to the agent's actions and provides feedback in the form of rewards or penalties.
3. A state space that defines the possible states of the environment.
4. An action space that defines the possible actions the agent can take in each state.
5. A reward function defines the reward or penalty that the agent receives for each action taken in a given state.

The goal of the agent is to learn a policy that maps states to actions in a way that maximizes the expected cumulative reward over time. The agent learns by exploring the environment and adjusting its policy based on the feedback it receives from the environment. This process is known as the exploration-exploitation trade-off, where the agent must balance the desire to explore new actions and states with the need to exploit its current knowledge to maximize the expected reward (Choi et al., 2020; Jovel & Greiner, 2021; Kober et al., 2013).

One of the challenges of reinforcement learning is the trade-off between exploration and exploitation. If the agent only exploits its current policy, it may get stuck in a suboptimal policy. On the other hand, if the agent only explores, it may take too long to learn an optimal policy. Researchers are constantly developing new algorithms and techniques to improve the exploration-exploitation trade-off in reinforcement learning (Kober et al., 2013).

In summary, reinforcement learning is a type of machine learning where an agent learns to take actions in an environment to maximize a reward signal. The agent learns through trial-and-error interactions with the environment, and the goal is to learn a policy that maximizes the expected cumulative reward over time. Reinforcement learning has many practical applications in various fields and is a powerful tool for developing autonomous systems that can learn and adapt to their environments.

12.3.4 Semi-Supervised Learning

Semi-supervised learning is a type of machine learning technique that falls between supervised learning and unsupervised learning. In supervised learning, the algorithm is trained using a labeled dataset, whereas, in unsupervised learning, the algorithm is trained on an unlabeled dataset without any predefined outputs. In semi-supervised learning, the algorithm is trained on a combination of both labeled and unlabeled data to improve its performance (Choi et al., 2020; Gao et al., 2018).

The main motivation behind semi-supervised learning is to address the limitations of supervised learning, which often requires a large amount of labeled data to achieve good performance (Antti & Harri, 2017; Gao et al., 2018). In many real-world scenarios, labeled data can be expensive, time-consuming, or even impossible to obtain, whereas unlabeled data is abundant and easy to obtain. By leveraging unlabeled data, semi-supervised learning algorithms can improve the accuracy of their predictions while reducing the amount of labeled data needed (Antti & Harri, 2017).

Semi-supervised learning has been applied to a wide range of applications, including natural language processing, computer vision, and speech recognition. Overall, semi-supervised learning is a promising approach that can leverage abundant unlabeled data to improve the performance of machine learning algorithms.

12.3.5 Internet of Things (IoT)

The Internet of Things (IoT) refers to the interconnectivity of physical objects, devices, vehicles, and other items embedded with sensors, software, and network connectivity that enable them to collect and exchange data. The IoT creates a network of connected devices that can communicate with each other and with humans, providing new opportunities for data-driven decision-making and automation (Al-Fuqaha et al., 2015; Atzori et al., 2010; Roman et al., 2013). IoT technology has the potential to revolutionize industries such as healthcare, transportation, and agriculture by enabling real-time monitoring and control, predictive maintenance, and improved resource utilization (Al-Fuqaha et al., 2015). However, it also raises concerns about data privacy, security, and interoperability (Roman et al., 2013). Figure 12.2 shows the basic components that come together to form an Internet of Things (Sadhu et al., 2022).

The IoT encompasses various components that work together to enable connectivity, data collection, processing, and communication. Here are some key components used in IoT systems (Al-Fuqaha et al., 2015; Atzori et al., 2010; Zanella et al., 2014):

1. Sensors and actuators: Sensors are devices that detect and measure physical parameters such as temperature, humidity, motion, and light. Actuators, on the other hand, are devices that perform actions or control physical processes based on received instructions. These components play crucial roles in collecting data from the environment and enabling interaction with the physical world.
2. Connectivity technologies: IoT devices require connectivity to exchange data and communicate with other devices or systems. Various wireless and wired communication technologies are used, including Wi-Fi, Bluetooth, Zigbee, NFC (near-field communication), and cellular networks (3G, 4G, and 5G). These technologies enable seamless data transmission and network integration.
3. Data processing and analytics: IoT systems generate massive amounts of data that need to be processed and analyzed to derive meaningful insights. Data processing components include edge computing devices, cloud computing platforms, and analytics algorithms. Edge computing enables data processing and analysis at the edge of the network, closer to the data source, reducing latency and bandwidth requirements. Cloud platforms provide scalable storage and computing resources for data processing and analytics.
4. Applications and services: IoT applications and services provide value to end-users by utilizing the collected and processed data. These can range from smart home automation and industrial monitoring to healthcare systems and environmental monitoring. IoT applications often leverage machine learning, artificial intelligence, and data visualization techniques to deliver actionable insights and enhanced user experiences.

The IoT has also been described as a paradigm shift in computing, with the potential to transform how we interact with the world around us. The IoT has the potential to enable new applications such

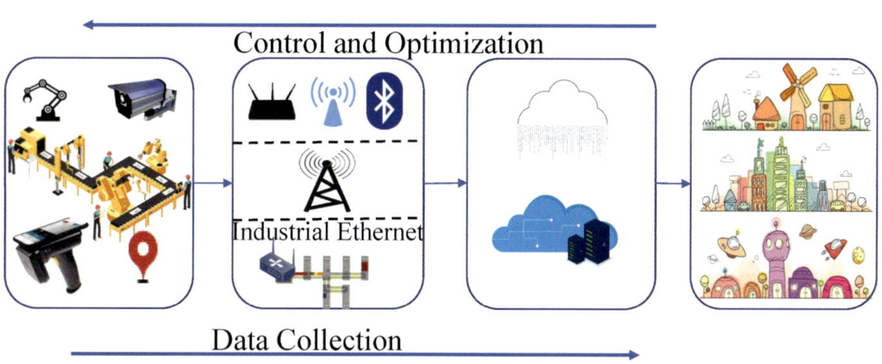

FIGURE 12.2 Primary technological components of IoT. Adapted from Zanella et al., (2014).

as smart homes, smart cities, and smart grids, which can improve energy efficiency, reduce waste, and enhance the quality of life (Atzori et al., 2010; Gubbi et al., 2013; Zanella et al., 2014).

Overall, the IoT represents a major technological trend that is likely to have significant impacts on society in the coming years. While there are challenges to be addressed, such as security and privacy concerns, the potential benefits of the IoT make it an area of significant interest for researchers and practitioners alike.

12.3.6 Blockchain Technology

Blockchain is a decentralized digital ledger technology that allows multiple participants to maintain a shared and synchronized record of transactions or data securely and transparently. At its core, a blockchain consists of a chain of blocks, where each block contains a set of transactions or data. These blocks are linked together sequentially, forming a chain-like structure. Each block contains a unique identifier called a cryptographic hash, which is generated based on the data within the block. Additionally, each block typically contains a reference to the previous block in the chain, creating a chronological order of transactions (Andreas, 2014; Don & Alex, 2016).

The decentralized nature of blockchain means that there is no central authority or intermediary controlling the entire system. Instead, multiple participants, known as nodes, maintain copies of the blockchain and collaborate to validate and record new transactions. This distributed network of nodes ensures that no single entity has complete control or can manipulate the data stored on the blockchain (Andreas, 2014; Don & Alex, 2016). Figure 12.3 describes how a basic blockchain concept works (Zubaydi et al., 2019).

Blockchain technology offers several advantages, including transparency, security, and decentralization. It enables trust among participants by providing a tamper-resistant and auditable record of transactions (Andreas, 2014; Don & Alex, 2016). It has the potential to streamline and automate processes, reduce costs, and eliminate the need for intermediaries in various industries. Blockchain technology was initially introduced as the underlying technology behind the cryptocurrency Bitcoin (Andreas, 2014; Don & Alex, 2016) but has since gained recognition for its potential applications in various industries, including finance, supply chain management, healthcare, and even tax administration.

12.4 MACHINE LEARNING, IOT, AND BLOCKCHAIN IN FOOD WASTE MANAGEMENT: APPLICATIONS AND CASE STUDIES

Machine learning, IoT, and blockchain technologies possess significant potential in transforming various industries, including food waste management through timely alerts and autonomous modes. These technologies can be harnessed in several applications to create more practical, transparent, secure, and sustainable solutions for eliminating the issue of food waste. Table 12.2 sums up the potential applications of machine learning, IoT, and blockchain technologies in food waste management.

12.4.1 Principles and Application of Machine Learning in Food Waste Management

12.4.1.1 Predicting Waste Generation Patterns

Predictive analysis emerges as a crucial application of AI in managing food waste. With the ability to examine past and present data, AI technologies can anticipate future waste generation patterns, thereby enabling better planning and resource distribution. An example of this application is seen in the strategies employed by several tech companies that deploy AI-driven systems to mitigate food waste in commercial environments. These systems, such as those seen in the UK-based firm named Winnow (www.winnowsolutions.com), utilize machine learning to discern trends and project future

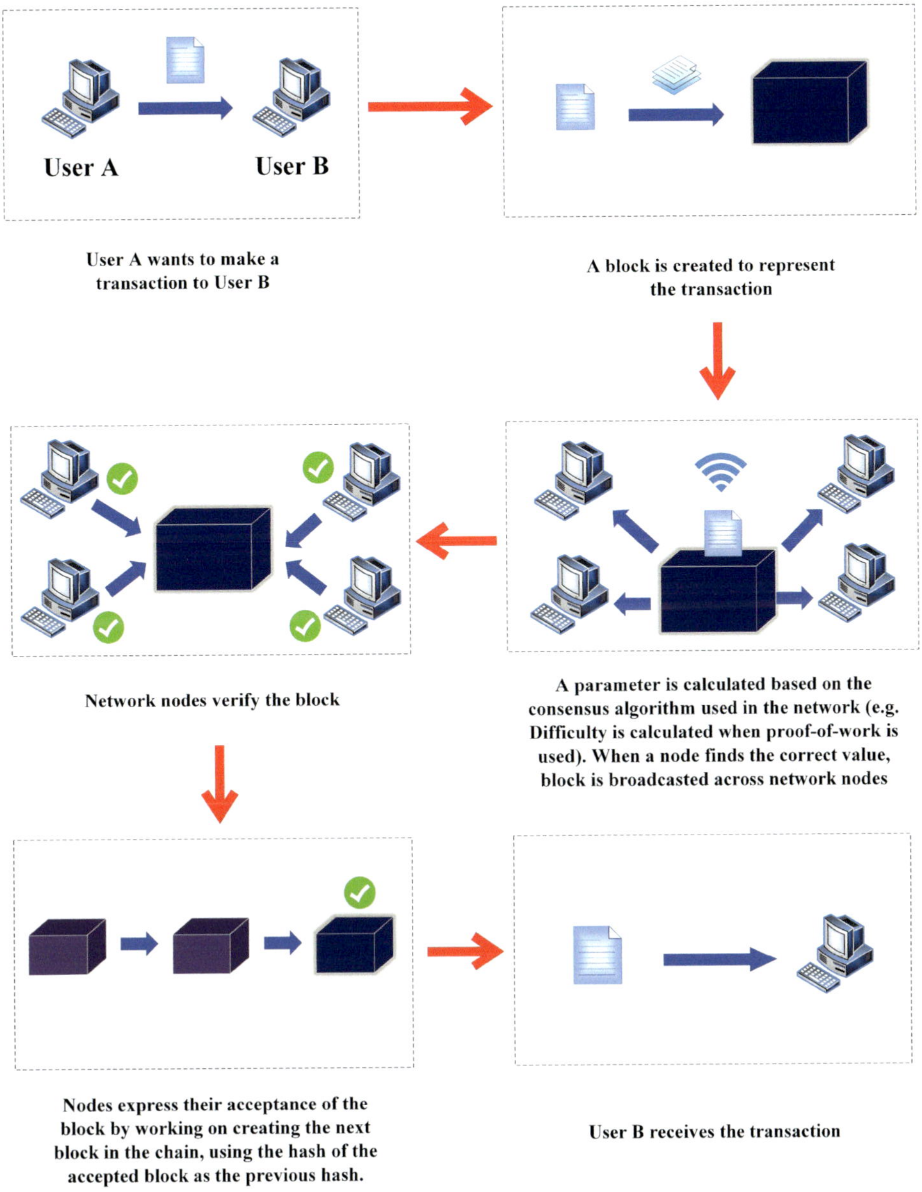

FIGURE 12.3 The working concept of blockchain technology. Adapted from Zubaydi et al., (2019).

waste. Consequently, commercial kitchens can refine their purchasing and production methods to more accurately reflect expected demand, leading to notable reductions in waste and corresponding financial savings (www.winnowsolutions.com).

12.4.1.2 Image Recognition for Waste Classification

Utilizing machine learning in image recognition is a transformative approach for discerning various types of food waste, which aids in facilitating correct segregation and disposal practices. In this context, certain innovative practices involve using such technology to categorize and track food waste, leading to more efficient management. One example is an initiative by a Spanish startup, VEnvirotech, that employs ML-based image recognition to sort organic waste for subsequent

TABLE 12.2
Summary of Technology Principles and Example Application

Technology	Principles Applied to Food Waste Management	Examples
Artificial intelligence/ machine learning	Prediction	Predicting waste generation patterns.
	Image recognition	Waste identification and classification.
	Decision support	When and what to buy to avoid over-purchasing.
	Learning from experience	Learning consumer behavior to improve shelf life.
	Optimization	Forecasting demand and supply to avoid overproduction.
Blockchain	Data decentralization	Food waste data democratization.
	Data transparency	Food supply chain transparency ensures waste generation traceability.
	Data security and integrity	Real-time access to unified, reliable data on food and food waste.
	Data immutability	Consumer engagement for responsible purchasing and food redistribution.
Internet of Things	Real-time data monitoring and collection	Real-time monitoring and collection of data on food – temperature, humidity, expiry date, etc.
		Tracking food products in the food supply chain can help identify waste-generation points for action.
		Consumer engagement in food waste reduction.

transformation into bioplastics through a bacterial conversion process (www.venvirotech.com). By accurately classifying the waste via image recognition, process optimization occurs, resulting in heightened efficiency, increased recycling rates, and decreased environmental impact.

12.4.1.3 Decision Support: When and What to Purchase

Artificial intelligence and machine learning are transforming the landscape of decision-making, especially in the realm of food waste management. By harnessing large volumes of data, these technologies offer key insights and informed suggestions to stakeholders, thus enhancing strategy formulation and policy decisions. A noteworthy illustration of this is the work of Wasteless, an Israeli startup (www.wasteless.com). Wasteless employs AI and ML algorithms to establish a "dynamic pricing model" for supermarkets. By assimilating and analyzing a plethora of data, including sales history and real-time inventory information, this tool generates variable product prices based on projected expiry dates. The model aims to incentivize consumers to buy products that are closer to their expiration, thus effectively reducing food waste. The instance of Wasteless provides a testament to the transformative potential of AI and ML in food waste management, facilitating evidence-based decision-making and augmenting overall waste management tactics (www.wasteless.com).

12.4.1.4 Learning from Experience: Learning Consumer Behavior

Artificial intelligence, with its ability to learn from experience and discern patterns, plays a transformative role in managing waste effectively. AI-powered platforms analyze diverse data sets – from sales, stock levels, and delivery schedules to external factors such as seasonal influences and weather patterns. Processing this vast array of data, AI reveals underlying patterns that guide strategic decisions about product ordering and inventory management. By closely aligning operations with actual consumer behavior and demand, retailers can reduce food waste significantly while simultaneously improving their profitability. This illustrates the potential of AI in shaping sustainable operations in the retail sector, making it an instrumental tool in devising successful waste reduction strategies.

12.4.1.5 Optimizing Forecasting Supply and Demand

Effective food waste management critically depends on accurate demand forecasting and inventory optimization, both areas where AI can provide significant value. An example of this can be found in companies like Crisp, which have designed sophisticated data platforms that employ AI to predict demand for products and optimize inventory levels (www.gocrisp.com). These platforms analyze a variety of data, including sales figures and market trends, to provide accurate demand forecasts. Food retailers as well as producers can leverage these platforms to align their production and inventory with actual demand, preventing overproduction and reducing the amount of food that goes to waste. The application of AI in this context not only aids in reducing food waste but also helps improve resource use efficiency and sustainability in the food industry.

12.4.1.6 Sustainability

The overarching principle guiding the application of AI and ML in food waste management is sustainability. By reducing food waste and enhancing resource-use efficiency, we can move significantly towards a more sustainable food ecosystem. Technologies such as AI and ML, when used wisely and responsibly, can make an invaluable contribution to global sustainability efforts.

12.4.1.7 Ethical Considerations

As we incorporate AI/ML in food waste management, ethical considerations, including data privacy and the responsible use of these technologies, must be at the forefront. AI and ML systems rely on large amounts of data, raising concerns about data privacy and security. Additionally, the decision-making processes of AI and ML systems can sometimes be opaque, leading to potential accountability issues. Therefore, it is essential to use these technologies responsibly and transparently, with appropriate regulatory oversight.

12.4.1.8 Case Studies of Machine Learning in Food Waste Management

Decoding Consumer Behavior with Machine Learning: A Study of Afresh Technologies

San Francisco-based company, Afresh Technologies, is a pioneer in leveraging AI and ML technologies to tackle the pressing issue of food waste, particularly in the fresh food retail sector (www.afresh.com). By understanding the unique complexities of this sector – from the perishability of products to fluctuating consumer demand – Afresh harnesses the power of AI to optimize operations, minimize waste, and enhance profitability. The company's innovative approach provides a valuable lens into the potential of AI in food waste management, demonstrating how data analytics, machine learning, and deep industry knowledge can revolutionize waste reduction strategies. This exploration of Afresh Technologies' work not only aids in understanding the present state of AI's role in managing food waste but also illuminates potential future developments in this crucial area (www.afresh.com).

The exploration of Afresh Technologies' contributions provides a compelling illustration of the potential of AI. It showcases how a powerful blend of data analytics, machine learning, and a keen understanding of the industry can revolutionize the way we approach and address food waste. Recognizing the inherently dynamic and unpredictable nature of the fresh food sector, Afresh Technologies has developed a unique approach that considers the multifaceted factors influencing food demand and waste. The company's platform harnesses diverse sets of data, including sales records, current stock levels, delivery schedules, and seasonal trends. It even integrates more unpredictable variables, such as external factors like weather patterns. This robust data collection ensures a comprehensive understanding of the complex operational dynamics of the fresh food retail sector.

Afresh employs advanced machine learning algorithms that are designed to make sense of this complex data. These algorithms process this information, discern patterns, and generate accurate demand forecasts, guiding retailers to make more informed, optimal ordering decisions. In doing so, the platform helps retail businesses reduce their surplus and subsequently their waste, while maintaining profitability. This is a vital step towards creating a more sustainable food supply chain, addressing the widespread issue of food waste, and enhancing the overall efficiency of fresh food retail operations.

Afresh's AI platform has a transformative role in the realm of food waste management, particularly within the fresh food retail sector. The platform's core strength lies in its capability to intelligently analyze and understand consumer purchasing patterns, a complex task requiring precise, data-driven insights. By leveraging machine learning algorithms, Afresh interprets buying trends, seasonal fluctuations, and unique buying behaviors, giving retailers a nuanced understanding of consumer demand (www.afresh.com). This ability to accurately anticipate consumer demand allows retailers to align their operations closely with actual market requirements. The result is a significant decrease in unsold fresh food items, which would typically end up as waste. By minimizing discrepancies between supply and demand, Afresh's solution tackles one of the main causes of food waste at the retail level: overstocking.

Furthermore, the platform's powerful predictive capabilities equip retailers with the tools to proactively manage their inventories. In the volatile fresh food retail environment, having an agile inventory management system is paramount for reducing waste. Afresh's platform can forecast stock requirements based on a multitude of factors, ensuring an appropriate level of stock at any given time. By intelligently managing inventory levels, the platform helps retailers avoid both overstocking, which leads to waste, and understocking, which could result in lost sales and unsatisfied customers. Afresh's AI platform represents a significant stride towards efficient, sustainable food waste management in the retail sector (www.afresh.com).

While Afresh's platform offers an innovative solution for food waste management, it is not without its challenges. The accuracy and efficiency of the platform largely depend on the quality of data input, highlighting the need for comprehensive and accurate data collection practices. Moreover, the implementation and integration of such a platform within a retailer's existing operations may require substantial resources and technical proficiency. Looking ahead, it would be beneficial for Afresh to continue refining its platform, focusing on ease of integration and user-friendliness to encourage wider adoption.

Harnessing Machine Learning to Combat Food Waste: An Analysis of Winnow Solutions

The burgeoning waste problem worldwide necessitates innovative approaches to waste management, particularly in the realm of food waste. A significant part of the solution lies in the ability to forecast future waste generation trends. By leveraging historical and current data, it is possible to predict waste volumes, types, and patterns, leading to more strategic planning and efficient resource allocation. Winnow Solutions, a technology company based in the UK, is a frontrunner in this space, leveraging machine learning to cut food waste in commercial kitchens (www.winnowsolutions.com). This section provides an in-depth analysis of Winnow Solutions' approach to reducing food waste.

Winnow Solutions offers a data-driven approach to food waste management. The company has developed an artificial intelligence tool, Winnow Vision, which employs machine learning to identify and measure food waste automatically in commercial kitchens (www.winnowsolutions.com). These environments often generate a significant amount of food waste due to factors like overproduction, mismanagement of inventory, and lack of tracking and accountability mechanisms. This is where predictive analytics, a powerful application of artificial intelligence, can play a transformative role. The Winnow Vision system works by capturing data about the type and quantity of food that is being discarded. This is accomplished through a smart scale placed under the waste bin and a camera above it that takes photos of the discarded food. The AI-powered system then identifies the type of food and measures its weight. Over time, the system learns to recognize different types of food waste more accurately, enhancing its predictive capabilities.

The data collected by Winnow Vision is translated into financial terms to show the real cost of the waste being generated, which is a powerful incentive for kitchens to minimize waste. With access to predictive insights, chefs and kitchen managers can adjust their procurement and production strategies to align more closely with anticipated demand, significantly reducing waste. In addition to cutting down on waste, these AI-driven solutions can contribute to considerable cost savings. According to Winnow, their technology has helped customers save $42 million per year by reducing food waste. The system is being used by several major organizations worldwide, including IKEA and the Armani Hotel Dubai.

Winnow's approach illustrates how AI can help us navigate the complex issue of food waste. By using AI to understand and predict waste patterns, we can turn a traditionally reactive process into a proactive one. As more businesses adopt similar technologies, we can look forward to a future where waste is significantly reduced and resources are used more sustainably. AI's predictive ability is not limited to commercial kitchens or food waste. It has vast potential to transform waste management across multiple sectors and waste types. As our capabilities to collect and analyze waste data continue to grow, we can expect AI to play an increasingly central role in waste management globally. By monitoring food discarded in commercial kitchens, Winnow Vision provides actionable insights to chefs and kitchen staff. This information helps them identify waste patterns, understand the cost implications of waste, and make informed decisions to reduce food waste. Using a camera, a set of smart scales, and the same type of machine learning technology found in autonomous vehicles, Winnow Vision "learns" to recognize different foods being thrown away. The system then captures the weight and cost of this waste, providing an analysis that can highlight opportunities for menu changes, portion control, and other waste reduction strategies. A prime example of Winnow Vision's efficacy is its application in IKEA's kitchens. The technology helped IKEA achieve its target of reducing food waste in its food operations by 50% by the end of 2020. IKEA used the Winnow Vision system to monitor its food waste, using the data generated to make adjustments that significantly reduced waste.

However, implementing Winnow Solutions' technology is not without challenges. For one, it requires kitchen staff to incorporate new procedures into their routines, which can meet resistance. Moreover, the technology requires an upfront investment, which may be a barrier for smaller businesses. Despite these hurdles, the results delivered by Winnow Solutions are promising, and its innovative approach showcases the potential of machine learning in food waste reduction. Looking ahead, the wider adoption of technologies like Winnow Vision could be a game-changer in the quest for more sustainable food systems.

12.4.2 Principles and Applications of Blockchain in Food Waste Management

12.4.2.1 Transparency and Traceability

Blockchain technology is a powerful tool for improving transparency and traceability in the food supply chain, thereby aiding food waste management. Certain initiatives harness this technology to offer end-to-end supply chain transparency. Such systems log and validate data about food production, processing, and distribution, making it available to both businesses and consumers. This empowers users to access trustworthy information about food origins, handling, and quality, bolstering waste reduction initiatives. The elevated transparency enabled by blockchain technology assists in pinpointing the sources of food waste in the supply chain, paving the way for targeted interventions to minimize waste. Systems like the Food Trust initiative by IBM are exemplary instances of this application of blockchain technology.

12.4.2.2 Real-Time Data Access and Sharing

Blockchain technology offers the ability to present a unified, real-time view of data to all participants in a supply chain. This capability is being championed by various food tech enterprises that are implementing blockchain to deliver real-time information about the condition and location of food products across the supply chain. Such a system allows stakeholders, from farmers to retailers, to access up-to-the-minute details about food products. This capability facilitates accurate demand forecasting, thereby preventing overproduction or over-ordering that can contribute to food waste. Such applications of blockchain demonstrate its potential in promoting more efficient and sustainable food supply chains. Ripe.io is a food-tech startup pioneering this frontier (Bhusal, 2021; technology).

12.4.2.3 Consumer Engagement

Blockchain technology has the potential to significantly influence consumer behavior in food waste management through its inherent qualities of transparency, traceability, and immutability. It can increase consumer awareness and education about the journey of food products and the implications of food waste, promoting more responsible purchasing decisions. Blockchain can also facilitate efficient food redistribution to reduce waste and create a reward system to incentivize responsible behavior. Additionally, it provides a feedback loop for businesses to improve waste management. However, leveraging blockchain for consumer behavior change requires a comprehensive approach that includes education, policy development, and infrastructure support, alongside the technology implementation.

12.4.2.4 Waste Tracking and Management

Blockchain's immutable record-keeping can be used to track food waste from generation to disposal. The use of blockchain for waste tracking and management can significantly enhance waste disposal efficiency. By recording how much waste is produced at each stage of the supply chain, where it goes, and how it is treated or disposed of, blockchain can help optimize waste management strategies. This application is similar to the plastic waste tracking model adopted by Plastic Bank, suggesting a possible pathway for food waste tracking. Plastic Bank (plasticbank.com), though primarily focused on plastic waste, presents a model for food waste tracking. By tracking waste and providing incentives for waste collection, the platform enables efficient waste management, thus preventing unnecessary landfilling.

12.4.2.5 Decentralization: Democratizing Food Waste Data

Decentralization in blockchain technology, which involves distributing data across a network of computers rather than having it controlled by a central authority, can contribute significantly to food waste management. This decentralization enhances transparency and trust, which can promote responsible practices and reduce food waste. It facilitates efficient redistribution of surplus food and improved traceability in the food supply chain. Blockchain can also be used to incentivize sustainable practices via decentralized applications or smart contracts. Additionally, the technology's data security and integrity can help prevent fraud and misinformation, contributing to waste reduction. However, the effectiveness of blockchain will depend on its adoption, the establishment of standards, regulatory support, and integration with other waste reduction efforts.

12.4.2.6 Case Studies of Blockchain in Food Waste Management

IBM Food Trust: A Blockchain-Based Platform for Food Traceability

In the quest to enhance food traceability, IBM has developed the Food Trust, a blockchain-based system designed to promote transparency and accountability in the food supply chain. This analysis aims to delve into the workings of IBM Food Trust and explore its implications for food waste management. Launched in 2018, IBM Food Trust aims to create a more transparent and efficient global food supply chain (www.ibm.com/products/supply-chain-intelligence-suite/food-trust). It achieves this by leveraging blockchain technology's unique properties, including immutability, decentralization, transparency, and security. Through this platform, stakeholders involved in the food supply chain, including farmers, processors, retailers, and consumers, can access and share information.

This includes data about food production, processing, transportation, and expiration, all of which are recorded on the blockchain.

IBM Food Trust offers significant potential for food waste management by enhancing supply chain transparency and traceability. By providing a clear and accurate view of food's journey from farm to fork, this system allows all stakeholders to identify waste hotspots in the supply chain. For example, if a particular stage in the supply chain consistently leads to spoilage due to delays, this issue can be identified and rectified. Moreover, by recording expiration dates and communicating them across the supply chain, the Food Trust can help ensure that food products are efficiently routed to market and sold before they spoil. This has the potential to significantly reduce food waste at the retail level.

Despite its potential, IBM Food Trust also faces challenges. The platform's effectiveness depends on widespread adoption by all stakeholders in the food supply chain. Encouraging such adoption may require addressing concerns about data privacy and the costs associated with implementing blockchain technology. Furthermore, while the Food Trust can enhance traceability, it does not directly address other causes of food waste, such as overproduction and over-purchasing by consumers. Therefore, it should ideally be part of a larger, multi-pronged strategy to address food waste.

Ripe.io: Transforming the Digital Food Supply Chain with Blockchain

Another leading proponent of the use of blockchain technology is Ripe.io (www.ripe.io), a company that leverages blockchain technology to transform the digital food supply chain. This section will analyze the role and implications of Ripe.io in the context of food waste management. Ripe.io utilizes blockchain technology and IoT to create a transparent and trustworthy food supply chain. The platform enables stakeholders to track and trace the journey of food products from farm to fork, thus offering insights into various factors including ripeness, freshness, and nutritional information.

The transparency provided by Ripe.io holds significant promise for reducing food waste. By providing stakeholders with the ability to track a product's journey in real time, potential wastage points within the supply chain can be identified and rectified. This tracking can also help ensure the quality of products, reducing the likelihood of waste due to spoilage. Additionally, Ripe.io's ability to provide real-time data on product ripeness and freshness can help retailers and consumers make more informed decisions, thus minimizing the chances of food waste due to overstocking or purchasing products with a short remaining shelf life.

Ripe.io has been involved in several collaborations to demonstrate the practical benefits of its blockchain solution. For instance, in a partnership with Ward's Berry Farm, Ripe.io was able to provide real-time data regarding the growing conditions of various crops, allowing stakeholders to make informed decisions that could prevent potential food waste. In another collaboration with the Dairy Farmers of America, Ripe.io's blockchain platform helped to enhance transparency in the dairy supply chain, promoting accountability and potentially reducing waste.

Despite its potential, the application of blockchain technology in food waste management does face some challenges. These include data privacy issues, the need for industry-wide standards and protocols, and the high initial setup costs that may deter smaller stakeholders. Addressing these challenges will be crucial for the wider adoption of solutions like Ripe.io.

12.4.3 Principles and Applications of the Internet of Things in Food Waste Management

12.4.3.1 Real-time Monitoring and Data Collection

The principle of real-time monitoring is especially relevant in the context of food waste management, allowing for the collection and analysis of data in real time. Sensors embedded in devices can detect and report conditions such as temperature, humidity, and expiration dates, thereby identifying potential food waste hotspots. One such application is seen in smart refrigerators. These appliances can track their contents, alert users when items are about to expire, and even suggest recipes to use up ingredients. For example, the Samsung Family Hub Refrigerator employs IoT to track food items and alert users about upcoming expiration dates (www.samsung.com/us/explore/family-hub-refrigerator/overview/). A UK-based firm known for its smart kitchen technology created a product, the FridgeCam (smarter.am/products/smarter-fridgecam), which captures images of the refrigerator's contents once the door closes and uses AI to identify items, track expiry dates, and predict when certain items are running low (Dekoninck & Barbaccia, 2019).

12.4.3.2 Optimized Logistics and Supply Chain Management

IoT can also play a vital role in enhancing supply chain efficiency. By tracking food products throughout the supply chain, IoT can help identify inefficiencies and optimize logistics. This could include refining transport schedules, storage conditions, and retail stocking practices. One case in point is the use of IoT by Zest Labs (www.zestlabs.com). The company uses IoT sensors and analytics to monitor fresh produce throughout the supply chain, thereby reducing waste through improved handling and transportation.

12.4.3.3 Predictive Analytics and Demand Forecasting

IoT, when combined with artificial intelligence and machine learning, can enable predictive analytics. This can help forecast future waste generation and demand, resulting in more efficient planning and resource allocation. Winnow Solutions (www.winnowsolutions.com) uses IoT and AI to reduce food waste in commercial kitchens. Their systems analyze trends and predict future waste, allowing kitchens to adjust their food preparation accordingly.

12.4.3.4 Consumer Engagement and Behavior Change

IoT can also facilitate consumer engagement in waste reduction efforts. Smart devices can provide personalized feedback and nudge consumers about their waste habits, helping them make more sustainable choices. A good example is the smart waste bin from Bin-e (www.bine.world). The device uses image recognition and IoT to identify, categorize, and manage waste, even providing users with data about their waste generation.

12.4.3.5 Case Studies of the Internet of Things in Food Waste Management

An Examination of Samsung's Family Hub Refrigerator: The Intersection of IoT and Food Waste Management

The application of IoT in domestic appliances has fostered the creation of innovative solutions for managing food waste. The Family Hub refrigerator, developed by Samsung, is a prime example of this approach. This section will analyze the features of the Family Hub refrigerator and evaluate its potential role in reducing household food waste. The Samsung Family Hub refrigerator (www.samsung.com/us/explore/family-hub-refrigerator/overview/) is a smart appliance equipped with IoT capabilities. It boasts a wide range of features designed to optimize food storage, reduce waste, and enhance the overall kitchen experience (Dekoninck & Barbaccia, 2019).

A central element of its food management system is the ability to keep track of what's inside the refrigerator. Using internal cameras and image recognition, the Family Hub can identify food items and monitor their expiry dates. Users can check their refrigerator's contents in real-time through a smartphone app, even while they're away from home.

The Family Hub refrigerator's potential for reducing food waste lies in its ability to increase consumer awareness about food storage and expiration. By providing real-time updates about food items and their expiry dates, the Family Hub can remind users to consume food items before they spoil. Furthermore, this technology can help curb overbuying – a significant contributor to food waste. By allowing users to see what they already have at home while they're grocery shopping, the Family Hub can prevent unnecessary purchases and encourage more mindful consumption.

While the Family Hub refrigerator offers exciting possibilities for waste reduction, it also presents a few challenges. Its effectiveness is largely reliant on users actively engaging with its features. Therefore, user-friendliness and seamless integration into daily routines are critical factors for success. Furthermore, the adoption of such smart appliances is currently limited by their high cost, making them inaccessible to a substantial segment of consumers. Efforts to make such technology more affordable will be crucial in expanding its impact.

Additionally, while the Family Hub can facilitate more efficient food management at the consumer level, it doesn't directly address food waste issues upstream in the supply chain. Thus, it represents one piece of the larger puzzle of comprehensive food waste reduction. Samsung's Family Hub refrigerator provides a glimpse into the future of smart home technology's role in managing food waste. By leveraging IoT capabilities, it empowers consumers with the necessary tools to monitor their food consumption and reduce waste. Despite facing challenges related to user engagement and affordability, smart appliances like the Family Hub highlight the potential of technology in contributing to a more sustainable food system.

An Analysis of Zest Labs' Approach to Reducing Waste in the Fresh Food Supply Chain

In the ongoing quest to combat food waste, technological innovation plays a crucial role. A case in point is Zest Labs, an ag-tech company, which has developed IoT-based solutions to optimize the fresh food supply chain and minimize waste. This section provides an analysis of Zest Labs' contributions to food waste management. Zest Labs leverages IoT and data analytics to address inefficiencies in the fresh food supply chain. The company's flagship solution, Zest Fresh, uses IoT-enabled sensors to monitor temperature and humidity throughout the supply chain, providing real-time data about the freshness of produce (www.zestlabs.com).

By providing actionable insights into the freshness of food items, Zest Fresh enables more informed decision-making. This visibility allows suppliers to prioritize the sale and distribution of produce based on its freshness, thereby reducing spoilage and waste. For instance, a batch of produce experiencing higher-than-average temperatures during transport could be identified and prioritized for sale at the retail level. This approach, known as first-expired-first-out (FEFO) inventory management, contrasts with the traditional first-in-first-out (FIFO) method, which does not account for variations in freshness and often results in unnecessary waste. Furthermore, by identifying issues in the supply chain, such as temperature fluctuations, Zest Fresh can help address these problems at their source. This preventive approach is key to achieving long-term reductions in food waste.

Despite its potential, Zest Labs' approach faces several challenges. These include the need for extensive cooperation across the supply chain, from growers to retailers, which may be complicated by disparate interests and logistical hurdles. Additionally, while Zest Fresh can dramatically reduce waste in the fresh produce sector, its impact on other types of food waste, such as processed or packaged foods, may be limited. As such, Zest Labs' solutions should be seen as part of a broader suite of tools needed to tackle food waste comprehensively. Zest Labs' use of IoT and data analytics offers a promising avenue for reducing food waste in the fresh food supply chain. By enhancing visibility and enabling more informed decision-making, tools like Zest Fresh can significantly decrease food spoilage and waste. However, these tools must be implemented as part of a multifaceted, cooperative approach to food waste reduction across the entire food system.

12.5 OBSTACLES, OPPORTUNITIES, AND FUTURE OUTLOOK

As highlighted above, AI tools can be leveraged for the efficient management of food waste through conceptual arguments with an empirical analysis (Tonini et al., 2018; Wen et al., 2018). Gathering the mentioned technologies together can promote the effectiveness of eliminating undesired food losses and could be a promising solution for food waste management. It is important to mention that adopting these technologies for full potential utilization requires concrete collaboration between stakeholders like businesses, policymakers, and technology developers (Corrado et al., 2019).

Having a number of advantages and great potential, it is also worth noting that the technologies may face several challenges and obstacles (Morone et al., 2019). The obstacles and challenges are listed in Table 12.3 and are further discussed in the subsequent subsections.

12.5.1 Obstacles

12.5.1.1 Data Inconsistency and Quality

In order to analyze and provide meaningful insights, the algorithms of machine learning need a large amount of precise and dependable data. However, inconsistent and unreliable data can exacerbate the lack of a standardized format and make it hard to interpret results. Therefore, data availability and the collection of quality data are among the crucial challenges that necessitate efforts to overcome these hurdles, and they should be addressed by collecting reliable data from multiple sources and ensuring their accuracy and quality (Corrado et al., 2019; Xue et al., 2017).

12.5.1.2 Cost and Affordability

The development and implementation of safe and sustainable methods that provide feasible technologies for converting food waste into value-added products or efficient food waste management may promote the circular economy. However, the integration of these modern approaches, including blockchain, IoT, and ML, may generate cost and affordability issues (Said et al., 2023). Providing industrial-scale application through these instruments has operational barriers, and these challenges lie in areas such as food waste collection, suitable machine learning methods, installation, skilled labor, internet connection, and process compatibility. Each of the cost items itself depends on the size of the process; i.e., when the size of the data increases, so do the computational expenses (Awad & Khanna, 2015). In order to overcome these hindrances, finding cost-effective solutions and providing public-private partnerships for open-source technologies are key parameters.

12.5.1.3 Privacy and Security

Privacy and security are crucial aspects of implementing IoT devices, particularly for collecting and sharing substantial data related to the food sector and consumer behaviors (Lei et al., 2022). Keeping these data private and secure through a blockchain is essential for gaining trust and ensuring government regulations (Feng et al., 2018). In protecting these sensitive data in the IoT system, some precautions and methods should be adopted, such as encryption to protect from unauthorized access, secure data sharing protocols, secure software and firmware updates, data consistency without third-party accession, network shielding, and protection through up-to-date industry-specific standards. Thus, it fosters confidence among stakeholders like suppliers, consumers, and policymakers, and leverages the benefits of computational technologies (Nair & Dorai, 2021).

TABLE 12.3
Summary of the Obstacles and Opportunities of Food Waste Management through Machine Learning, IoT, and Blockchain

Obstacles	Opportunities
Data inconsistency and quality	Real-time monitoring and intervention
Cost and affordability	Efficient inventory management
Privacy and security	Supply chain transparency
Behavioral and cultural factors	Environmental and social impact
Infrastructure and connectivity	Redistribution and sharing platforms
Collaboration and standardization	Consumer awareness and engagement

12.5.1.4 Behavioral and Cultural Factors

Behavioral and cultural factors are another huge obstacle for the wider application area of machine learning, IoT, and blockchain. The main reason for that hindrance lies in the lack of awareness, education, and other cultural norms. Shifting behavior and mindsets for successfully reducing food wastage can be achieved through education starting from an early age, awareness campaigns, incentives, and regulations from policymakers. In that aspect, consumers, suppliers, and all the stakeholders should have a straightforward collaboration, and they can adopt their crucial role in tackling food waste (Al-Obadi et al., 2022; Jones-Garcia et al., 2022).

12.5.1.5 Infrastructure and Connectivity

The achievement of globalized food waste management depends on the presence of a robust and scalable infrastructure for a wide range of data collection and transmission. However, although it seems achievable in developed or high-income countries, there can be a lack of infrastructure and limited availability of reliable connections in rural areas or developing countries (Pardini et al., 2019). This can directly affect the real-time monitoring and efficient data transfer between IoT systems. The lack of infrastructure can be exacerbated by ineffective physical deployment and functioning of IoT devices, resulting in a hindrance to the adoption of IoT systems, particularly in remote or underserved areas (Anagnostopoulos et al., 2017).

12.5.1.6 Collaboration and Standardization

To successfully address food waste management, it is critical to foster cooperation among all groups of responsible stakeholders, including producers, suppliers, retailers, and consumers. However, due to the diversities, ensuring strong coordination and standardized sharing practices between the various entities and sectors can be impractical. Therefore, having common frameworks may facilitate a guideline for stakeholders in aligning required practices. This can be achievable through standardized terminology, applications, and methodologies. Principally, collaboration and standardization are decisive barriers to effective food waste management, and they can be eliminated by having a common goal and working together with stakeholders, aligning their efforts, and sharing information seamlessly (Lai et al., 2020; Sharma et al., 2020). The potential usage of machine learning, IoT, and blockchain in food waste management can be harnessed by addressing each challenge mentioned above for a more sustainable food strategy of the future.

12.5.2 Opportunities

12.5.2.1 Real-time Monitoring and Intervention

Leveraging IoT devices in the food supply chain for efficient food wastage management can enable real-time monitoring of critical parameters affecting spoilage and shelf life, such as temperature, humidity, and quality. Along the food supply chain, this facilitates proactive intervention when deviations or unprecedented issues are detected (Wen et al., 2018). In order to eliminate these unfavorable conditions that can affect food quality, these technologies can establish suitable storage conditions through adjustment, redirecting shipments, or implementing corrective actions. Thus, real-time monitoring can prevent food spoilage, extend shelf life, and minimize waste through corrective measures and fine-tuning solutions (Cappelletti et al., 2022).

12.5.2.2 Efficient Inventory Management

Inventory management, particularly in the food industry due to the perishable products and limited shelf life, is a critical aspect of running a successful enterprise. Machine learning algorithms and other software tools have the power to follow the food lifetime and optimize the inventory management process. Using a large amount of available data with high-accuracy forecasting enables them to align their inventory with market necessities. On the other hand, one of the critical benefits of effective inventory management is eliminating overproduction, which causes unnecessary costs and exacerbates food wastage (Liegeard & Manning, 2020). Therefore, implementing these technologies into food waste management can be a cost-saving approach and ensure a steady fresh food supply.

12.5.2.3 Supply Chain Transparency

In today's globalized world, ensuring transparency is crucial for gaining stakeholders' trust and reaching accountability goals. In this aspect, by using blockchain technology with its decentralized and immutable ledger, it can be possible to record every transaction and operation transparently and securely (Saurabh & Dey, 2021). Thus, each step of the journey for the food, from preparation to disposal, can be documented via distinguished verification. This would facilitate tracing the origin of food, which can often be challenging and time-consuming for stakeholders to pinpoint and figure out the preparation and certifications. In summary, these technologies possess great potential for improving transparency and trustworthiness in the food industry and waste management processes (Ada et al., 2021; Park & Li, 2021).

12.5.2.4 Environmental and Social Impact

One of the most important environmental impacts of effective food waste management is the mitigation of greenhouse gas emissions by reducing the amount of potential food waste. When food wastes are landfilled, the anaerobic digestion process releases methane, one of the greenhouse gases responsible for causing climate change. Through an efficient management process, it (i) minimizes the waste amount, (ii) provides more efficient landfill area usage, (iii) eliminates the excess amount of water, energy, and land usage, and (iv) contributes to a sustainable and circular economy (Chandan et al., 2023; Morrow & Zarrebini, 2019).

12.5.2.5 Redistribution and Sharing Platforms

As blockchain technology has found a number of application areas in the last couple of years, it is gaining further attention for promoting a sustainable environment and future. Within that scope, it is critical to harness the potential of blockchain in food waste management by adopting decentralized food-sharing platforms to identify surplus food simply for sending to the individuals that require it. Thus, excess food can be used and won't be considered waste anymore, and these technologies can reduce the amount of waste by addressing food insecurity and building stronger communities (Centobelli et al., 2022; Patil et al., 2023).

12.5.2.6 Consumer Awareness and Engagement

Implementing advanced technologies for food waste management can create a consumer profile with high awareness by empowering them about the importance of reducing waste for a more sustainable future (Dey et al., 2022). IoT devices and machine learning technology can facilitate the consumer in gaining a sense of responsibility and promoting behavioral change toward reducing waste through varied applications, i.e., smart labels can provide detailed information about the product lifecycle, expiration dates, proper storage guidelines, and potential environmental effects (Hamsagayathri et al., 2020; Khan et al., 2020). This can be a journey to create a society that fosters a high sense of responsibility, has sustainable consumption patterns, and is conscious about food waste management.

12.5.3 Future Outlook

12.5.3.1 A Circular Food Supply Chain Ecosystem Approach

The bifurcation of the food system into upstream and downstream stages, encompassing production and distribution versus consumption and waste management respectively, inhibits the efficient handling of food waste. Evidence from several case studies indicates that digital initiatives predominantly concentrate on either the upstream or downstream facets of the food system. Nevertheless, we propose that adopting circular ecosystem approaches can significantly aid in food waste prevention by integrating all stages of the food cycle and enhancing food waste management (Liaros, 2021).

A circular economy ecosystem is an organizational construct that autonomously organizes and collectively produces a sustainable ecosystem as a result of system-level interaction. Its defining traits include cyclical processes of recycling, reusing, and minimizing resource use in the economy (Ranta et al., 2020). The food system is a multifaceted network where cooperation is indispensable for effective food delivery to consumers. Coordinating activities within this ecosystem entails a complex interplay among organizations, actors, resources, and the food itself. An inherent challenge faced by ecosystems is the aptitude for, and processing of, information (Wang, 2021).

Therefore, digital representation and monitoring play an instrumental role in ensuring comprehensive oversight and no omissions. This can be realized through a digitally-enabled circular food ecosystem encompassing all players within a specific food system. The integration of technologies such as the IoT, machine learning, and blockchain into a circular ecosystem facilitates collaboration between actors, thereby promoting a circular economy, resource reuse, and waste prevention.

The primary strategy to address food waste should be to reincorporate it back into the food system. This can be achieved by aligning with all partners in the food supply chain and the food waste management stakeholders, both downstream and upstream, within a digitally-enabled circular ecosystem. As depicted in Figure 12.4, every component of the food system should be interconnected within a circular food ecosystem that manages its waste using circular principles (www.ellenmacarthurfoundation.org), such as eliminating waste and pollution, maintaining products and materials at their highest value, and rejuvenating nature. These principles provide a roadmap toward devising an optimal waste prevention system within the food ecosystem. Present digital technologies, however, within the linear economic system are limited in their scope, catering only to consumption while neglecting waste handling.

The waste management system acknowledges the inherent value of waste and investigates diverse methods for its utilization. Nonetheless, the production and distribution sectors of the food system lack a mechanism to recover the waste or channel it to suitable partners who may find a use for it. As a result, the food system must proactively search for partners willing to acquire waste and by-products during production and distribution. Additionally, consumers should be informed about the final destination of their waste, whether it becomes a biological nutrient for the biosphere, a technical nutrient for re-manufacturing (McDonough & Braungart, 2002), or ends up in a landfill.

Several companies, such as Winnow, IBM Food Trust, and Plastic Bank, offer specialized waste monitoring services. However, these systems lack interoperability and fail to offer a comprehensive ecosystem approach to food waste management, primarily due to their disconnected nature. We suggest that food waste management should commence even before production, by forging connections with consumers. The leverage of data science, artificial intelligence, and blockchain-enabled applications can be instrumental in engaging all stakeholders, encompassing producers, consumers, and waste disposers. The unique attributes of digital technologies, such as programmability and re-combinability (Yoo et al., 2010; Yoo et al., 2012), facilitate their application to circular economy ecosystems designed for efficient resource management.

Additionally, connecting production and consumption through the IoT across the production line, extending to consumers' kitchens and refrigerators, is essential. IBM's Food Trust has made substantial progress in facilitating blockchain connectivity for all food producers, but it currently lacks the inclusion of waste and consumption aspects. We recommend an approach that aligns with the cradle-to-cradle philosophy within a farm-to-farm framework (McDonough & Braungart, 2002). Each section of the food system can implement machine learning, IoT, and connect to a blockchain-powered supply chain using the underlying circular principles (MacArthur, 2013; McDonough & Braungart, 2002).

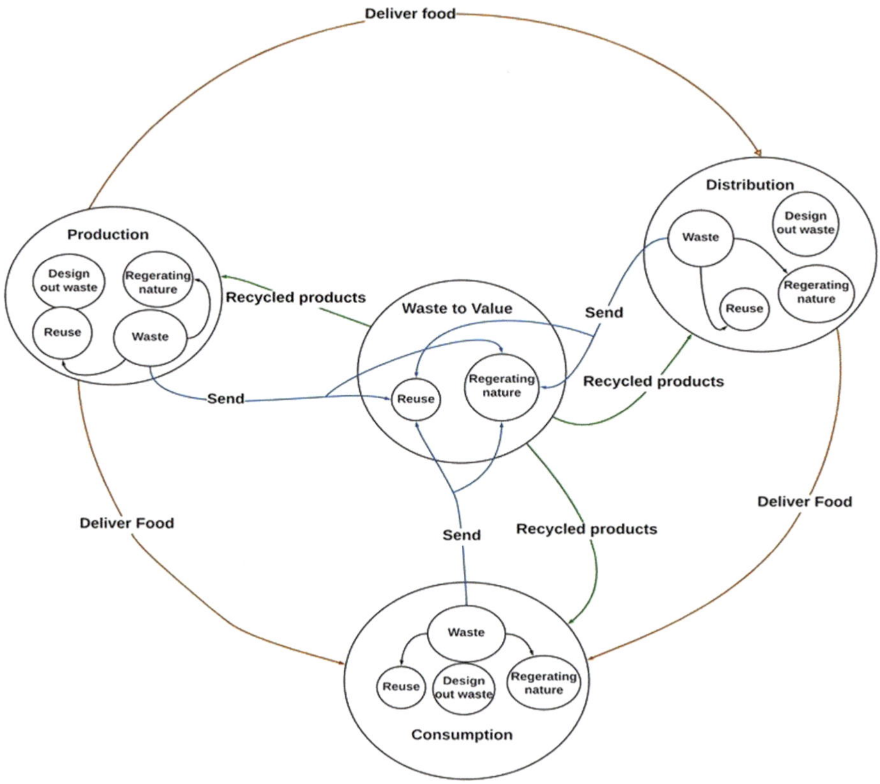

FIGURE 12.4 Circular food waste management ecosystem.

12.5.3.2 Digitally-Enabled Circular Food Supply Chain Ecosystem

The Ellen MacArthur Foundation (www.ellenmacarthurfoundation.org) presents their butterfly model that comprises two cycles: the biological cycle and the technical cycle. The biological cycle emphasizes the reintroduction of products and materials into the biosphere through non-toxic, restorative loops, highlighting the concept's core principles of resource regeneration and reuse. In contrast, the technical nutrient cycle facilitates quality enhancements through a process called upcycling. Upcycling involves transforming materials into new ones with improved quality and enhanced functionality, promoting resource efficiency and waste reduction. The food system contains both biological and technical cycles that must be captured and separated in order to ensure resource recovery and conservation. The Ellen MacArthur Foundation provides principles (www.ellenmacarthurfoundation.org) that can be applied to the selection of digital technologies for waste prevention and recovery.

Figure 12.5 proposes a digitalized circular food waste management ecosystem, leveraging AI/ML, blockchain, and IoT technologies. The proposed system brings together all food stakeholders, including those in production, distribution, consumption, and waste recovery. In this ecosystem, stakeholders contribute food and food-related waste data to a common platform using IoT tools. These tools monitor the food supply chain in real-time, identifying resource usage and inefficiencies. The data collected forms the basis for AI/ML analysis, providing insights to optimize processes and reduce waste. Thus, IoT and AI together enable real-time adjustments and informed decision-making throughout the entire food cycle, affecting aspects such as demand and supply forecasting, consumer behavior, waste generation patterns, and retail operations. The ecosystem platform is underpinned by a blockchain, which provides transparency and traceability of data

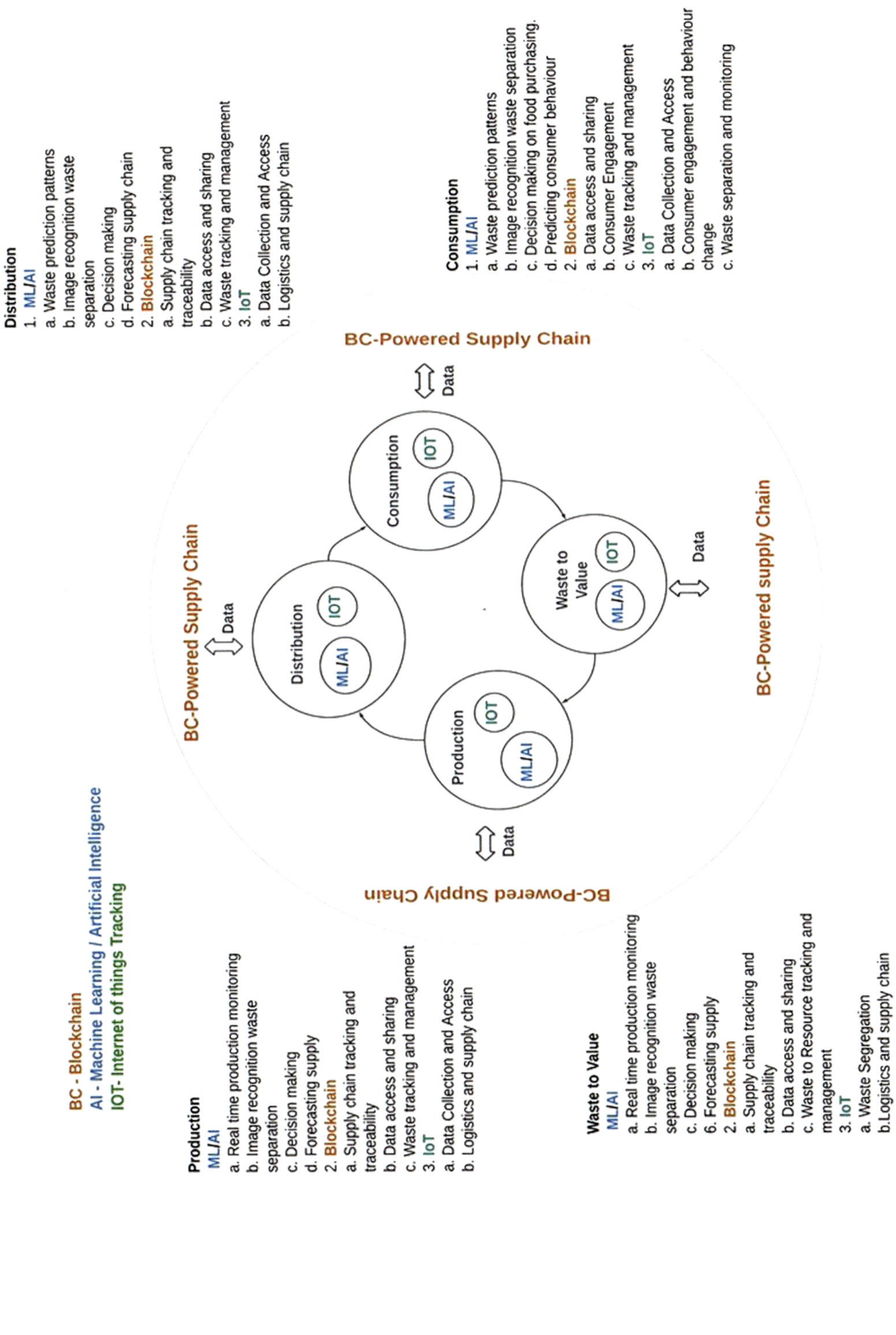

FIGURE 12.5 Proposed digitalization of the circular food waste management ecosystem.

along the food supply chain. It allows stakeholders to pinpoint where food waste is being generated and ensure that waste, which cannot be reused, is appropriately managed. Blockchain assures secure, real-time data access, maintains data integrity, and facilitates collaborative opportunities for waste prevention and recovery among stakeholders.

New players, referred to as "Waste Value Creators" (WVCs), are introduced into the supply chain. WVCs create value from food waste, serving it back to farmers (in a farm-to-farm model) or consumers. They can track, trace, and collect food waste, further enhancing the system's effectiveness. By implementing such a blockchain-powered supply chain and leveraging IoT and AI technologies, we can realize an ideal future where waste is minimized and resources are efficiently utilized. This approach not only regenerates the natural ecosystem but also necessitates the collective efforts of all stakeholders involved in the food system, integrating innovative technological solutions into their operations. This way, the proposed digitalized circular food waste management ecosystem ensures sustainability, resource optimization, and waste reduction.

12.6 CONCLUSIONS AND PERSPECTIVES

In conclusion, implementing machine learning, the Internet of Things, and blockchain technologies presents opportunities for overcoming the intricate challenges of food waste management. The cooperation of these technologies not only enhances real-time monitoring and data analysis across the entire food supply chain but also establishes a transparent and immutable record of transactions within the entire chain. As mentioned above, the opportunities of this innovative approach can enable stakeholders to identify inefficiencies, minimize losses, and make informed decisions to optimize resource allocation. As we navigate into a future defined by sustainability and resource conservation, harnessing the power of tools such as predictive analytics, intelligent sensors, and decentralized ledgers is expected to increase. Ultimately, this will potentially pave the way for a more efficient global food system.

ACKNOWLEDGEMENTS

The authors gratefully acknowledge FORMAS (Project number: 2021–02458) and Cost Action (CA20133) Cross Border Transfer and Development of Sustainable Resource Recovery Strategies Towards Zero Waste (FULLRECO4US) for their support.

REFERENCES

Ada, N., Kazancoglu, Y., Sezer, M.D., Ede-Senturk, C., Ozer, I., Ram, M. 2021. Analyzing Barriers of Circular Food Supply Chains and Proposing Industry 4.0 Solutions. *Sustainability*, **13**(12), 6812.

Al-Fuqaha, A., Guizani, M., Mohammadi, M., Aledhari, M., Ayyash, M. 2015. Internet of Things: A Survey on Enabling Technologies, Protocols, and Applications. *IEEE Communications Surveys & Tutorials*, **17**(4), 2347–2376.

Al-Obadi, M., Ayad, H., Pokharel, S., Ayari, M.A. 2022. Perspectives on Food Waste Management: Prevention and Social Innovations. *Sustainable Production and Consumption*, **31**, 190–208.

Al-Rumaihi, A., McKay, G., Mackey, H.R., Al-Ansari, T. 2020. Environmental Impact Assessment of Food Waste Management Using Two Composting Techniques. *Sustainability*, **12**(4), 1595.

Anagnostopoulos, T., Zaslavsky, A., Kolomvatsos, K., Medvedev, A., Amirian, P., Morley, J., Hadjieftymiades, S. 2017. Challenges and Opportunities of Waste Management in IoT-Enabled Smart Cities: A Survey. *IEEE Transactions on Sustainable Computing*, **2**(3), 275–289.

Andreas, M.A. 2014. *Mastering Bitcoin: Unlocking Digital Crypto-Currencies*. O'Reilly Media, Inc.

Antti, T., Harri, V. 2017. Mean Teachers Are Better Role Models: Weight-Averaged Consistency Targets Improve Semi-Supervised Deep Learning Results. in: *Proceedings of the 31st International Conference on Neural Information Processing Systems*, Curran Associates Inc. Long Beach, California, USA.

Atzori, L., Iera, A., Morabito, G. 2010. The Internet of Things: A survey. *Computer Networks*, **54**(15), 2787–2805.

Awad, M., Khanna, R. 2015. Support Vector Regression. in: *Efficient Learning Machines: Theories, Concepts, and Applications for Engineers and System Designers*, (Eds.) M. Awad, R. Khanna, Apress. Berkeley, CA, pp. 67–80.

Bengio, Y., Courville, A.C., Vincent, P. 2012. Unsupervised Feature Learning and Deep Learning: A Review and New Perspectives. *ArXiv*, abs/1206.5538.

Bhusal, C.S. 2021. Blockchain Technology in Agriculture: A Case Study of Blockchain Start-Up Companies. *International Journal of Computer Science & Information Technology*, **13**(5). https://doi.org/10.5121/ijcsit.2021.13503

Buzby, J.C., Farah-Wells, H., Hyman, J. 2014. *The Estimated Amount, Value, and Calories of Postharvest Food Losses at the Retail and Consumer Levels in the United States*. USDA-ERS Economic Information.

Cappelletti, F., Papetti, A., Rossi, M., Germani, M. 2022. Smart strategies for household food waste management. *Procedia Computer Science*, **200**, 887–895.

Centobelli, P., Cerchione, R., Vecchio, P.D., Oropallo, E., Secundo, G. 2022. Blockchain technology for bridging trust, traceability and transparency in circular supply chain. *Information & Management*, **59**(7), 103508.

Chandan, A., John, M., Potdar, V. 2023. Achieving UN SDGs in Food Supply Chain Using Blockchain Technology. *Sustainability*, **15**(3), 2109.

Choi, R.Y., Coyner, A.S., Kalpathy-Cramer, J., Chiang, M.F., Campbell, J.P. 2020. Introduction to Machine Learning, Neural Networks, and Deep Learning. *Translational Vision Science & Technology*, **9**(2), 14.

Corrado, S., Caldeira, C., Eriksson, M., Hanssen, O.J., Hauser, H.-E., van Holsteijn, F., Liu, G., Östergren, K., Parry, A., Secondi, L., Stenmarck, Å., Sala, S. 2019. Food waste accounting methodologies: Challenges, opportunities, and further advancements. *Global Food Security*, **20**, 93–100.

Dekoninck, E., Barbaccia, F. 2019. Streamlined Assessment to Assist in the Design of Internet-of-Things (IOT) Enabled Products: A Case Study of the Smart Fridge. *Proceedings of the Design Society: International Conference on Engineering Design*, **1**(1), 3721–3730.

Dey, S., Saha, S., Singh, A.K., McDonald-Maier, K. 2022. SmartNoshWaste: Using Blockchain, Machine Learning, Cloud Computing and QR Code to Reduce Food Waste in Decentralized Web 3.0 Enabled Smart Cities. *Smart Cities*, **5**(1), 162–176.

Dladla, I., Machete, F., Shale, K. 2016. A -Review of Factors Associated with Indiscriminate Dumping of Waste in Eleven African Countries. *African Journal of Science, Technology, Innovation and Development*, **8**(5–6), 475–481.

Don, T., Alex, T. 2016. *Blockchain Revolution: How the Technology Behind Bitcoin Is Changing Money, Business, and the World*. Portfolio.

Eriksson, M., Spångberg, J. 2017. Carbon Footprint and Energy Use of Food Waste Management Options for Fresh Fruit and Vegetables from Supermarkets. *Waste Management*, **60**, 786–799.

Eriksson, M., Strid, I., Hansson, P.-A. 2015. Carbon Footprint of Food Waste Management Options in the Waste Hierarchy – A Swedish Case Study. *Journal of Cleaner Production*, **93**, 115–125.

Feng, L., Zhang, H., Lou, L., Chen, Y. 2018. A Blockchain-Based Collocation Storage Architecture for Data Security Process Platform of WSN. *2018 IEEE 22nd International Conference on Computer Supported Cooperative Work in Design ((CSCWD))*, 9–11 May 2018. pp. 75–80.

Gao, F., Yang, Y., Wang, J., Sun, J., Yang, E., Zhou, H. 2018. A Deep Convolutional Generative Adversarial Networks (DCGANs)-Based Semi-Supervised Method for Object Recognition in Synthetic Aperture Radar (SAR) Images. *Remote Sensing*, **10**(6), 846.

Girotto, F., Alibardi, L., Cossu, R. 2015. Food Waste Generation and Industrial Uses: A Review. *Waste Management*, **45**, 32–41.

Grycová, B., Koutník, I., Pryszcz, A. 2016. Pyrolysis Process for the Treatment of Food Waste. *Bioresource Technology*, **218**, 1203–1207.

Gubbi, J., Buyya, R., Marusic, S., Palaniswami, M. 2013. Internet of Things (IoT): A Vision, Architectural Elements, and Future Directions. *Future Generation Computer Systems*, **29**(7), 1645–1660.

Gustavsson, J., Cederberg, C., Sonesson, U., Otterdijk, R., Meybeck, A. 2011. *Global Food Losses and Food Waste- Extent, Causes and Prevention*. FAO.

Hamsagayathri, P., Rajakumari, K., Ramya, P., Shoukath Ali, K. 2020. *Blockchain, Big Data and Machine Learning*. CRC Press.

Hebrok, M., Boks, C. 2017. Household Food Waste: Drivers and Potential Intervention Points for Design – An Extensive Review. *Journal of Cleaner Production*, **151**, 380–392.

Holm-Nielsen, J.B., Al Seadi, T., Oleskowicz-Popiel, P. 2009. The Future of Anaerobic Digestion and Biogas Utilization. *Bioresource Technology*, **100**(22), 5478–5484.

Jones-Garcia, E., Bakalis, S., Flintham, M. 2022. Consumer Behaviour and Food Waste: Understanding and Mitigating Waste with a Technology Probe. *Foods*, **11**(14), 2048.

Jovel, J., Greiner, R. 2021. An Introduction to Machine Learning Approaches for Biomedical Research. *Frontiers in Medicine*, **8**. https://doi.org/10.3389/fmed.2021.771607

Julian, P., Mark, B., Sarah, M. 2010. Food Waste within Food Supply Chains: Quantification and Potential for Change to 2050. *Philosophical Transactions of the Royal Society,* 365, 3065–3081.

Khan, P.W., Byun, Y.-C., Park, N. 2020. IoT-Blockchain Enabled Optimized Provenance System for Food Industry 4.0 Using Advanced Deep Learning. *Sensors*, **20**(10), 2990.

Kinach, L., Parizeau, K., Fraser, E.D.G. 2020. Do Food Donation Tax Credits for Farmers Address Food Loss/Waste and Food Insecurity? A Case Study from Ontario. *Agric Human Values*, **37**(2), 383–396.

Kober, J., Bagnell, J.A., Peters, J. 2013. Reinforcement Learning in Robotics: A Survey. *The International Journal of Robotics Research*, **32**(11), 1238–1274.

Krause, M.J., W. Chickering, G., Townsend, T.G., Reinhart, D.R. 2016. Critical review of the methane generation potential of municipal solid waste. *Critical Reviews in Environmental Science and Technology*, **46**(13), 1117–1182.

Lai, C.S., Jia, Y., Dong, Z., Wang, D., Tao, Y., Lai, Q.H., Wong, R.T.K., Zobaa, A.F., Wu, R., Lai, L.L. 2020. A Review of Technical Standards for Smart Cities. *Clean Technologies*, **2**(3), 290–310.

Lee, J.-Y., Cheon, J.-Y., Kwon, H.-P., Yoon, H.-S., Lee, S.-S., Kim, J.-H., Park, J.-K., Kim, C.-G. 2006. Attenuation of Landfill Leachate at Two Uncontrolled Landfills. *Environmental Geology*, **51**(4), 581–593.

Lee, S.-H., Choi, K.-I., Osako, M., Dong, J.-I. 2007. Evaluation of Environmental Burdens Caused by Changes of Food Waste Management Systems in Seoul, Korea. *Science of The Total Environment*, **387**(1), 42–53.

Lei, M., Xu, L., Liu, T., Liu, S., Sun, C. 2022. Integration of Privacy Protection and Blockchain-Based Food Safety Traceability: Potential and Challenges. *Foods*, **11**(15), 2262.

Liaros, S. 2021. Circular Food Futures: What Will They Look Like? *Circular Economy and Sustainability*, **1**(4), 1193–1206.

Liegeard, J., Manning, L. 2020. Use of Intelligent Applications to Reduce Household Food Waste. *Critical Reviews in Food Science and Nutrition*, **60**(6), 1048–1061.

Liu, X., Shi, Y., Kong, L., Tong, L., Cao, H., Zhou, H., Lv, Y. 2022. Long-Term Application of Bio-Compost Increased Soil Microbial Community Diversity and Altered Its Composition and Network. *Microorganisms*, **10**(2), 462.

MacArthur, E. 2013. Towards the circular economy, economic and business rationale for an accelerated transition. Ellen MacArthur Foundation: Cowes, UK, 1.

McDonough, W., Braungart, M. 2002. *Cradle to Cradle: Remaking the Way We Make Things*. North Point Press.

Mohammed, M., Ozbay, I., Karademir, A., Isleyen, M. 2017. Pre-treatment and Utilization of Food Waste as Energy Source by Bio-drying Process. *Energy Procedia*, **128**, 100–107.

Mokrane, S., Buonocore, E., Capone, R., Franzese, P.P. 2023. Exploring the Global Scientific Literature on Food Waste and Loss. *Sustainability*, **15**(6), 4757.

Morone, P., Koutinas, A., Gathergood, N., Arshadi, M., Matharu, A. 2019. Food Waste: Challenges and Opportunities for Enhancing the Emerging Bio-economy. *Journal of Cleaner Production*, **221**, 10–16.

Morrow, M.J., Zarrebini, M. 2019. Blockchain and the Tokenization of the Individual: Societal Implications. *Future Internet*, **11**(10), 220.

Nair, P.R., Dorai, D.R. 2021. Evaluation of Performance and Security of Proof of Work and Proof of Stake Using Blockchain. *2021 Third International Conference on Intelligent Communication Technologies and Virtual Mobile Networks (ICICV)*, 4–6 Feb. 2021. pp. 279–283.

Ojha, S., Bußler, S., Schlüter, O.K. 2020. Food Waste Valorisation and Circular Economy Concepts in Insect Production and Processing. *Waste Management*, **118**, 600–609.

Pardini, K., Rodrigues, J.J.P.C., Kozlov, S.A., Kumar, N., Furtado, V. 2019. IoT-Based Solid Waste Management Solutions: A Survey. *Journal of Sensor and Actuator Networks*, **8**(1), 5.

Park, A., Li, H. 2021. The Effect of Blockchain Technology on Supply Chain Sustainability Performances. *Sustainability*, **13**(4), 1726.

Patil, S., Nikam, O., Nair, S., Raut, A., Lobo, V.B. 2023. Sustainable Food Waste Management and Tracking System Using Blockchain. *2023 International Conference on Advancement in Computation & Computer Technologies (InCACCT)*, 5–6 May 2023. pp. 848–853.

Paul van der, W., Jason, A.G. 2017. A Systematic Review of Food Losses and Food Waste Generation in Developed Countries. *Proceedings of the Institution of Civil Engineers - Waste and Resource Management*, **170**(2), 66–77.

Ranta, V., Keränen, J., Aarikka-Stenroos, L. 2020. How B2B suppliers Articulate Customer Value Propositions in the Circular Economy: Four Innovation-Driven Value Creation Logics. *Industrial Marketing Management*, **87**, 291–305.

Roman, R., Zhou, J., Lopez, J. 2013. On the Features and Challenges of Security and Privacy in Distributed Internet of Things. *Computer Networks*, **57**(10), 2266–2279.

Sadhu, P.K., Yanambaka, V.P., Abdelgawad, A. 2022. Internet of Things: Security and Solutions Survey. *Sensors*, **22**(19), 7433.

Said, Z., Sharma, P., Thi Bich Nhuong, Q., Bora, B.J., Lichtfouse, E., Khalid, H.M., Luque, R., Nguyen, X.P., Hoang, A.T. 2023. Intelligent Approaches for Sustainable Management and Valorisation of Food Waste. *Bioresource Technology*, **377**, 128952.

San Martin, D., Ramos, S., Zufía, J. 2016. Valorisation of Food Waste to Produce New Raw Materials for Animal Feed. *Food Chemistry*, **198**, 68–74.

Saurabh, S., Dey, K. 2021. Blockchain Technology Adoption, Architecture, and Sustainable Agri-Food Supply Chains. *Journal of Cleaner Production*, **284**, 124731.

Shafiee-Jood, M., Cai, X. 2016. Reducing Food Loss and Waste to Enhance Food Security and Environmental Sustainability. *Environmental Science & Technology*, **50**(16), 8432–8443.

Sharma, M., Joshi, S., Kannan, D., Govindan, K., Singh, R., Purohit, H.C. 2020. Internet of Things (IoT) Adoption Barriers of Smart Cities' Waste Management: An Indian Context. *Journal of Cleaner Production*, **270**, 122047.

Slorach, P.C., Jeswani, H.K., Cuéllar-Franca, R., Azapagic, A. 2019. Energy Demand and Carbon Footprint of Treating Household Food Waste Compared to Its Prevention. *Energy Procedia*, **161**, 17–23.

Thi, N.B.D., Kumar, G., Lin, C.-Y. 2015. An Overview of Food Waste Management in Developing Countries: Current Status and Future Perspective. *Journal of Environmental Management*, **157**, 220–229.

Tonini, D., Albizzati, P.F., Astrup, T.F. 2018. Environmental Impacts of Food Waste: Learnings and Challenges from a Case Study on UK. *Waste Management*, **76**, 744–766.

Truong, L., Morash, D., Liu, Y., King, A. 2019. Food Waste in Animal Feed with a Focus on Use for Broilers. *International Journal of Recycling of Organic Waste in Agriculture*, **8**(4), 417–429.

Wainaina, S., Lukitawesa, Kumar Awasthi, M., Taherzadeh, M.J. 2019. Bioengineering of Anaerobic Digestion for Volatile Fatty Acids, Hydrogen or Methane Production: A Critical Review. *Bioengineered*, **10**(1), 437–458.

Wainaina, S., Taherzadeh, M.J. 2023. Automation and Artificial Intelligence in Filamentous Fungi-Based Bioprocesses: A Review. *Bioresource Technology*, **369**, 128421.

Wang, P. 2021. Connecting the Parts with the Whole: Toward an Information Ecology Theory of Digital Innovation Ecosystems. *MIS Quarterly*, **45**(1), 397–422.

Wei, Y., Li, J., Shi, D., Liu, G., Zhao, Y., Shimaoka, T. 2017. Environmental Challenges Impeding the Composting of Biodegradable Municipal Solid Waste: A Critical Review. *Resources, Conservation and Recycling*, **122**, 51–65.

Wen, Z., Hu, S., De Clercq, D., Beck, M.B., Zhang, H., Zhang, H., Fei, F., Liu, J. 2018. Design, Implementation, and Evaluation of an Internet of Things (IoT) Network System for Restaurant Food Waste Management. *Waste Management*, **73**, 26–38.

Xue, L., Liu, G., Parfitt, J., Liu, X., Van Herpen, E., Stenmarck, Å., O'Connor, C., Östergren, K., Cheng, S. 2017. Missing Food, Missing Data? A Critical Review of Global Food Losses and Food Waste Data. *Environmental Science & Technology*, **51**(12), 6618–6633.

Yoo, Y., Boland Jr R.J., Lyytinen, K. and Majchrzak, A., 2012. Organizing for innovation in the digitized world. *Organization science*, **23**(5), 1398–1408.

Yoo, Y., Henfridsson, O., Lyytinen, K. 2010. Research commentary—the new organizing logic of digital innovation: an agenda for information systems research. *Information systems research*, **21**(4), 724–735.

Zaki, M.J., Meira, J.W. 2014. *Data Mining and Analysis: Fundamental Concepts and Algorithms*. Cambridge University Press.

Zanella, A., Bui, N., Castellani, A., Vangelista, L., Zorzi, M. 2014. Internet of Things for Smart Cities. *IEEE Internet of Things Journal*, **1**(1), 22–32.

Zubaydi, H.D., Chong, Y.-W., Ko, K., Hanshi, S.M., Karuppayah, S. 2019. A Review on the Role of Blockchain Technology in the Healthcare Domain. *Electronics*, **8**(6), 679.

Index